Microbiology Monographs

Volume 33

Series Editor

Alexander Steinbüchel
Münster, Germany

The Springer book series ***Microbiology Monographs*** presents carefully refereed volumes on selected microbiological topics. Microbiology is a still rapidly expanding field with significant impact on many areas of basic and applied science. The growth in knowledge of microbial physiology, cell structure, biotechnological capabilities and other aspects of microorganisms is increasing dramatically even in the face of the breakthroughs that already have been made.

Reflecting these most recent achievements, the series' wide scope encompasses such topics as inclusions in prokaryotes, predatory prokaryotes, magnetoreception and magnetosomes in bacteria, uncultivable microorganisms, microbial endosymbionts, bacterial resistance, extremophilic microorganisms, analyses of genome sequences and structures, microorganisms as cell factories for chemicals and fuels, metabolic engineering, gene transfer and expression systems and distinct physiological groups of bacteria.

The volume editors are well-known experts in their particular fields, and each volume offers 10 to 20 comprehensive review articles covering all relevant aspects of the topic in focus.

More information about this series at http://www.springer.com/series/7171

Dietmar Schlosser
Editor

Laccases in Bioremediation and Waste Valorisation

Springer

Editor
Dietmar Schlosser
Department of Environmental Microbiology
Helmholtz Centre for Environmental Research – UFZ
Leipzig, Sachsen, Germany

ISSN 1862-5576 ISSN 1862-5584 (electronic)
Microbiology Monographs
ISBN 978-3-030-47905-3 ISBN 978-3-030-47906-0 (eBook)
https://doi.org/10.1007/978-3-030-47906-0

This Springer imprint is published by the registered company Springer Nature Switzerland AG.
The registered company address is: Gewerbestrasse 11, 6330 Cham, Switzerland

Preface

Laccases (EC 1.10.3.2; p-diphenol:dioxygen oxidoreductase) are ubiquitous multicopper oxidases occurring in fungi, bacteria, plants, and insects. The variable catalytic properties of the numerous and diverse members of this enzyme class share the common characteristic of generating organic radical species as primary oxidation products, which can give rise to both degradative and synthetic processes. In the different groups of their producing organisms, laccases are implicated in various biological functions such as lignin decomposition, detoxification of hazardous compounds, morphogenesis, microbial interactions, pathogenesis, and lignin biosynthesis. The considerable substrate promiscuity of laccases has resulted in a steadily increasing range of different potential biotechnological applications. Among these are bioremediation, especially with respect to treatment, detoxification, and decolourization of various types of wastewaters and process effluents, and waste valorization approaches. Here, substantial research and technology development efforts have been made during the last few decades.

This *Microbiology Monographs* volume aims to present the current knowledge with regard to fundamental characteristics of laccases and most recent advances in the application of these enzymes in the fields of bioremediation and waste valorization. The first three chapters (Mäkelä et al.: Fungal laccases and their potential in bioremediation applications; Martins et al.: Bacterial laccases: some recent advances and applications; and Giacobbe et al.: Old enzymes at the forefront of lignocellulosic waste valorisation) provide general and comprehensive introductions of fungal and bacterial laccases (i.e., the two most important enzyme groups from the application viewpoint) and their practical applicability for bioremediation and lignocellulosic waste valorization. The subsequent chapters deal with possible combinations of laccases and further potentially cooperating enzymes (Haghbeen and Schlosser: Laccases in the context of potentially cooperating enzymes), in-depth insights into laccase immobilization for wastewater treatment (Vaithyanathan et al.: Immobilized laccase—a promising bioremediation tool for the removal of organic contaminants in wastewater; and Ariste and Cabana: Challenges in applying cross-linked laccase aggregates in bioremediation of emerging contaminants from municipal wastewater), and environmental biosensor applications of laccases (Parra-Arroyo et al.: Laccase-assisted cues: state-of-the-art analytical modalities for detection, quantification, and redefining "removal" of environmentally related contaminants of high concern). Finally, the quest for enzymes with improved and better fitting properties is addressed in two chapters dealing with laccase engineering by directed and computational evolution (Mateljak et al.: Laccase engineering by directed and computational evolution) and novel enzymes from extreme environments (Pérez-Llano et al.: Laccases from extremophiles), respectively.

We would like to express our gratitude to all authors for their substantial efforts and their enthusiastic engagement in preparing the individual chapters. Also, I (Dietmar Schlosser) am grateful to Alexander Steinbüchel (the Series Editor of *Microbiology Monographs*) and Springer Nature for giving me the opportunity to edit this book. The help and advice of Bibhuti Bhusan Sharma, the Production Editor in charge of the book project, and of the Associate Editor Markus Späth during all phases of the writing and publishing process is further gratefully acknowledged. We hope very much that this *Microbiology Monographs* volume, which tackles the fascinating field of laccases, will be appreciated by the readers.

Leipzig, Germany, Dietmar Schlosser
Münster, Germany Alexander Steinbüchel
April 2020

Contents

Fungal Laccases and Their Potential in Bioremediation Applications

Miia R. Mäkelä, Marja Tuomela, Annele Hatakka, and Kristiina Hildén

Contents

Abstract Basidiomycete white rot and litter-decomposing fungi secrete unspecific oxidative enzymes for degradation of aromatic polymer lignin. Recalcitrant organopollutants with structural similarities to lignin can also be degraded by these enzymes, i.e., laccases and class II heme peroxidases. Laccases belong to a superfamily of multicopper oxidases, which have been characterized in basidiomycete and ascomycete fungal species. Laccases catalyze one-electron transfer reactions from phenolic and low-redox-potential compounds with the concomitant reduction of molecular oxygen to water. In the presence of small molecular weight mediator compounds, the substrate spectrum of laccases expands to non-phenolic molecules and larger organic polymers. Fungal laccases have broad substrate range, and therefore they have attracted attention as "green catalysts" in different areas of biotechnology including bioremediation. Applications for fungal laccases are found in wastewater treatment, detoxification, and decolorization of industrial effluents as well as in bioremediation of contaminated soils. In this chapter, we describe the properties of fungal laccases, their reactions with mediator compounds, and their recombinant production in different host organisms. We also discuss the potential

M. R. Mäkelä · M. Tuomela · A. Hatakka · K. Hildén (✉)
Department of Microbiology, University of Helsinki, Helsinki, Finland
e-mail: kristiina.s.hilden@helsinki.fi

D. Schlosser (ed.), *Laccases in Bioremediation and Waste Valorisation*,
Microbiology Monographs 33, https://doi.org/10.1007/978-3-030-47906-0_1

and challenges as well as the recent trends of the use of fungal laccases in bioremediation of wastewaters and soils contaminated with various xenobiotics.

1 Introduction

Bioremediation has been acknowledged to be environmentally friendly and an economically viable way to treat toxic and recalcitrant, usually xenobiotic compounds. The conversion of different pollutants in soil, compost, or effluents by microbial activities has been recognized. The characteristics of bacterial species, such as their ability to grow in various environments and to use several types of compounds as nutrient sources, make them suitable for several applications in bioremediation. By contrast, fungi as eukaryotic organisms grow on milder conditions, such as in more narrow temperature range, and usually require aerobic conditions, although fungi inhabiting extreme environments also exist. However, many saprotrophic wood- and litter-decomposing fungi produce high redox potential, unspecific extracellular oxidative enzymes that are targeted to degrade the most recalcitrant natural polymer, aromatic lignin (Hatakka 2005). Therefore, interest in bioremediation by fungi has focused mainly on aromatic compounds, which are often large molecules impossible to be degraded intracellularly by bacteria (Tuomela and Hatakka 2011). These compounds remain totally or partially non-degraded, e.g., in activated sludge wastewater treatment plants, and thus end up in the water system. In soil, xenobiotics are often in aged form and typically bound to humic substances (Bollag and Myers 1992). It is proposed that fungi degrade these pollutants predominantly with their lignin-modifying enzymes, i.e., laccases and lignin-modifying peroxidases, and therefore bioremediation studies have focused on white rot and litter-decomposing fungi (Qayyum et al. 2009).

Besides cellulose and hemicellulose, lignin is one of the three main polymers in the cell walls of the lignified plants. Wood-degrading white rot fungi and soil litter-decomposing fungi have a key role in the carbon cycle on Earth, since they are the only organisms which are able to efficiently degrade and even mineralize lignin (Hatakka 2005). Lignin degradation by white rot fungi has been intensively studied for decades, mostly for biotechnical applications in the pulp and paper industry, in the treatment of effluents from pulp mills or textile industry, and in soil bioremediation.

Bioremediation of wastewaters has focused on textile dyes, personal care products and pharmaceuticals (PCPPs), and endocrine disrupters (endocrine-disrupting chemicals, EDCs), i.e., compounds that mimic natural hormones (Wesenberg et al. 2002; Solís et al. 2012; Auriol et al. 2007; Husain and Qayyum 2013). The polluting compounds in soil systems are polyaromatic hydrocarbons (PAHs), halogenated compounds (chloro- or bromophenols, polychlorinated biphenyls (PCBs), and polychlorinated dibenzo-*p*-dioxins and polychlorinated dibenzofurans (PCDD/F)), various agrochemicals used as pesticides or herbicides, and nitroaromatics, such as trinitrotoluene (TNT) used in explosives (Steffen and Tuomela 2011). These compounds originate typically from wood treatment, gas works, and military activities. For all these contaminant groups, degradation by white rot fungi has been demonstrated, and several studies with fungal oxidative enzymes have also been performed in vitro.

Enzymes as environmentally friendly industrial biocatalysts can contribute considerably to bioremediation processes. Due to their ability to unspecifically oxidize a broad range of substrate molecules by reducing molecular oxygen to water, fungal laccases are attractive enzymes for the treatment of water and soil systems contaminated with various organopollutants. In addition, the common ability of fungi from different ecophysiological groups to produce several laccase isoenzymes together with the availability of several commercially produced fungal laccases has allowed testing of these enzymes for degradation of xenobiotic compounds in different bioremediation systems (e.g., Asif et al. 2017; Bilal et al. 2017; Herkommerová et al. 2018; Barrios-Estrada et al. 2018; Becker et al. 2016; Singh et al. 2015). The increasing number of fungal whole genome sequences has provided a wealth of laccase candidates (de Vries et al. 2017; Floudas et al. 2012; Floudas et al. 2015; Riley et al. 2014). Furthermore, the recent advancements in genetic engineering and genome editing techniques will allow more economically feasible recombinant production of fungal laccases as well as improvement of their biochemical properties towards bioremediation applications (Kun et al. 2019).

2 Fungi

More than 90% of all wood-rotting basidiomycetes cause white rot type of wood decay (Eriksson et al. 1990; Hatakka and Hammel 2010). White rot fungi are a heterogeneous group of species, which are more commonly found on angiosperm than on gymnosperm tree species in nature (Eriksson et al. 1990). After white rot fungi have degraded lignin from wood cell walls, the wood is characteristically white or yellowish and has soft, fiber-like structure. Litter-decomposing basidiomycete fungi live in soil environments in the humus layer of forests and grasslands decaying, e.g., dead leaves and needles. Their decay process results in the formation of so-called white rot humus. Lignin modification system of litter-decomposing species has shown that some species are capable of lignin degradation, although to a lesser extent than wood-rotting fungi (Steffen et al. 2000). They also possess similar oxidative enzymes, i.e., laccases and class II heme peroxidases, as wood-rotting white rot species (Lundell et al. 2014). Since litter-decomposing fungi are able to colonize and dwell in soil and obviously successfully compete with other soil microbes, these fungi and their enzymes are considered to be ideal for treatment of contaminated soils.

Wood-rotting brown rot fungi are basidiomycetes that mainly decompose the cellulose and hemicellulose components of wood and modify lignin only to a limited extent (Eriksson et al. 1990). Brown-rotted wood is dark, shrinks, and typically breaks into brick-shaped or cubical fragments that easily break down into brown powder. Only a small proportion, roughly 7%, of the known wood-degrading basidiomycete species belongs to brown rot group, which occurs most frequently on gymnosperm wood (Eriksson et al. 1990). Phylogenetic analyses have suggested that most brown rot fungi evolved repeatedly from white rot species, most likely by the selective loss of some biodegradation mechanisms (Floudas et al. 2012). Brown rot fungi lack class II heme peroxidases, but laccase-encoding genes are present in almost all studied brown rot species.

Ascomycota is a large and diverse phylum of fungi, including yeasts and filamentous fungi, and they can be found in a wide range of biotopes. Saprobic ascomycetes, which are able to degrade plant biomass, are also the main recombinant enzyme-producing species in industry (Mäkelä et al. 2014). Ascomycetous soft rot fungi are mostly soil-inhabiting species with intermediate capabilities to degrade wood. They typically attack wood in wet environments resulting in brownish, soft, and sponge-like wood, which cracks when dry. Soft rot-causing fungi depolymerize cellulose and hemicelluloses rapidly and are able to convert lignin to some extent (Liers et al. 2006; Shary et al. 2007). Although these fungi do not produce heme containing lignin-modifying peroxidases, they possess genes encoding multicopper oxidases (MCOs) and laccases, the biological role of which is not clear (Levasseur et al. 2010). Laccase production has been also described in non-wood-degrading ascomycete species such as in thermophilic *Myceliophthora thermophila* and plant pathogenic *Melanocarpus albomyces* (Kiiskinen et al. 2002) and *Magnaporthe grisea* fungi (Iyer and Chattoo 2003).

3 Structural Characteristics and Reaction Mechanisms of Fungal Laccases

Laccases (EC 1.10.3.2. diphenol:oxygen oxidoreductase) are four-copper-containing metalloenzymes, which belong to Auxiliary Activities (AA) 1_1 family in the Carbohydrate-Active enZymes (CAZy) database (www.cazy.org). Most white rot fungi produce laccases (Baldrian 2006), but they are widely distributed in nature and also present in other fungi and higher plants, and laccase-like multicopper oxidases (MCOs) are found even in bacteria and archaea.

Laccases and MCOs display a highly conserved three-dimensional structure, which consists of four substrate binding loops and four highly conserved signature sequences with amino acids that act as copper ligands. Fungal MCO superfamily forms a phylogenetically divergent group of so-called blue oxidases, which can be phylogenetically divided into subfamilies including sensu stricto laccases, Fe^{2+}-oxidizing Fet3-type ferroxidases, pigment-synthesizing laccase-like enzymes, and putative ascorbate oxidases (Kües and Rühl 2011). Fungal laccases are thought to participate, e.g., in lignocellulose decay, morphogenesis, plant-pathogen interactions, and stress defense (Baldrian 2006; Thurston 1994). However, some white fungi lack genes encoding sensu stricto laccases.

Laccases have broad substrate specificity including phenols, polyphenols, aromatic amines, thiols, and heterocyclic compounds as well as inorganic ions (Solomon et al. 1996; Xu et al. 1996).

They catalyze four consecutive one-electron transfer reactions from various substrates to corresponding radicals utilizing molecular oxygen as the oxidant (Baldrian 2006; Thurston 1994). The electrons withdrawn from the reducing organic substrates are transferred to molecular oxygen with concomitant production of water according to the overall reaction:

$$4\ \text{benzenediol} + O_2 \leftrightarrow 4\ \text{benzosemiquinone} + 2\ H_2O$$

The crystal structures of the ascomycete laccases (f.ex. *M. albomyces*, *Thielavia arenaria*) and several basidiomycetes laccases (f.ex. *Coprinus cinerea*, *Trametes versicolor*, and *Lentinus tigrinus*) have been solved and summarized in Hakulinen and Rouvinen (2015). All structures exhibit a similar three cupredoxin domain architecture. The copper atoms located in two centers (T1, T2/T3) play a key role in the reduction of oxygen to water. The T1 site contains the mononuclear Cu1, while Cu2, Cu3a, and Cu3b are located in the T2/T3 trinuclear center (Morozova et al. 2007a, b; Solomon et al. 1996). The substrate binding site is located in a negatively charged cavity near to the T1 site (Fig. 1) (Hilden et al. 2013). Cu1 is the primary electron acceptor site in a laccase-catalyzed oxidation. Highly conserved amino acid residues (ten His and one Cys) coordinate the copper atoms, and the electrons from T1 Cu are sequentially transferred to the trinuclear cluster, where oxygen is reduced to water. In the resting form, laccase is oxidized having all four coppers in the 2+ state (Solomon et al. 1996). A substrate donates four electrons, which are transferred one at a time to trinuclear copper cluster resulting in reduced form of the enzyme. The fully reduced trinuclear center reacts with oxygen generating two oxygen-bound intermediates (Augustine et al. 2010). First, the two-electron reduced species, i.e., peroxide intermediate, is formed followed by the four-electron reduced species, i.e., native intermediate. The catalytic cycle is terminated when the native intermediate is decayed to the fully oxidized resting form.

More than 100 fungal laccases have been purified and characterized (www.brenda-enzymes.org). Typically, fungal laccases are globular extracellular enzymes (Baldrian 2006). They have acidic p*I*s and pH optimum between 2 and 7. Laccases from white rot fungi usually have molecular masses of 60–80 kDa, and they are glycosylated. In the case of extracellular laccases, glycosylation is needed for secretion as well as thermal stability, conformational stability, and protection from proteolysis (Morozova et al. 2007a, b).

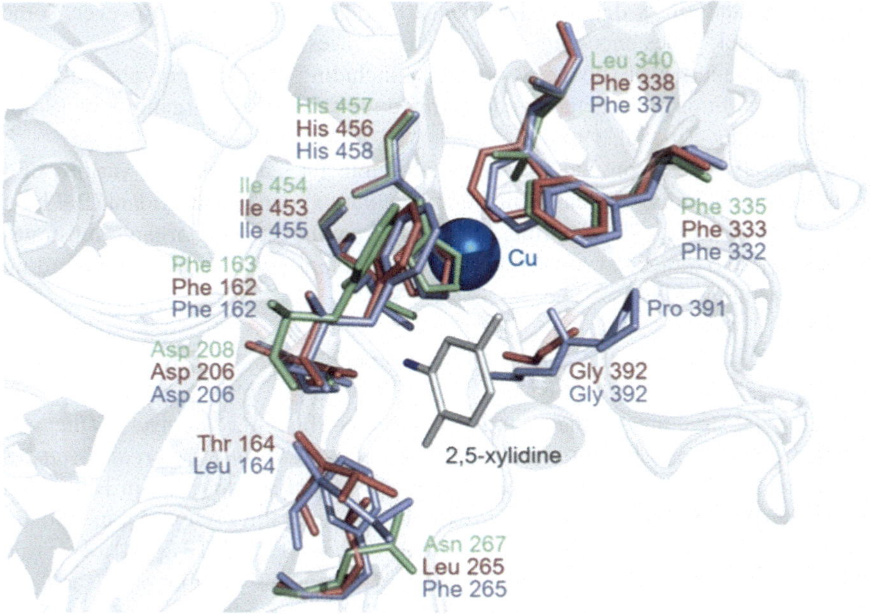

Fig. 1 Superimposition of amino acid residues close to the T1-Cu active site of the modeled structure of the white rot fungus *Obba rivulosa* (syn. *Physisporinus rivulosus*) Lcc1 (green), *O. rivulosa* Lcc2 (red), and crystal structure of *T. versicolor* laccase (LacIIIb, PDB accession 1KYA [purple]) with the co-crystallized substrate analog 2,5-xylidine. Reprinted by permission from: Springer Nature, Applied Microbiology and Biotechnology, Heterologous expression and structural characterization of two low pH laccases from a biopulping white-rot fungus *Physisporinus rivulosus*, Hilden K, Mäkelä MR, Lundell T, Kuuskeri J, Chernykh A, Golovleva L, Archer DB, Hatakka A, © 2013

4 Laccase Mediators

Laccases are able to directly oxidize substrates with relatively low redox potentials such as phenols (0.5–1.0 V) (Bourbonnais et al. 1998; Tadesse et al. 2008). Non-phenolic lignin units with high redox potential (over 1.5 V) are above the oxidation capacity of fungal laccases and therefore cannot be directly oxidized. However, in the presence of diffusible electron carriers, so-called laccase redox mediator compounds, oxidation of larger molecules and even delignification are possible by laccase (Fig. 2) (Call and Mücke 1997).

Redox mediators are small molecules, which can act as electron shuttles between the oxidant and the target substrate. Oxidized mediator released from the active site of laccase is able to perform oxidative reactions on substrates, which have too high redox potential for laccases (Canas and Camarero 2010). Thus, mediators can expand the substrate repertoire of laccases by oxidation of non-phenolic substrates as well. The ideal mediator is a good substrate for laccase and stable in its oxidized form. Preferably, the oxidation capacity is high without inhibition or degradation of

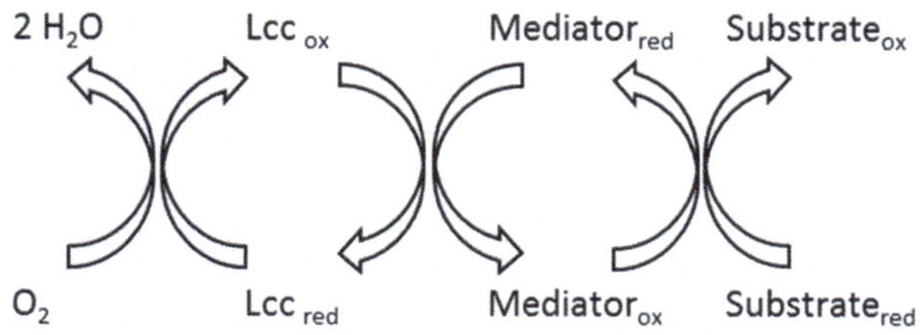

Fig. 2 Laccase-catalyzed oxidation of substrate in the presence of low molecular weight mediator compound. Reprinted from Biochimica et Biophysica Acta 1379 (3), Bourbonnais R, Leech D, Paice MG, Electrochemical analysis of the interactions of laccase mediators with lignin model compounds, 381–390, © 1998, with permission from Elsevier

laccase. In addition, the mediator should in ideal case perform many redox cycles without degradation, and it should not produce unwanted side reactions.

Redox mediators can be classified as natural and synthetic mediators. Although over 100 redox mediators have been reported, most of them have been proven inefficient due to the instability of their intermediates resulting in incomplete redox cycles and poor substrate oxidation (Bourbonnais et al. 1998; Díaz González et al. 2009; González Arzola et al. 2009). Some of these compounds act as co-oxidants or activators of laccase rather than mediators and are consumed in the reaction. The common synthetic mediators are, for example, 2,2′-azino-bis-3-ethylthiazoline-6-sulfonate (ABTS), 1-hydroxybenzotriazole (1-HBT), N-hydroxypthalimide (HPI), violuric acid (VA), and 2,2,6,6-tetra methyl-1-piperidinyloxy free radical (TEMPO) (Christopher et al. 2014). Despite the oxidative efficiency, they are needed in large amounts for industrial-scale oxidation processes that raises the costs. In addition, many mediators are possible environmental hazards themselves. For that reason identification of other alternatives, so-called natural mediators, has been in focus in order to substitute synthetic compounds (Cho et al. 2004; Camarero et al. 2005, 2007). Natural compounds from plants or industrial by-products have been searched to find an ideal redox mediator, and lignocellulose-derived syringyl-type plant phenolics seem to be promising substrates for laccases. They are potentially inexpensive and available, e.g., in industrial hardwood pulping waste (Camarero et al. 2005; Canas and Camarero 2010).

Synthetic laccase mediators are widely investigated for applications such as boosting bleaching of pulp in the pulp and paper industry, bleaching of textiles, and enhancing the efficiency in the degradation of environmental contaminants. In bleaching of paper pulp, the most frequently studied mediators are ABTS, HBT, and VA (Johannes and Majcherczyk 2000). For the degradation of contaminants such as PAHs, herbicides, or recalcitrant dyes, various natural mediators have been screened with laccase (Johannes and Majcherczyk 2000; Cañas et al. 2007; Maruyama et al. 2006; Camarero et al. 2005). The highest degradation was achieved with 4-hydroxybenzoic acid (HBA), which enhanced the degradation of PAH by laccase almost as much as HBT (Johannes and Majcherczyk 2000; Maruyama et al. 2006). However, the degradation degree of a urea herbicide dymron ((1-R′,Rdimethylbenzyl)-3-(4-methylphenyl) urea) was much lower with HBA compared to synthetic mediators (Maruyama et al. 2006). In screening of 44 natural mediators, none was as efficient as ABTS for decolorization of dyes by *Pycnoporus cinnabarinus* laccase (Camarero et al. 2005). Nevertheless, *p*-coumaric acid was able to enhance PAH degradation even more than ABTS (Cañas et al. 2007). Laccase mediators suggested for bioremediation are comprehensively listed in Majeau et al. (2010).

Mediators may also prevent polymerization of reactive products, as in addition to degradation laccases catalyze polymerization of contaminants (Leontievsky et al. 2001). In many cases, polymerization detoxifies the contaminants, but the handling of the formed polymers may cause an additional effort, especially if a large amount of polymeric material is produced.

So far, there has been limited utilization of laccase-mediator system in industrial applications (Christopher et al. 2014). The drawbacks of using laccase redox mediators are their high cost in the amounts required in large-scale oxidation processes or insufficient environmental safety (Morozova et al. 2007b). Also, the major problem in the use of synthetic mediators is the instability of intermediates formed during enzymatic reactions. This results in incomplete redox cycles or poor substrate oxidation.

5 Recombinant Production of Fungal Laccases

Laccases are excellent catalysts for industrial processes, since they are relatively robust and can be used in the presence of various organic solvents. Several reviews have summarized their role in biotechnological, chemical, and industrial applications that show the great interest in laccases (Ba and Vinoth Kumar 2017; Mayer and Staples 2002; Riva 2006; Rodriguez Couto and Toca Herrera 2006; Senthivelan et al. 2016). In industrial processes, large quantities of enzyme are required. Optimization of expression conditions and enzyme properties are therefore essential for increasing market demand. Improved characteristics of laccase, including increased activity or better stability, can be achieved by protein engineering, i.e., DNA "shuffling" methods, directed evolution, or site-directed or random mutagenesis (Bulter et al. 2003; Mate and Alcalde 2015; Mateljak et al. 2019).

Various fungal laccases have been produced recombinantly in eukaryotic hosts such as in fungi (*Pichia pastoris*, *Saccharomyces cerevisiae*, *Aspergillus niger*) and plants (de Wilde et al. 2008; Wang et al. 2004; Sonoki et al. 2005). However, the activity of fungal laccases produced in plants has been lower than in those produced in fungal hosts (Sakamoto et al. 2008; Nakagawa et al. 2010). Commercial laccase preparations originating from both ascomycete and basidiomycete species are available. As examples, commercially produced laccases from the basidiomycete white rot species are those from *Cerrena unicolor* (Jena Bioscience), *Agaricus bisporus*, and *T. versicolor* (ASA Spezialenzyme GmbH and Sigma-Aldrich). The laccase from the thermophilic ascomycete *M. thermophila* has been commercialized by

Novozymes (Mate and Alcalde 2017), and the ascomycete *Aspergillus* sp. laccase is sold by Sigma-Aldrich.

P. pastoris and *S. cerevisiae* are the most commonly used yeast hosts for heterologous expression of laccases due to the various production strains and vectors available for these species. Other yeast species such as *Kluyveromyces lactis*, *Pichia methanolica*, and *Yarrowia lipolytica* have been also employed as hosts for laccase production (Table 1). *K. lactis*, *S. cerevisiae*, and *Y. lipolytica* are generally regarded as safe (GRAS) organisms, which are beneficial when producing proteins for food or medical applications. High activity levels, up to 7200 U/l, have been achieved with laccase from the ascomycete species *Cyathus bulleri* produced in *P. pastoris* (Garg et al. 2012). The protein yield of most laccases is below 50 mg/l when produced in yeast hosts. The highest yields have been reported for laccases from the plant pathogen species *Botrytis aclada* and *Moniliophthora roreri* produced in *P. pastoris* (Kittl et al. 2012).

A. niger is a well-known filamentous ascomycete host that produces high levels of recombinant enzymes for many industrial applications (Cairns et al. 2018). While overall expression of basidiomycete laccases in *Aspergillus* species is successful, the production levels may vary significantly (Table 1). The production of laccases has reached the activity levels in the range of 592–774,000 U/l (Téllez-Jurado et al. 2006; Hatamoto et al. 1999). The potential of *A. niger* as a host for industrial enzyme production has been shown by expression of *T. versicolor* laccases ending to much higher yield compared to expression of these genes in *P. pastoris* (Bohlin et al. 2006). Together with *Aspergillus* species, *T. reesei* is the most commonly used filamentous fungus for industrial production of heterologous and homologous proteins. The production levels reported for *M. albomyces* laccase expression in *T. reesei*, from 200 to 920 mg/l, are among the highest recombinant laccase yields (Becker et al. 2016; Kiiskinen et al. 2004).

6 Wastewater Treatment

Numerous studies have focused on bioremediation of wastewaters by fungal laccases. Particularly, degradation and conversion of industrially used dye compounds, personal care products and pharmaceuticals (PCPPs), and endocrine-disrupting chemicals (EDCs) have gained significant interest (Wesenberg et al. 2002; Solís et al. 2012; Auriol et al. 2007; Husain and Qayyum 2013) due to their possible harmful impacts on, e.g., aquatic ecosystems and public health (Asif et al. 2017).

In general, purified fungal laccases have been shown to more effectively remove phenolic than non-phenolic compounds from aqueous systems (Asif et al. 2017). However, because of the different physicochemical properties of the non-phenolic compounds, laccases can degrade notably high levels of, for example, ibuprofen, a common anti-inflammatory drug with non-phenolic benzene ring structure. It should be noted that better degradation of the pollutants has often been achieved with fungal

Table 1 Recombinant fungal laccases and their production yields in yeasts and filamentous fungi

Laccase (GenBank accession no)	Host	Source of the native protein	Production level (mg/l)	References
AJ571698	*Saccharomyces cerevisiae*	*Melanocarpus albomyces*	3–7	Andberg et al. (2009)
AY397783		*Trametes* sp. C30	2	Klonowska et al. (2005)
JN559771	*Pichia pastoris*	*Botrytis aclada*	495	Kittl et al. (2012)
FJ473385		*Ganoderma lucidum*	6	You et al. (2014)
KY111767		*Moniliophthora roreri*	1000	Bronikowski et al. (2017)
AF297228		*Pleurotus sajor-caju*	4.85	Soden et al. (2002)
AF170093		*Pycnoporus cinnabarinus*	8	Otterbein et al. (2000)
AJ294820, AM292415		*Trametes trogii*	17–286	Colao et al. (2006, 2009)
X84683, Y18012		*Trametes versicolor*	4.9–100	Bohlin et al. (2006), Huang et al. (2016)
AF388910		*Trametes* sp. AH28-2	4	Hong et al. (2006)
JF719545		*Y. lipolytica*	4.9	Kalyani et al. (2015)
AF025481	*Kluyveromyces lactis*	*P. cinnabarinus*	5	Lambertz et al. (2016)
AJ005018, Z34848		*Pleurotus ostreatus*	1.6–1.9	Piscitelli et al. (2005)
AF170093	*Yarrowia lipolytica*	*P. cinnabarinus*	20	Madzak et al. (2005)
AF414109		*T. versicolor*	2.5	Jolivalt et al. (2005)
AF170093	*Aspergillus niger*	*P. cinnabarinus*	70	Record et al. (2002)
EF446161		*Phanerochaete flavido-alba*	30	Benghazi et al. (2014)
CAA06291, AJ005018		*P. ostreatus*	13	Macellaro et al. (2014)
AY397783		*Trametes* sp. C30	840	Mekmouche et al. (2014)
AAD30964	*Aspergillus oryzae*	*Coprinopsis cinerea*	135	Yaver et al. (1999)
T10922		*Myceliophthora thermophila*	19	Berka et al. (1997)
AF170093		*P. cinnabarinus*	80	Sigoillot et al. (2004)
CAE00180	*Trichoderma reesei*	*M. albomyces*	200–920	Kiiskinen et al. (2004), Andberg et al. (2009)
CAA36379		*Phlebia radiata*	19.5	Saloheimo and Niku-Paavola (1991)
U44851, U44431		*T. versicolor*	800–1000	Baker and White (2001)

crude enzyme preparations (Nguyen et al. 2014a, b, 2015; Lloret et al. 2010), which may be due to the presence of natural mediator compounds in these mixtures as well as the combinatorial effect of several fungal oxidative enzymes. In addition, the degradation of non-phenolic compounds and the repertoire of compounds affected by fungal laccases can be improved by including laccase mediator compounds to the reaction system (Nguyen et al. 2016; Ashe et al. 2016; Margot et al. 2015). Typically, the major obstacle in using the laccase mediators is their high production cost, thus often making the enzymatic treatments economically unfeasible.

Due to their unspecific catalytic action, laccases have high potential for detoxification and decolorization of industrial effluents contaminated with synthetic dyes of different chemical structures. These complex aromatic compounds originating from textile, food, pharmaceutical, printing, paper, and chemical industries include compounds such as azo dyes, triphenylmethane, indigo, and anthraquinone dyes (Legerská et al. 2016). Only industrial textile dyeing processes have been approximated to discharge 280,000 tons of xenobiotic dye compounds as wastewater annually (Jin et al. 2007). In addition to their toxicity and potentially carcinogenic or mutagenic effects, these compounds can cause increased biochemical and chemical oxygen demand levels in wastewaters.

High-redox-potential fungal laccases are especially suitable for degradation of recalcitrant azo dyes that represent up to 70% of industrially used dyes (Wang et al. 2013). Notably, satisfactory degradation of these dyes can be achieved even without addition of laccase mediator compounds. For example, purified *T. versicolor* laccase has been reported to efficiently decolorize azo dyes amaranth and Remazol brilliant blue R without mediators (Champagne and Ramsay 2005). Complete decolorization of azo and also anthraquinone dyes by crude and purified *C. unicolor* laccase has been observed without a mediator when sufficient amount of enzyme was used and when incubation time was extended (Michniewicz et al. 2008). The litter-decomposing fungus *Podoscypha elegans* was shown to decolorize five azo dyes commonly used in textile industry by producing remarkable laccase and low manganese peroxidase activity under the studied conditions (Pramanik and Chaudhuri 2018). However, aromatic amines that are reduced intermediates of azo dyes are more toxic than the original dye compounds (Gomi et al. 2011). Therefore, in addition to the degree of decolorization, it is crucial to determine the nature and toxicity of the resulting compounds.

Immobilization often improves the reusability and biochemical properties of fungal laccases (Fernández-Fernández et al. 2013; Bilal et al. 2017). Immobilization of purified *T. hirsuta* laccase has been shown to enhance the efficiency of the degradation of triarylmethane, indigo, azo, and anthraquinone dyes by increasing thermal stability and inhibitor tolerance of the enzyme (Abadulla et al. 2000). Approximately 90% of the azo dye amaranth was decolorized with immobilized laccase from the ascomycete fungus *C. bulleri* in the presence of laccase mediator (Chhabra et al. 2015). Use of nanotechnology-based support materials for laccase immobilization offers several advantages: degradation of contaminant is even faster and more efficient than with free laccase or other laccase immobilization materials (Dai et al. 2011; Ji et al. 2016; Koyani and Vazquez-Duhalt 2016). Also, recombinant laccases have been tested for decolorization and detoxification of dyes and textile industry wastewaters. The laccases from the ascomycete *M. thermophila* and the basidiomycete *T. trogii*, produced in *S. cerevisiae*, were able to decolorize triphenylmethane, azo, and anthraquinone dyes as well as textile industry wastewater samples (Herkommerová et al. 2018). In most cases, the decolorization rates were increased when HBT mediator was added to the reaction mixtures. It is noteworthy that some dye compounds such as anthraquinone can act as laccase mediator for degradation of azo dyes (Zeng et al. 2012; Senthivelan et al. 2016).

During the last decades, there has been increasing concern about EDC molecules that disturb the endocrine system of different organisms. These include synthetic compounds used in plastic industry, such as bisphenols and phthalates, antibacterial agent triclosan (TCS), and nonionic surfactant biodegradation metabolite nonylphenol (NP) among many others. EDCs have been detected in treated wastewater effluents, in groundwater, as well as in drinking water (Snyder 2008), and extensive research has been concentrated on their removal by fungal enzymes (reviewed by Barrios-Estrada et al. 2018).

Both native and recombinant fungal laccases have been shown to be potent candidates for removal of EDCs with and without mediators as well as in immobilized systems. Degradation of bisphenol A (2,2-bis(4-hydroxyphenyl)propane; BPA) has been demonstrated with several white rot species from the genus *Trametes* (e.g., Ji et al. 2009; Fukuda et al. 2001, 2004; Spina et al. 2015; Hongyan et al. 2019), which are generally known as good laccase producers. The litter-decomposing fungi *Stropharia coronilla* and *Stropharia rugosoannulata* also produced laccases during conversion of BPA (Kabiersch et al. 2011). As an example of ascomycete laccases, the commercial *Aspergillus oryzae* laccase with syringaldehyde as a mediator removed over 95% of BPA when continuous enzymatic membrane reactor was applied (Nguyen et al. 2014a). Also laccases from the white rot fungus *Pycnoporus sanguineus* are efficient degraders of common EDCs NP and TCS as shown by the two purified thermostable laccases as well as the crude culture supernatant from this fungus (Ramírez-Cavazos et al. 2014). Often, the fungal laccases that are effective against BPA also remove NP and TCS (Cabana et al. 2007; Spina et al. 2015; Dai et al. 2015). Even a field-scale experiment has been performed with a commercial laccase from the ascomycete fungus *Thielavia* sp. (AB Enzymes, Germany), in which biologically purified domestic and industrial wastewater containing BPA was further treated with laccase and 66% removal of BPA was achieved (Gasser et al. 2014).

One of the current global challenges is prevention of the spreading of the antimicrobial resistance. Therefore, efficient removal and degradation of antibiotics from wastewaters, ranging from municipal wastewaters to those from hospitals and pharmaceutical industry, is crucial. In recent years, there has been a rapid increase in utilization of laccases for removal of antibiotics. The target compounds in these studies include penicillins, tetracyclines, sulfonamides, quinolones, trimethoprim, sulfamethoxazole, and tetracycline (reviewed by Yang et al. 2017). Several studies have been conducted with commercially available *T. versicolor* laccase (Sigma-Aldrich, Fluka), and in general, mediator compounds have been observed to enhance the degradation of the antibiotics (Becker et al. 2016; Kumar and Cabana 2016; Margot et al. 2015; Rodriguez-Rodriguez et al. 2012; Llorca et al. 2015; de Cazes et al. 2014; Ashrafi et al. 2016; Ding et al. 2016). Also the commercial Novozym 51003 laccase from the ascomycete fungus *M. thermophila* has been tested for the degradation of sulfamethoxazole when added to municipal wastewater sample in environmentally relevant concentration (Nguyen et al. 2014b). *T. versicolor* laccase, immobilized to enzymatic membrane reactor, with syringaldehyde as a mediator was reported to degrade over 50% of 32 out of 38 tested antibiotics (Becker et al. 2016). Despite of the effectiveness of this laccase system towards a wide range of antibiotics, a time-dependent increase of toxicity was observed, indicating that improvement of the system to further degradation of the transformation products is needed. Similarly, increased toxicity of the effluent was detected when syringaldehyde was added as a mediator for *M. thermophila* laccase reaction for the degradation of sulfamethoxazole, but granular activated carbon was reported to reduce the toxicity (Nguyen et al. 2014b).

7 Bioremediation of Soil

Biological treatment of contaminated soils is often more effective compared to chemical and physical remediation technologies (Sharma et al. 2018; Chen et al. 2015). Fungal laccases have been shown to be able to remove a wide range of organopollutants also in soil systems (Durán and Esposito 2000; Field et al. 1993). Typical organopollutants in soils are polyaromatic hydrocarbons (PAHs), halogenated compounds (chloro- or bromophenols, polychlorinated biphenyls (PCBs), polychlorinated dibenzo-*p*-dioxins, polychlorinated dibenzofurans (PCDD/Fs)), various agrochemicals used as pesticides or herbicides, and nitroaromatics, such as 2,4,6-trinitrotoluene (TNT) used in explosives, but also antibiotics and other PCPPs as well as EDCs are present in soil environments (Yang et al. 2017; Tanaka et al. 2001).

Commercial *T. versicolor* laccase was tested for remediation of PAH-contaminated shoal soil with three types of laccase-carrying electrospun fibrous membranes fabricated via emulsion electrospinning (Dai et al. 2011). This system showed highly efficient removal of phenanthrene, fluoranthene, benz[a]anthracene, and benzo[a]pyrene from soil in the aqueous solution with greatly shorter half-lives compared to the treatments with free laccase or membrane adsorption. Soil oxygen concentration and pH were shown to have significant influence on degradation of the insecticide dichlorodiphenyltrichloroethane (DDT) when a purified laccase from the white rot fungus *Panus conchatus* was tested, as increased oxygen concentration and pH range 2.5–4.5 in soil enhanced the effect of the laccase treatment (Zhao and Yi 2010). Co-remediation of DDT-contaminated soil with the white rot fungus *Flammulina velutipes* and its purified laccase was shown to be more rapid and efficient compared to the use of only the fungus or laccase enzyme (Fan et al. 2013). In addition, no significant difference was observed between sterilized and non-sterilized soils, suggesting that the indigenous microbes of the soil were not able to remove DDT. *T. versicolor* was shown to be a good candidate for the degradation of the chlorinated herbicide atrazine in non-sterile calcareous clay soil with low moisture and organic matter content (Bastos and Magan 2009).

Laccases can also couple soil pollutants, such as chlorinated phenols and anilines, to soil organic matter, thus lowering the biological availability and toxicity of the xenobiotics (Ahn et al. 2002). Free and immobilized *T. villosa* laccase was shown to efficiently transform 2,4-dichlorophenol (2,4-DCP) and bind it to soil organic matter when test soils were amended with this compound (Ahn et al. 2002). Laccases from *T. versicolor* and *Rhizoctonia praticola* have also been shown to couple various chlorinated phenols and aromatic amines to phenolic humic compounds (Bollag 1992). Experiments with the whole fungus *T. versicolor* in the degradation of ^{14}C-labelled pentachlorophenol (PCP) in soil indicated that part of the ^{14}C-label was transiently bound to soil organic matter, but this part was apparently later attacked and mineralized by the fungus (Tuomela et al. 1998). Commercial *T. versicolor* laccase showed potential in treatment of sandy loam soil contaminated with sulfadimethoxine, a commonly used antibiotic in animal feeding that is released into agricultural lands by manure application (Singh et al. 2015). Significant reduction in the extractable fraction of sulfadimethoxine was detected when laccase and laccase-mediator systems were applied, indicating that the contaminant was bound to the soil. Furthermore, enhanced transformation was observed when the organic matter content in soil was increased by addition of peat. *T. versicolor* laccase was also tested for simultaneous removal of 14 different antibiotics (Ding et al. 2016). While sulfonamides and tetracyclines were mainly removed by laccase oxidation, laccase-mediated adsorption of quinolones to soil was reported.

Treatment of contaminated soil by fungi and their enzymes is more challenging than that of the aqueous environments and liquid systems. This is because enzymes often bind to soil organic matter or clay making their extraction and activity measurements difficult or even impossible, thus greatly hampering the evaluation of the enzyme performance. Most of the bioremediation studies have been performed on artificially contaminated soils that are spiked with organic pollutants (Pointing 2001; Šašek 2003). However, use of fungal treatment under non-sterile and in situ conditions is important to make the experiments potentially transferable to a field scale (Šašek 2003). Inoculation of laccase-producing fungi to the contaminated soil seems to be the most feasible bioremediation method, because the use of isolated or heterologously produced enzymes is not economically feasible in large scale (Viswanath et al. 2014). For this, pine bark has been shown to be a promising co-substrate for fungal propagation into soil (Valentín et al. 2009, 2010). As an example of successful field-scale experiment, bioremediation of soil contaminated with chlorophenols, creosote, TNT, petroleum hydrocarbons, organochlorine compound lindane, or even polychlorinated dioxins has been performed with ligninolytic fungi (Steffen and Tuomela 2011).

8 Conclusions

Fungal laccases are promising biocatalysts in various biotechnology fields, and their suitability for bioremediation applications have been studied already for 20 years. This has led to several field- or pilot-scale treatments with laccase-producing fungi or their enzymes. Although bioremediation of soil has been proven challenging due to, e.g., heterogeneity of soil matrix, fungal laccases have shown potential in removing a wide range of organopollutants also in soil systems. Due to the less complicated liquid matrix, treatment of wastewaters has been demonstrated to be more feasible approach and seem to be the closest to a commercial application by fungi and their laccases. Especially the development of methods for decolorization of textile dyes and removal of EDCs and PCPPs by tertiary treatment has received wide attention. In addition, an increased amount of research has been dedicated to removal of antibiotics from wastewaters by fungal laccases in order to prevent the spreading of the antimicrobial resistance, which is one of the major global concerns at the moment. High production costs of recombinantly produced laccases often prevent their use in large-scale bioremediation applications. However, the recent developments in laccase immobilization studies by using nanoparticles have offered promising results with, e.g., improved operational lifetime and stabilization of laccases, thus possibly enabling the process scale use of recombinant fungal laccases for bioremediation.

In conclusion, the application potential of fungal laccases in environmental field is considerable. This is further emphasized by recent trends of so-called green remediation and in situ treatment technologies in environmental field, which may give an advantage to biological methods, particularly fungal technologies.

Acknowledgments The Academy of Finland grants no. 308284 and no. 297847 to MRM and KH are acknowledged, respectively. The European Union's Horizon 2020 research and innovation program under grant agreement No 720918 is also acknowledged.

References

Abadulla E, Tzanov T, Costa S, Robra K-H, Cavaco-Paulo A, Gübitz GM (2000) Decolorization and detoxification of textile dyes with a laccase from *Trametes hirsuta*. Appl Environ Microbiol 66(8):3357. https://doi.org/10.1128/AEM.66.8.3357-3362.2000

Ahn MY, Dec J, Kim JE, Bollag JM (2002) Treatment of 2,4-dichlorophenol polluted soil with free and immobilized laccase. J Environ Qual 31(5):1509–1515. https://doi.org/10.2134/jeq2002.1509

Andberg M, Hakulinen N, Auer S, Saloheimo M, Koivula A, Rouvinen J, Kruus K (2009) Essential role of the C-terminus in *Melanocarpus albomyces* laccase for enzyme production, catalytic properties and structure. FEBS J 276(21):6285–6300. https://doi.org/10.1111/j.1742-4658.2009.07336.x

Ashe B, Nguyen LN, Hai FI, Lee D-J, van de Merwe JP, Leusch FDL, Price WE, Nghiem LD (2016) Impacts of redox-mediator type on trace organic contaminants degradation by laccase:

degradation efficiency, laccase stability and effluent toxicity. Int Biodeter Biodegr 113:169–176. https://doi.org/10.1016/j.ibiod.2016.04.027

Ashrafi SD, Nasseri S, Alimohammadi M, Mahvi AH, Faramarzi MA (2016) Optimization of the enzymatic elimination of flumequine by laccase-mediated system using response surface methodology. Desalin Water Treat 57(31):14478–14487. https://doi.org/10.1080/19443994.2015.1063462

Asif MB, Hai FI, Singh L, Price WE, Nghiem LD (2017) Degradation of pharmaceuticals and personal care products by white-rot fungi—a critical review. Curr Pollut Rep 3(2):88–103. https://doi.org/10.1007/s40726-017-0049-5

Augustine AJ, Kjaergaard C, Qayyum M, Ziegler L, Kosman DJ, Hodgson KO, Hedman B, Solomon EI (2010) Systematic perturbation of the trinuclear copper cluster in the multicopper oxidases: the role of active site asymmetry in its reduction of O_2 to H_2O. J Am Chem Soc 132 (17):6057–6067. https://doi.org/10.1021/ja909143d

Auriol M, Filali-Meknassi Y, Tyagi RD, Adams CD (2007) Laccase-catalyzed conversion of natural and synthetic hormones from a municipal wastewater. Water Res 41(15):3281–3288. https://doi.org/10.1016/j.watres.2007.05.008

Ba S, Vinoth Kumar V (2017) Recent developments in the use of tyrosinase and laccase in environmental applications. Crit Rev Biotechnol 37(7):819–832. https://doi.org/10.1080/07388551.2016.1261081

Baker CJO, White TC (2001) Expression of laccase I and IV genes from *Trametes versicolor* in *Trichoderma reesei*. ACS Symp Ser Am Chem Soc 785. https://doi.org/10.1021/bk-2001-0785.ch026

Baldrian P (2006) Fungal laccases – occurrence and properties. FEMS Microbiol Rev 30:215–242. https://doi.org/10.1111/j.1574-4976.2005.00010.x

Barrios-Estrada C, de Jesus R-AM, Munoz-Gutierrez BD, Iqbal HMN, Kannan S, Parra-Saldivar R (2018) Emergent contaminants: endocrine disruptors and their laccase-assisted degradation – a review. Sci Total Environ 612:1516–1531. https://doi.org/10.1016/j.scitotenv.2017.09.013

Bastos AC, Magan N (2009) *Trametes versicolor*: potential for atrazine bioremediation in calcareous clay soil, under low water availability conditions. Int Biodeter Biodegr 63(4):389–394. https://doi.org/10.1016/j.ibiod.2008.09.010

Becker D, Varela Della Giustina S, Rodriguez-Mozaz S, Schoevaart R, Barcelo D, de Cazes M, Belleville MP, Sanchez-Marcano J, de Gunzburg J, Couillerot O, Volker J, Oehlmann J, Wagner M (2016) Removal of antibiotics in wastewater by enzymatic treatment with fungal laccase – degradation of compounds does not always eliminate toxicity. Bioresour Technol 219:500–509. https://doi.org/10.1016/j.biortech.2016.08.004

Benghazi L, Record E, Suarez A, Gomez-Vidal JA, Martinez J, de la Rubia T (2014) Production of the *Phanerochaete flavido-alba* laccase in *Aspergillus niger* for synthetic dyes decolorization and biotransformation. World J Microbiol Biotechnol 30(1):201–211. https://doi.org/10.1007/s11274-013-1440-z

Berka RM, Schneider P, Golightly EJ, Brown SH, Madden M, Brown KM, Halkier T, Mondorf K, Xu F (1997) Characterization of the gene encoding an extracellular laccase of *Myceliophthora thermophila* and analysis of the recombinant enzyme expressed in *Aspergillus oryzae*. Appl Environ Microbiol 63(8):3151

Bilal M, Asgher M, Parra-Saldivar R, Hu H, Wang W, Zhang X, Iqbal HMN (2017) Immobilized ligninolytic enzymes: an innovative and environmental responsive technology to tackle dye-based industrial pollutants – a review. Sci Total Environ 576:646–659. https://doi.org/10.1016/j.scitotenv.2016.10.137

Bohlin C, Jönsson LJ, Roth R, van Zyl WH (2006) Heterologous expression of *Trametes versicolor* laccase in *Pichia pastoris* and *Aspergillus niger*. In: McMillan JD, Adney WS, Mielenz JR, Klasson KT (eds) Twenty-seventh symposium on biotechnology for fuels and chemicals. Humana Press, Totowa, NJ, pp 195–214. https://doi.org/10.1007/978-1-59745-268-7_15

Bollag JM (1992) Decontaminating soil with enzymes. Environ Sci Technol 26(10):1876–1881. https://doi.org/10.1021/es00034a002

Bollag J-M, Myers C (1992) Detoxification of aquatic and terrestrial sites through binding of pollutants to humic substances. Sci Total Environ 117–118:357–366. https://doi.org/10.1016/0048-9697(92)90102-X

Bourbonnais R, Leech D, Paice MG (1998) Electrochemical analysis of the interactions of laccase mediators with lignin model compounds. Biochim Biophys Acta 1379(3):381–390. https://doi.org/10.1016/S0304-4165(97)00117-7

Bronikowski A, Hagedoorn P-L, Koschorreck K, Urlacher VB (2017) Expression of a new laccase from *Moniliophthora roreri* at high levels in *Pichia pastoris* and its potential application in micropollutant degradation. AMB Express 7(1):73–73. https://doi.org/10.1186/s13568-017-0368-3

Bulter T, Alcalde M, Sieber V, Meinhold P, Schlachtbauer C, Arnold FH (2003) Functional expression of a fungal laccase in *Saccharomyces cerevisiae* by directed evolution. Appl Environ Microbiol 69(2):987. https://doi.org/10.1128/AEM.69.2.987-995.2003

Cabana H, Jiwan JL, Rozenberg R, Elisashvili V, Penninckx M, Agathos SN, Jones JP (2007) Elimination of endocrine disrupting chemicals nonylphenol and bisphenol A and personal care product ingredient triclosan using enzyme preparation from the white rot fungus *Coriolopsis polyzona*. Chemosphere 67(4):770–778. https://doi.org/10.1016/j.chemosphere.2006.10.037

Cairns TC, Nai C, Meyer V (2018) How a fungus shapes biotechnology: 100 years of *Aspergillus niger* research. Fungal Biol Biotechnol 5(1):13. https://doi.org/10.1186/s40694-018-0054-5

Call HP, Mücke I (1997) History, overview and applications of mediated ligninolytic systems, especially laccase-mediator-systems (Lignozym®-process). J Biotechnol 53:163–202. https://doi.org/10.1016/S0168-1656(97)01683-0

Camarero S, Ibarra D, Martínez MJ, Martínez ÁT (2005) Lignin-derived compounds as efficient laccase mediators for decolorization of different types of recalcitrant dyes. Appl Environ Microbiol 71(4):1775–1784. https://doi.org/10.1128/AEM.71.4.1775-1784.2005

Camarero S, Ibarra D, Martínez ÁT, Romero J, Gutiérrez A, del Río JC (2007) Paper pulp delignification using laccase and natural mediators. Enzyme Microb Technol 40(5):1264–1271. https://doi.org/10.1016/j.enzmictec.2006.09.016

Canas AI, Camarero S (2010) Laccases and their natural mediators: biotechnological tools for sustainable eco-friendly processes. Biotechnol Adv 28(6):694–705. https://doi.org/10.1016/j.biotechadv.2010.05.002

Cañas AI, Alcalde M, Plou F, Martínez MJ, Martínez ÁT, Camarero S (2007) Transformation of polycyclic aromatic hydrocarbons by laccase is strongly enhanced by phenolic compounds present in soil. Environ Sci Technol 41(8):2964–2971. https://doi.org/10.1021/es062328j

Champagne PP, Ramsay JA (2005) Contribution of manganese peroxidase and laccase to dye decoloration by *Trametes versicolor*. Appl Microbiol Biotechnol 69(3):276–285. https://doi.org/10.1007/s00253-005-1964-8

Chen M, Xu P, Zeng G, Yang C, Huang D, Zhang J (2015) Bioremediation of soils contaminated with polycyclic aromatic hydrocarbons, petroleum, pesticides, chlorophenols and heavy metals by composting: applications, microbes and future research needs. Biotechnol Adv 33(6 Pt 1):745–755. https://doi.org/10.1016/j.biotechadv.2015.05.003

Chhabra M, Mishra S, Sreekrishnan TR (2015) Immobilized laccase mediated dye decolorization and transformation pathway of azo dye acid red 27. J Environ Health Sci Eng 13:38. https://doi.org/10.1186/s40201-015-0192-0

Cho N-S, Shin W, Jeong S-W, Leonowicz A (2004) Degradation of lignosulfonate by fungal laccase with low molecular mediators. Bull Korean Chem Soc 25:1551–1554. https://doi.org/10.5012/bkcs.2004.25.10.1551

Christopher LP, Yao B, Ji Y (2014) Lignin biodegradation with laccase-mediator systems. Front Energy Res 2(12). https://doi.org/10.3389/fenrg.2014.00012

Colao MC, Lupino S, Garzillo AM, Buonocore V, Ruzzi M (2006) Heterologous expression of lcc1 gene from *Trametes trogii* in *Pichia pastoris* and characterization of the recombinant enzyme. Microb Cell Fact 5(1):31. https://doi.org/10.1186/1475-2859-5-31

Colao MC, Caporale C, Silvestri F, Ruzzi M, Buonocore V (2009) Modeling the 3-D structure of a recombinant laccase from *Trametes trogii* active at a pH close to neutrality. Protein J 28 (9–10):375–383. https://doi.org/10.1007/s10930-009-9204-1

Dai Y, Yin L, Niu J (2011) Laccase-carrying electrospun fibrous membranes for adsorption and degradation of PAHs in shoal soils. Env Sci Technol 45(24):10611–10618. https://doi.org/10.1021/es203286e

Dai Y, Song Y, Wang S, Yuan Y (2015) Treatment of halogenated phenolic compounds by sequential tri-metal reduction and laccase-catalytic oxidation. Water Res 71:64–73. https://doi.org/10.1016/j.watres.2014.12.047

de Cazes M, Belleville MP, Petit E, Llorca M, Rodríguez-Mozaz S, de Gunzburg J, Barceló D, Sanchez-Marcano J (2014) Design and optimization of an enzymatic membrane reactor for tetracycline degradation. Catal Today 236:146–152. https://doi.org/10.1016/j.cattod.2014.02.051

de Vries RP, Riley R, Wiebenga A, Aguilar-Osorio G, Amillis S, Uchima CA, Anderluh G, Asadollahi M, Askin M, Barry K, Battaglia E, Bayram Ö, Benocci T, Braus-Stromeyer SA, Caldana C, Cánovas D, Cerqueira GC, Chen F, Chen W, Choi C, Clum A, dos Santos RAC, Damásio ARdL, Diallinas G, Emri T, Fekete E, Flipphi M, Freyberg S, Gallo A, Gournas C, Habgood R, Hainaut M, Harispe ML, Henrissat B, Hildén KS, Hope R, Hossain A, Karabika E, Karaffa L, Karányi Z, Kraševec N, Kuo A, Kusch H, LaButti K, Lagendijk EL, Lapidus A, Levasseur A, Lindquist E, Lipzen A, Logrieco AF, MacCabe A, Mäkelä MR, Malavazi I, Melin P, Meyer V, Mielnichuk N, Miskei M, Molnár ÁP, Mulé G, Ngan CY, Orejas M, Orosz E, Ouedraogo JP, Overkamp KM, Park H-S, Perrone G, Piumi F, Punt PJ, Ram AFJ, Ramón A, Rauscher S, Record E, Riaño-Pachón DM, Robert V, Röhrig J, Ruller R, Salamov A, Salih NS, Samson RA, Sándor E, Sanguinetti M, Schütze T, Sepčić K, Shelest E, Sherlock G, Sophianopoulou V, Squina FM, Sun H, Susca A, Todd RB, Tsang A, Unkles SE, van de Wiele N, van Rossen-Uffink D, Oliveira JVdC, Vesth TC, Visser J, Yu J-H, Zhou M, Andersen MR, Archer DB, Baker SE, Benoit I, Brakhage AA, Braus GH, Fischer R, Frisvad JC, Goldman GH, Houbraken J, Oakley B, Pócsi I, Scazzocchio C, Seiboth B, vanKuyk PA, Wortman J, Dyer PS, Grigoriev IV (2017) Comparative genomics reveals high biological diversity and specific adaptations in the industrially and medically important fungal genus *Aspergillus*. Genome Biol 18(1):28. https://doi.org/10.1186/s13059-017-1151-0

de Wilde C, Uzan E, Zhou Z, Kruus K, Andberg M, Buchert J, Record E, Asther M, Lomascolo A (2008) Transgenic rice as a novel production system for *Melanocarpus* and *Pycnoporus* laccases. Transgenic Res 17(4):515–527. https://doi.org/10.1007/s11248-007-9124-9

Díaz González M, Vidal T, Tzanov T (2009) Electrochemical study of phenolic compounds as enhancers in laccase-catalyzed oxidative reactions. Electroanalysis 21(20):2249–2257. https://doi.org/10.1002/elan.200904678

Ding H, Wu Y, Zou B, Lou Q, Zhang W, Zhong J, Lu L, Dai G (2016) Simultaneous removal and degradation characteristics of sulfonamide, tetracycline, and quinolone antibiotics by laccase-mediated oxidation coupled with soil adsorption. J Hazard Mater 307:350–358. https://doi.org/10.1016/j.jhazmat.2015.12.062

Durán N, Esposito E (2000) Potential applications of oxidative enzymes and phenoloxidase-like compounds in wastewater and soil treatment: a review. Appl Catal B Environ 28(2):83–99. https://doi.org/10.1016/S0926-3373(00)00168-5

Eriksson K-EL, Blanchette RA, Ander P (1990) Microbial and enzymatic degradation of wood and wood components. Springer, Berlin

Fan B, Zhao Y, Mo G, Ma W, Wu J (2013) Co-remediation of DDT-contaminated soil using white rot fungi and laccase extract from white rot fungi. J Soil Sediment 13(7):1232–1245. https://doi.org/10.1007/s11368-013-0705-3

Fernández-Fernández M, Sanromán MÁ, Moldes D (2013) Recent developments and applications of immobilized laccase. Biotechnol Adv 31(8):1808–1825. https://doi.org/10.1016/j.biotechadv.2012.02.013

Field JA, de Jong E, Feijoo-Costa G, de Bont JAM (1993) Screening for ligninolytic fungi applicable to the biodegradation of xenobiotics. Trends Biotechnol 11(2):44–49. https://doi.org/10.1016/0167-7799(93)90121-O

Floudas D, Binder M, Riley R, Barry K, Blanchette RA, Henrissat B, Martínez AT, Otillar R, Spatafora JW, Yadav JS, Aerts A, Benoit I, Boyd A, Carlson A, Copeland A, Coutinho PM, de Vries RP, Ferreira P, Findley K, Foster B, Gaskell J, Glotzer D, Górecki P, Heitman J, Hesse C, Hori C, Igarashi K, Jurgens JA, Kallen N, Kersten P, Kohler A, Kües U, Kumar TKA, Kuo A, LaButti K, Larrondo LF, Lindquist E, Ling A, Lombard V, Lucas S, Lundell T, Martin R, McLaughlin DJ, Morgenstern I, Morin E, Murat C, Nagy LG, Nolan M, Ohm RA, Patyshakuliyeva A, Rokas A, Ruiz-Dueñas FJ, Sabat G, Salamov A, Samejima M, Schmutz J, Slot JC, St. John F, Stenlid J, Sun H, Sun S, Syed K, Tsang A, Wiebenga A, Young D, Pisabarro A, Eastwood DC, Martin F, Cullen D, Grigoriev IV, Hibbett DS (2012) The paleozoic origin of enzymatic lignin decomposition reconstructed from 31 fungal genomes. Science 336 (6089):1715–1719. https://doi.org/10.1126/science.1221748

Floudas D, Held BW, Riley R, Nagy LG, Koehler G, Ransdell AS, Younus H, Chow J, Chiniquy J, Lipzen A, Tritt A, Sun H, Haridas S, LaButti K, Ohm RA, Kües U, Blanchette RA, Grigoriev IV, Minto RE, Hibbett DS (2015) Evolution of novel wood decay mechanisms in Agaricales revealed by the genome sequences of *Fistulina hepatica* and *Cylindrobasidium torrendii*. Fungal Genet Biol 76:78–92. https://doi.org/10.1016/j.fgb.2015.02.002

Fukuda T, Uchida H, Takashima Y, Uwajima T, Kawabata T, Suzuki M (2001) Degradation of bisphenol A by purified laccase from *Trametes villosa*. Biochem Biophys Res Commun 284 (3):704–706. https://doi.org/10.1006/bbrc.2001.5021

Fukuda T, Uchida H, Suzuki M, Miyamoto H, Morinaga H, Nawata H, Uwajima T (2004) Transformation products of bisphenol A by a recombinant *Trametes villosa* laccase and their estrogenic activity. J Chem Technol Biotechnol 79(11):1212–1218. https://doi.org/10.1002/jctb.1115

Garg N, Bieler N, Kenzom T, Chhabra M, Ansorge-Schumacher M, Mishra S (2012) Cloning, sequence analysis, expression of *Cyathus bulleri* laccase in *Pichia pastoris* and characterization of recombinant laccase. BMC Biotechnol 12(1):75. https://doi.org/10.1186/1472-6750-12-75

Gasser CA, Yu L, Svojitka J, Wintgens T, Ammann EM, Shahgaldian P, Corvini PF, Hommes G (2014) Advanced enzymatic elimination of phenolic contaminants in wastewater: a nano approach at field scale. Appl Microbiol Biotechnol 98(7):3305–3316. https://doi.org/10.1007/s00253-013-5414-8

Gomi N, Yoshida S, Matsumoto K, Okudomi M, Konno H, Hisabori T, Sugano Y (2011) Degradation of the synthetic dye amaranth by the fungus *Bjerkandera adusta* Dec 1: inference of the degradation pathway from an analysis of decolorized products. Biodegradation 22 (6):1239–1245. https://doi.org/10.1007/s10532-011-9478-9

González Arzola K, Arévalo MC, Falcón MA (2009) Catalytic efficiency of natural and synthetic compounds used as laccase-mediators in oxidising veratryl alcohol and a kraft lignin, estimated by electrochemical analysis. Electrochim Acta 54(9):2621–2629. https://doi.org/10.1016/j.electacta.2008.10.059

Hakulinen N, Rouvinen J (2015) Three-dimensional structures of laccases. Cell Mol Life Sci 72 (5):857–868. https://doi.org/10.1007/s00018-014-1827-5

Hatakka A (2005) Biodegradation of lignin. In: Steinbuchel A (ed) Biopolymers online, vol 1. Weinheim, Wiley-VCH, pp 129–179. https://doi.org/10.1002/3527600035.bpol1005

Hatakka A, Hammel K (2010) Fungal biodegradation of lignocelluloses. In: Hofrichter M (ed) The Mycota X, industrial applications, vol 2. Springer, Berlin, pp 319–340. https://doi.org/10.1007/978-3-642-11458-8_15

Hatamoto O, Sekine H, Nakano E, Abe K (1999) Cloning and expression of a cDNA encoding the laccase from *Schizophyllum commune*. Biosci Biotechnol Biochem 63(1):58–64. https://doi.org/10.1271/bbb.63.58

Herkommerová K, Dostál J, Pichová I (2018) Decolorization and detoxification of textile wastewaters by recombinant *Myceliophthora thermophila* and *Trametes trogii* laccases. 3. Biotech 8 (12):505. https://doi.org/10.1007/s13205-018-1525-3

Hilden K, Mäkelä MR, Lundell T, Kuuskeri J, Chernykh A, Golovleva L, Archer DB, Hatakka A (2013) Heterologous expression and structural characterization of two low pH laccases from a biopulping white-rot fungus *Physisporinus rivulosus*. Appl Microbiol Biotechnol 97 (4):1589–1599. https://doi.org/10.1007/s00253-012-4011-6

Hong Y, Xiao Y, Zhou H, Fang W, Zhang M, Wang J, Wu L, Yu Z (2006) Expression of a laccase cDNA from *Trametes* sp. AH28-2 in *Pichia pastoris* and mutagenesis of transformants by nitrogen ion implantation. FEMS Microbiol Lett 258(1):96–101. https://doi.org/10.1111/j.1574-6968.2006.00209.x

Hongyan L, Zexiong Z, Shiwei X, He X, Yinian Z, Haiyun L, Zhongsheng Y (2019) Study on transformation and degradation of bisphenol A by *Trametes versicolor* laccase and simulation of molecular docking. Chemosphere 224:743–750. https://doi.org/10.1016/j.chemosphere.2019.02.143

Huang L, Liu Y, Liu X, Ban L, Wang Y, Li M, Lu F (2016) Functional expression of *Trametes versicolor* thermotolerant laccase variant in *Pichia pastoris*. Biotechnol Biotech Eq 30 (2):261–269. https://doi.org/10.1080/13102818.2015.1134278

Husain Q, Qayyum S (2013) Biological and enzymatic treatment of bisphenol A and other endocrine disrupting compounds: a review. Crit Rev Biotechnol 33(3):260–292. https://doi.org/10.3109/07388551.2012.694409

Iyer G, Chattoo BB (2003) Purification and characterization of laccase from the rice blast fungus, *Magnaporthe grisea*. FEMS Microbiol Lett 227(1):121–126. https://doi.org/10.1016/s0378-1097(03)00658-x

Ji G, Zhang H, Huang F, Huang X (2009) Effects of nonionic surfactant Triton X-100 on the laccase-catalyzed conversion of bisphenol A. J Environ Sci 21(11):1486–1490. https://doi.org/10.1016/S1001-0742(08)62444-4

Ji C, Hou J, Wang K, Zhang Y, Chen V (2016) Biocatalytic degradation of carbamazepine with immobilized laccase-mediator membrane hybrid reactor. J Membrane Sci 502:11–20. https://doi.org/10.1016/j.memsci.2015.12.043

Jin X-C, Liu G-Q, Xu Z-H, Tao W-Y (2007) Decolorization of a dye industry effluent by *Aspergillus fumigatus* XC6. Appl Microbiol Biotechnol 74(1):239–243. https://doi.org/10.1007/s00253-006-0658-1

Johannes C, Majcherczyk A (2000) Natural mediators in the oxidation of polycyclic aromatic hydrocarbons by laccase mediator systems. Appl Environ Microbiol 66(2):524–528. https://doi.org/10.1128/AEM.66.2.524-528.2000

Jolivalt C, Madzak C, Brault A, Caminade E, Malosse C, Mougin C (2005) Expression of laccase IIIb from the white-rot fungus *Trametes versicolor* in the yeast *Yarrowia lipolytica* for environmental applications. Appl Microbiol Biotechnol 66(4):450–456. https://doi.org/10.1007/s00253-004-1717-0

Kabiersch G, Rajasärkka J, Ullrich R, Tuomela M, Hofrichter M, Virta M, Hatakka A, Steffen K (2011) Fate of bisphenol A during treatment with the litter-decomposing fungi *Stropharia rugosoannulata* and *Stropharia coronilla*. Chemosphere 83(3):226–232. https://doi.org/10.1016/j.chemosphere.2010.12.094

Kalyani D, Tiwari MK, Li J, Kim SC, Kalia VC, Kang YC, Lee J-K (2015) A highly efficient recombinant laccase from the yeast *Yarrowia lipolytica* and its application in the hydrolysis of biomass. PLoS One 10(3):e0120156–e0120156. https://doi.org/10.1371/journal.pone.0120156

Kiiskinen LL, Viikari L, Kruus K (2002) Purification and characterisation of a novel laccase from the ascomycete *Melanocarpus albomyces*. Appl Microbiol Biotechnol 59(2–3):198–204. https://doi.org/10.1007/s00253-002-1012-x

Kiiskinen L-L, Rättö M, Kruus K (2004) Screening for novel laccase-producing microbes. J Appl Microbiol 97:640–646. https://doi.org/10.1111/j.1365-2672.2004.02348.x

Kittl R, Mueangtoom K, Gonaus C, Khazaneh ST, Sygmund C, Haltrich D, Ludwig R (2012) A chloride tolerant laccase from the plant pathogen ascomycete *Botrytis aclada* expressed at high levels in *Pichia pastoris*. J Biotechnol 157(2):304–314. https://doi.org/10.1016/j.jbiotec.2011.11.021

Klonowska A, Gaudin C, Asso M, Fournel A, Réglier M, Tron T (2005) LAC3, a new low redox potential laccase from *Trametes* sp. strain C30 obtained as a recombinant protein in yeast. Enzyme Microb Technol 36(1):34–41. https://doi.org/10.1016/j.enzmictec.2004.03.022

Koyani RD, Vazquez-Duhalt R (2016) Laccase encapsulation in chitosan nanoparticles enhances the protein stability against microbial degradation. Environ Sci Pollut Res Int 23 (18):18850–18857. https://doi.org/10.1007/s11356-016-7072-8

Kües U, Rühl M (2011) Multiple multi-copper oxidase gene families in basidiomycetes – what for? Curr Genomics 12(2):72–94. https://doi.org/10.2174/138920211795564377

Kumar VV, Cabana H (2016) Towards high potential magnetic biocatalysts for on-demand elimination of pharmaceuticals. Bioresour Technol 200:81–89. https://doi.org/10.1016/j.biortech.2015.09.100

Kun RS, Gomes ACS, Hilden KS, Salazar Cerezo S, Mäkelä MR, de Vries RP (2019) Developments and opportunities in fungal strain engineering for the production of novel enzymes and enzyme cocktails for plant biomass degradation. Biotechnol Adv. https://doi.org/10.1016/j.biotechadv.2019.02.017

Lambertz C, Eckert C, Fisher R, Commandeur U (2016) *Kluyveromyces lactis* as an expression host for enzymes that degrade lignocellulosic biomass. Cell Chem Technol 50(3–4):385–389

Legerská B, Chmelová D, Ondrejovič M (2016) Degradation of synthetic dyes by laccases – a mini-review. Nova Biotechnol Chim 15(1):90. https://doi.org/10.1515/nbec-2016-0010

Leontievsky A, Myasoedova N, Baskunov B, Golovleva L, Bucke C, Evans C (2001) Transformation of 2,4,6-trichlorophenol by free and immobilized fungal laccase. Appl Microbiol Biotechnol 57(1):85–91. https://doi.org/10.1007/s002530100756

Levasseur A, Saloheimo M, Navarro D, Andberg M, Pontarotti P, Kruus K, Record E (2010) Exploring laccase-like multicopper oxidase genes from the ascomycete *Trichoderma reesei*: a functional, phylogenetic and evolutionary study. BMC Biochem 11(1):32. https://doi.org/10.1186/1471-2091-11-32

Liers C, Ullrich R, Steffen KT, Hatakka A, Hofrichter M (2006) Mineralization of ^{14}C-labelled synthetic lignin and extracellular enzyme activities of the wood-colonizing ascomycetes *Xylaria hypoxylon* and *Xylaria polymorpha*. Appl Microbiol Biotechnol 69(5):573–579. https://doi.org/10.1007/s00253-005-0010-1

Llorca M, Rodriguez-Mozaz S, Couillerot O, Panigoni K, de Gunzburg J, Bayer S, Czaja R, Barcelo D (2015) Identification of new transformation products during enzymatic treatment of tetracycline and erythromycin antibiotics at laboratory scale by an on-line turbulent flow liquid-chromatography coupled to a high resolution mass spectrometer LTQ-Orbitrap. Chemosphere 119:90–98. https://doi.org/10.1016/j.chemosphere.2014.05.072

Lloret L, Eibes G, Lú-Chau TA, Moreira MT, Feijoo G, Lema JM (2010) Laccase-catalyzed degradation of anti-inflammatories and estrogens. Biochem Eng J 51(3):124–131. https://doi.org/10.1016/j.bej.2010.06.005

Lundell TK, Mäkelä MR, de Vries RP, Hildén KS (2014) Chapter 11 – Genomics, lifestyles and future prospects of wood-decay and litter-decomposing basidiomycota. In: Martin FM (ed) Advances in botanical research, vol 70. Academic Press, pp 329–370. https://doi.org/10.1016/B978-0-12-397940-7.00011-2

Macellaro G, Baratto MC, Piscitelli A, Pezzella C, Fabrizi de Biani F, Palmese A, Piumi F, Record E, Basosi R, Sannia G (2014) Effective mutations in a high redox potential laccase from *Pleurotus ostreatus*. Appl Microbiol Biotechnol 98(11):4949–4961. https://doi.org/10.1007/s00253-013-5491-8

Madzak C, Otterbein L, Chamkha M, Moukha S, Asther M, Gaillardin C, Beckerich J-M (2005) Heterologous production of a laccase from the basidiomycete *Pycnoporus cinnabarinus* in the

dimorphic yeast *Yarrowia lipolytica*. FEMS Yeast Res 5(6–7):635–646. https://doi.org/10.1016/j.femsyr.2004.10.009
Majeau J-A, Brar SK, Tyagi RD (2010) Laccases for removal of recalcitrant and emerging pollutants. Bioresour Technol 101(7):2331–2350. https://doi.org/10.1016/j.biortech.2009.10.087
Mäkelä MR, Donofrio N, de Vries RP (2014) Plant biomass degradation by fungi. Fungal Genet Biol 72:2–9. https://doi.org/10.1016/j.fgb.2014.08.010
Margot J, Copin P-J, von Gunten U, Barry DA, Holliger C (2015) Sulfamethoxazole and isoproturon degradation and detoxification by a laccase-mediator system: influence of treatment conditions and mechanistic aspects. Biochem Eng J 103:47–59. https://doi.org/10.1016/j.bej.2015.06.008
Maruyama T, Komatsu C, Michizoe J, Ichinose H, Goto M (2006) Laccase-mediated oxidative degradation of the herbicide dymron. Biotechnol Prog 22(2):426–430. https://doi.org/10.1021/bp050251h
Mate DM, Alcalde M (2015) Laccase engineering: from rational design to directed evolution. Biotechnol Adv 33(1):25–40. https://doi.org/10.1016/j.biotechadv.2014.12.007
Mate DM, Alcalde M (2017) Laccase: a multi-purpose biocatalyst at the forefront of biotechnology. Microb Biotechnol 10(6):1457–1467. https://doi.org/10.1111/1751-7915.12422
Mateljak I, Monza E, Lucas MF, Guallar V, Aleksejeva O, Ludwig R, Leech D, Shleev S, Alcalde M (2019) Increasing redox potential, redox mediator activity, and stability in a fungal laccase by computer-guided mutagenesis and directed evolution. ACS Catalysis 9(5):4561–4572. https://doi.org/10.1021/acscatal.9b00531
Mayer AM, Staples RC (2002) Laccase: new functions for an old enzyme. Phytochemistry 60:551–565. https://doi.org/10.1016/S0031-9422(02)00171-1
Mekmouche Y, Zhou S, Cusano AM, Record E, Lomascolo A, Robert V, Simaan AJ, Rousselot-Pailley P, Ullah S, Chaspoul F, Tron T (2014) Gram-scale production of a basidiomycetous laccase in *Aspergillus niger*. J Biosci Bioeng 117(1):25–27. https://doi.org/10.1016/j.jbiosc.2013.06.013
Michniewicz A, Ledakowicz S, Ullrich R, Hofrichter M (2008) Kinetics of the enzymatic decolorization of textile dyes by laccase from *Cerrena unicolor*. Dyes Pigm 77(2):295–302. https://doi.org/10.1016/j.dyepig.2007.05.015
Morozova OV, Shumakovich GP, Gorbacheva MA, Shleev SV, Yaropolov AI (2007a) "Blue" laccases. Biochem (Mosc) 72(10):1136–1150. https://doi.org/10.1134/S0006297907100112
Morozova OV, Shumakovich GP, Shleev SV, Yaropolov AI (2007b) Laccase-mediator systems and their applications: a review. Appl Biochem Microbiol 43(5):523–535. https://doi.org/10.1134/S0003683807050055
Nakagawa Y, Sakamoto Y, Kikuchi S, Sato T, Yano A (2010) A chimeric laccase with hybrid properties of the parental *Lentinula edodes* laccases. Microbiol Res 165(5):392–401. https://doi.org/10.1016/j.micres.2009.08.006
Nguyen LN, Hai FI, Price WE, Leusch FDL, Roddick F, McAdam EJ, Magram SF, Nghiem LD (2014a) Continuous biotransformation of bisphenol A and diclofenac by laccase in an enzymatic membrane reactor. Int Biodeter Biodegr 95:25–32. https://doi.org/10.1016/j.ibiod.2014.05.017
Nguyen LN, Hai FI, Price WE, Leusch FDL, Roddick F, Ngo HH, Guo W, Magram SF, Nghiem LD (2014b) The effects of mediator and granular activated carbon addition on degradation of trace organic contaminants by an enzymatic membrane reactor. Bioresour Technol 167:169–177. https://doi.org/10.1016/j.biortech.2014.05.125
Nguyen LN, Hai FI, Price WE, Kang J, Leusch FDL, Roddick F, van de Merwe JP, Magram SF, Nghiem LD (2015) Degradation of a broad spectrum of trace organic contaminants by an enzymatic membrane reactor: complementary role of membrane retention and enzymatic degradation. Int Biodeter Biodegr 99:115–122. https://doi.org/10.1016/j.ibiod.2014.12.004
Nguyen LN, van de Merwe JP, Hai FI, Leusch FDL, Kang J, Price WE, Roddick F, Magram SF, Nghiem LD (2016) Laccase–syringaldehyde-mediated degradation of trace organic

contaminants in an enzymatic membrane reactor: removal efficiency and effluent toxicity. Bioresour Technol 200:477–484. https://doi.org/10.1016/j.biortech.2015.10.054

Otterbein L, Record E, Longhi S, Asther M, Moukha S (2000) Molecular cloning of the cDNA encoding laccase from *Pycnoporus cinnabarinus* I-937 and expression in *Pichia pastoris*. Eur J Biochem 267(6):1619–1625. https://doi.org/10.1046/j.1432-1327.2000.01166.x

Piscitelli A, Giardina P, Mazzoni C, Sannia G (2005) Recombinant expression of *Pleurotus ostreatus* laccases in *Kluyveromyces lactis* and *Saccharomyces cerevisiae*. Appl Microbiol Biotechnol 69(4):428–439. https://doi.org/10.1007/s00253-005-0004-z

Pointing SB (2001) Feasibility of bioremediation by white-rot fungi. Appl Microbiol Biotechnol 57 (1–2):20–33. https://doi.org/10.1007/s002530100745

Pramanik S, Chaudhuri S (2018) Laccase activity and azo dye decolorization potential of *Podoscypha elegans*. Mycobiology 46(1):79–83. https://doi.org/10.1080/12298093.2018.1454006

Qayyum H, Maroof H, Yasha K (2009) Remediation and treatment of organopollutants mediated by peroxidases: a review. Crit Rev Biotechnol 29(2):94–119. https://doi.org/10.1080/07388550802685306

Ramírez-Cavazos LI, Junghanns C, Ornelas-Soto N, Cárdenas-Chávez DL, Hernández-Luna C, Demarche P, Enaud E, García-Morales R, Agathos SN, Parra R (2014) Purification and characterization of two thermostable laccases from *Pycnoporus sanguineus* and potential role in degradation of endocrine disrupting chemicals. J Mol Catal B Enzym 108:32–42. https://doi.org/10.1016/j.molcatb.2014.06.006

Record E, Punt PJ, Chamkha M, Labat M, van den Hondel CAMJJ, Asther M (2002) Expression of the *Pycnoporus cinnabarinus* laccase gene in *Aspergillus niger* and characterization of the recombinant enzyme. FEBS J 269(2):602–609. https://doi.org/10.1046/j.0014-2956.2001.02690.x

Riley R, Salamov AA, Brown DW, Nagy LG, Floudas D, Held BW, Levasseur A, Lombard V, Morin E, Otillar R, Lindquist EA, Sun H, LaButti KM, Schmutz J, Jabbour D, Luo H, Baker SE, Pisabarro AG, Walton JD, Blanchette RA, Henrissat B, Martin F, Cullen D, Hibbett DS, Grigoriev IV (2014) Extensive sampling of basidiomycete genomes demonstrates inadequacy of the white-rot/brown-rot paradigm for wood decay fungi. Proc Natl Acad Sci USA 111 (27):9923. https://doi.org/10.1073/pnas.1400592111

Riva S (2006) Laccases: blue enzymes for green chemistry. Trends Biotechnol 24(5):219–226. https://doi.org/10.1016/j.tibtech.2006.03.006

Rodriguez Couto S, Toca Herrera JL (2006) Industrial and biotechnological applications of laccases: a review. Biotechnol Adv 24:500–513. https://doi.org/10.1016/j.biotechadv.2006.04.003

Rodriguez-Rodriguez CE, Garcia-Galan MA, Blanquez P, Diaz-Cruz MS, Barcelo D, Caminal G, Vicent T (2012) Continuous degradation of a mixture of sulfonamides by *Trametes versicolor* and identification of metabolites from sulfapyridine and sulfathiazole. J Hazard Mater 213–214:347–354. https://doi.org/10.1016/j.jhazmat.2012.02.008

Sakamoto Y, Nakade K, Yano A, Nakagawa Y, Hirano T, Irie T, Watanabe H, Nagai M, Sato T (2008) Heterologous expression of lcc1 from Lentinula edodes in tobacco BY-2 cells results in the production an active, secreted form of fungal laccase. Appl Microbiol Biotechnol 79 (6):971–980. https://doi.org/10.1007/s00253-008-1507-1

Saloheimo M, Niku-Paavola M-L (1991) Heterologous production of a ligninolytic enzyme: expression of the *Phlebia radiata* laccase gene in *Trichoderma reesei*. Bio/Technol 9 (10):987–990. https://doi.org/10.1038/nbt1091-987

Šašek V (2003) Why mycoremediations have not yet come into practice. In: Šašek V, Glaser JA, Baveye P (eds) The utilization of bioremediation to reduce soil contamination: problems and solutions. Springer Netherlands, Dordrecht, pp 247–266. https://doi.org/10.1007/978-94-010-0131-1_22

Senthivelan T, Kanagaraj J, Panda RC (2016) Recent trends in fungal laccase for various industrial applications: an eco-friendly approach – a review. Biotechnol Bioprocess Eng 21(1):19–38. https://doi.org/10.1007/s12257-015-0278-7

Sharma B, Dangi AK, Shukla P (2018) Contemporary enzyme based technologies for bioremediation: a review. J Environ Manage 210:10–22. https://doi.org/10.1016/j.jenvman.2017.12.075

Shary S, Ralph SA, Hammel KE (2007) New insights into the ligninolytic capability of a wood decay ascomycete. Appl Environ Microbiol 73(20):6691–6694. https://doi.org/10.1128/AEM.01361-07

Sigoillot C, Record E, Belle V, Robert JL, Levasseur A, Punt PJ, van den Hondel CAMJJ, Fournel A, Sigoillot JC, Asther M (2004) Natural and recombinant fungal laccases for paper pulp bleaching. Appl Microbiol Biotechnol 64(3):346–352. https://doi.org/10.1007/s00253-003-1468-3

Singh R, Sidhu SS, Zhang H, Huang Q (2015) Removal of sulfadimethoxine in soil mediated by extracellular oxidoreductases. Environ Sci Pollut Res Int 22(21):16868–16874. https://doi.org/10.1007/s11356-015-4893-9

Snyder SA (2008) Occurrence, treatment, and toxicological relevance of EDCs and pharmaceuticals in water. Ozone Sci Eng 30(1):65–69. https://doi.org/10.1080/01919510701799278

Soden DM, O'Callaghan J, Dobson ADW (2002) Molecular cloning of a laccase isozyme gene from *Pleurotus sajor-caju* and expression in the heterologous *Pichia pastoris* host. Microbiology 148(12):4003–4014. https://doi.org/10.1099/00221287-148-12-4003

Solís M, Solís A, Pérez HI, Manjarrez N, Flores M (2012) Microbial decolouration of azo dyes: a review. Process Biochem 47(12):1723–1748. https://doi.org/10.1016/j.procbio.2012.08.014

Solomon EI, Sundaram UM, Machonkin TE (1996) Multicopper oxidases and oxygenases. Chem Rev 96(7):2563–2606. https://doi.org/10.1021/cr950046o

Sonoki T, Kajita S, Ikeda S, Uesugi M, Tatsumi K, Katayama Y, Iimura Y (2005) Transgenic tobacco expressing fungal laccase promotes the detoxification of environmental pollutants. Appl Microbiol Biotechnol 67(1):138–142. https://doi.org/10.1007/s00253-004-1770-8

Spina F, Cordero C, Schilirò T, Sgorbini B, Pignata C, Gilli G, Bicchi C, Varese GC (2015) Removal of micropollutants by fungal laccases in model solution and municipal wastewater: evaluation of estrogenic activity and ecotoxicity. J Clean Prod 100:185–194. https://doi.org/10.1016/j.jclepro.2015.03.047

Steffen K, Tuomela M (2011) Fungal soil bioremediation: developments towards large-scale applications. In: Hofrichter M (ed) The Mycota X. Industrial applications. Springer, Berlin, pp 451–467. https://doi.org/10.1007/978-3-642-11458-8_22

Steffen KT, Hofrichter M, Hatakka A (2000) Mineralisation of ^{14}C-labelled synthetic lignin and ligninolytic enzyme activities of litter-decomposing basidiomycetous fungi. Appl Microbiol Biotechnol 54(6):819–825. https://doi.org/10.1007/s002530000473

Tadesse MA, DAnnibale A, Galli C, Gentili P, Sergi F (2008) An assessment of the relative contributions of redox and steric issues to laccase specificity towards putative substrates. Org Biomol Chem 6(5):868–878. https://doi.org/10.1039/b716002j

Tanaka T, Tonosaki T, Nose M, Tomidokoro N, Kadomura N, Fujii T, Taniguchi M (2001) Treatment of model soils contaminated with phenolic endocrine-disrupting chemicals with laccase from *Trametes* sp. in a rotating reactor. J Biosci Bioeng 92(4):312–316. https://doi.org/10.1016/S1389-1723(01)80232-2

Téllez-Jurado A, Arana-Cuenca A, González Becerra AE, Viniegra-González G, Loera O (2006) Expression of a heterologous laccase by *Aspergillus niger* cultured by solid-state and submerged fermentations. Enzyme Microb Technol 38(5):665–669. https://doi.org/10.1016/j.enzmictec.2005.07.021

Thurston CF (1994) The structure and function of fungal laccases. Microbiology 140:19–26

Tuomela M, Hatakka A (2011) 6.16 – Oxidative fungal enzymes for bioremediation. In: Moo-Young M (ed) Comprehensive biotechnology, 2nd edn. Academic Press, Burlington, pp 183–196. https://doi.org/10.1016/B978-0-08-088504-9.00370-6

Tuomela M, Lyytikäinen M, Oivanen P, Hatakka A (1998) Mineralization and conversion of pentachlorophenol (PCP) in soil inoculated with the white-rot fungus *Trametes versicolor*. Soil Biol Biochem 31(1):65–74. https://doi.org/10.1016/S0038-0717(98)00106-0

Valentín L, Kluczek-Turpeinen B, Oivanen P, Hatakka A, Steffen K, Tuomela M (2009) Evaluation of basidiomycetous fungi for pretreatment of contaminated soil. J Chem Technol Biotechnol 84 (6):851–858. https://doi.org/10.1002/jctb.2188

Valentín L, Kluczek-Turpeinen B, Willför S, Hemming J, Hatakka A, Steffen K, Tuomela M (2010) Scots pine (*Pinus sylvestris*) bark composition and degradation by fungi: potential substrate for bioremediation. Bioresour Technol 101(7):2203–2209. https://doi.org/10.1016/j.biortech.2009.11.052

Viswanath B, Rajesh B, Janardhan A, Kumar AP, Narasimha G (2014) Fungal laccases and their applications in bioremediation. Enzyme Res 2014:21. https://doi.org/10.1155/2014/163242

Wang G-D, Li Q-J, Luo B, Chen X-Y (2004) Ex planta phytoremediation of trichlorophenol and phenolic allelochemicals via an engineered secretory laccase. Nat Biotechnol 22(7):893–897. https://doi.org/10.1038/nbt982

Wang ZW, Liang JS, Liang Y (2013) Decolorization of Reactive Black 5 by a newly isolated bacterium *Bacillus* sp. YZU1. Int Biodeter Biodegr 76:41–48. https://doi.org/10.1016/j.ibiod.2012.06.023

Wesenberg D, Buchon F, Agathos SN (2002) Degradation of dye-containing textile effluent by the agaric white-rot fungus *Clitocybula dusenii*. Biotech Lett 24(12):989–993. https://doi.org/10.1023/a:1015660927710

Xu F, Shin W, Brown SH, Wahleithner JA, Sundaram UM, Solomon EI (1996) A study of a series of recombinant fungal laccases and bilirubin oxidase that exhibit significant differences in redox potential, substrate specificity, and stability. Biochim Biophys Acta 1292:303–311. https://doi.org/10.1016/0167-4838(95)00210-3

Yang J, Li W, Ng TB, Deng X, Lin J, Ye X (2017) Laccases: production, expression regulation, and applications in pharmaceutical biodegradation. Front Microbiol 8:832–832. https://doi.org/10.3389/fmicb.2017.00832

Yaver DS, Overjero MDC, Xu F, Nelson BA, Brown KM, Halkier T, Bernauer S, Brown SH, Kauppinen S (1999) Molecular characterization of laccase genes from the basidiomycete *Coprinus cinereus* and heterologous expression of the laccase Lcc1. Appl Environ Microbiol 65(11):4943

You L-F, Liu Z-M, Lin J-F, Guo L-Q, Huang X-L, Yang H-X (2014) Molecular cloning of a laccase gene from *Ganoderma lucidum* and heterologous expression in *Pichia pastoris*. J Basic Microbiol 54(S1):S134–S141. https://doi.org/10.1002/jobm.201200808

Zeng X, Cai Y, Liao X, Zeng X, Luo S, Zhang D (2012) Anthraquinone dye assisted the decolorization of azo dyes by a novel *Trametes trogii* laccase. Process Biochem 47 (1):160–163. https://doi.org/10.1016/j.procbio.2011.10.019

Zhao Y, Yi X (2010) Effects of soil oxygen conditions and soil pH on remediation of DDT-contaminated soil by laccase from white rot fungi. Int J Environ Res Public Health 7 (4):1612–1621. https://doi.org/10.3390/ijerph7041612

Bacterial Laccases: Some Recent Advances and Applications

Lígia O. Martins, Eduardo Pinho Melo, Antonio Sanchez-Amat, and Maria Paula Robalo

Contents

Abstract Laccases belong to the large family of multi-copper oxidases (MCOs) that couple the one-electron oxidation of substrates with the four-electron reduction of molecular oxygen to water. Because of their high relative non-specific oxidation capacity particularly on phenols and aromatic amines as well as the lack of requirement for expensive organic cofactors, they have found application in a large number of biotechnological fields. The vast majority of studies and applications were performed using fungal laccases, but bacterial laccases show interesting properties such as optimal temperature above 50 °C, optimal pH at the neutral to alkaline range, thermal and chemical stability and increased salt tolerance. Additionally, bacterial systems benefit from a wide range of molecular biology tools that facilitates their

L. O. Martins (✉)
Instituto de Tecnologia Química e Biológica António Xavier, Universidade Nova de Lisboa, Oeiras, Portugal
e-mail: lmartins@itqb.unl.pt

E. P. Melo
Centro de Ciências do Mar, Universidade do Algarve, Faro, Portugal

A. Sanchez-Amat
Department of Genetics and Microbiology, Universidad de Murcia, Murcia, Spain

M. P. Robalo
Área Departamental de Engenharia Química, ISEL – Instituto Superior de Engenharia de Lisboa, Instituto Politécnico de Lisboa, R. Conselheiro Emídio Navarro, Lisboa, Portugal

Centro de Química Estrutural, Complexo I, Instituto Superior Técnico, Universidade de Lisboa, Lisboa, Portugal

D. Schlosser (ed.), *Laccases in Bioremediation and Waste Valorisation*, Microbiology Monographs 33, https://doi.org/10.1007/978-3-030-47906-0_2

engineering and achievement of high yields of protein production and set-up of cost-effective bioprocesses. In this review we will provide up-to-date information on the distribution and putative physiological role of bacterial laccases and highlight their distinctive structural and biochemical properties, discuss the key role of copper in the biochemical properties, discuss thermostability determinants and, finally, review biotechnological applications with a focus on catalytic mechanisms on phenolics and aromatic amines.

1 Distribution and Physiological Role of Laccases in the Prokaryotic World

Laccases act on a surprisingly wide range of substrates, including diphenols, polyphenols, differently substituted phenols, diamines, aromatic amines, benzenethiols and even some inorganic compounds such as metal ions Mn(II), Cu(I) and Fe(II). Laccases are one of the oldest enzymes ever isolated from plant and fungal sources (Mayer and Staples 2002), and in the last two decades, an extensive number of laccases of prokaryotic origin have been identified and characterised; for recent reviews see Martins et al. (2015), Chandra and Chowdhary (2015), Tonin et al. (2016), Chauhan et al. (2017) and Guan et al. (2018). A phylogenetic analysis of these enzymes revealed that they can be clustered, and in the next sub-chapter, we will discuss the physiological relevance of proteins within the different groups (Fig. 1).

(A) Proteins involved in copper resistance. The wide distribution of laccases involved in copper resistance in the phylogenetic tree suggests that this is a physiological role commonly conserved in bacteria. Copper is a micronutrient required by living organisms that in excessive amounts exerts cellular toxicity. It is a redox active metal with two different oxidation states, Cu(I) and Cu(II). Bacteria have developed different mechanisms of copper tolerance that include the action of laccases in the periplasmic space playing a role in oxidising Cu(I) to the less toxic Cu(II) (Pérez et al. 2018). Several copper resistance-related laccases can be identified: **A.1. CueO.** *Escherichia coli* CueO, one of the best characterised bacterial laccases involved in copper resistance (Grass and Rensing 2001), is included in a branch including different enzymes from proteobacteria. CueO, in addition of oxidising Cu(I), also oxidises catecholate siderophores, avoiding that Cu(II) is reduced back to Cu(I) (Grass et al. 2004). *cueO* is regulated at the transcriptional level by CueR that also regulates the copper ATPase efflux CopA that works in coordination for copper detoxification (Outten et al. 2000; Rademacher and Masepohl 2012). **A.2. PcoA.** In addition to CueO, which is chromosomally encoded (Outten et al. 2001; Huffman et al. 2002), some *E. coli* strains possess a plasmid-encoded laccase involved in copper resistance, PcoA (Brown et al. 1995), that, similar to CueO, most likely exhibited Cu(I) oxidase activity in the periplasm

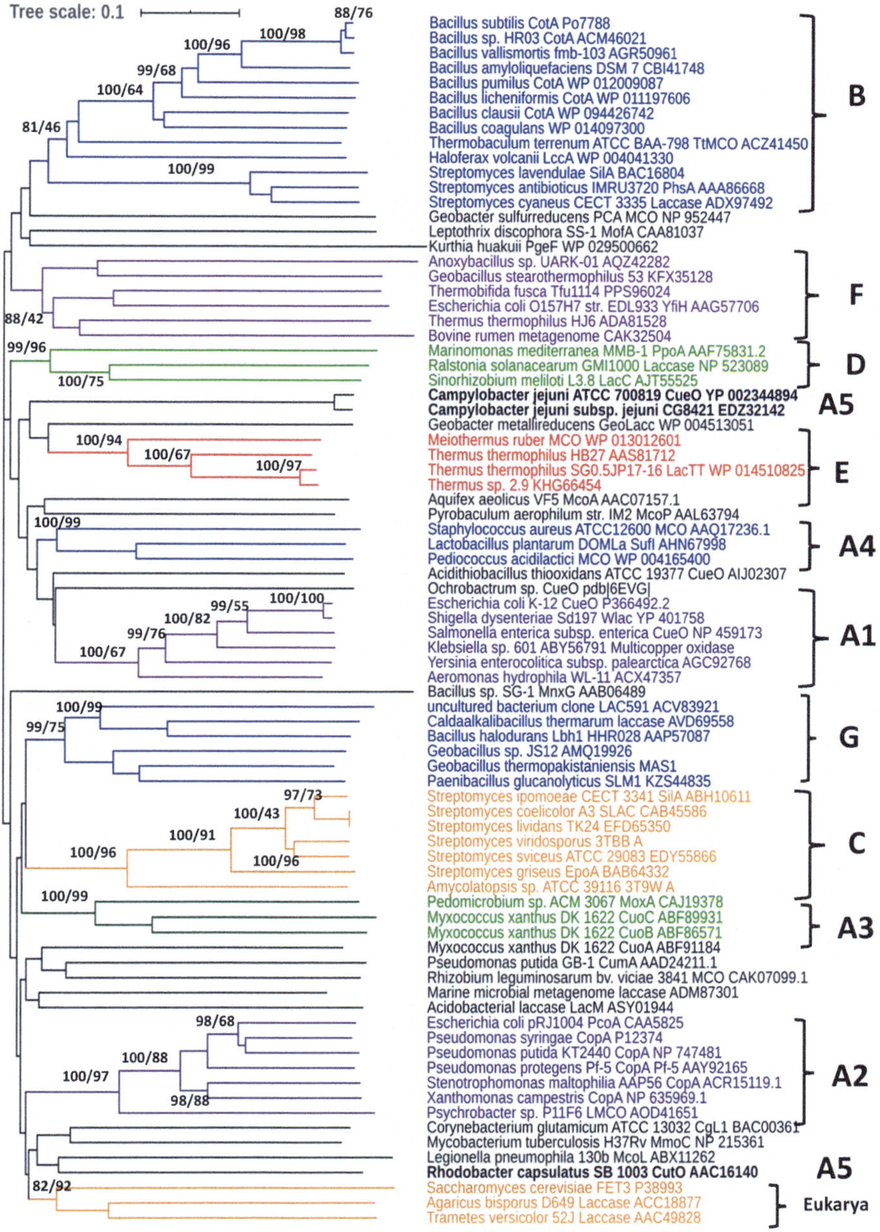

Fig. 1 Phylogenetic relationship of bacterial laccases with reported enzymatic activity; three eukaryotic enzymes were also included for comparison. The phylogenetic analysis was performed using the program MEGA X with sequences aligned by the program MUSCLE built into MEGA (Kumar et al. 2018). The tree using the Neighbour-Joining (NJ) method is drawn to scale, with branch lengths in the same units as those of the evolutionary distances used to infer the phylogenetic tree. The evolutionary distances were computed using the p-distance method and are in the units of the number of amino acid differences per site. The final display of the tree was obtained by using iTOL (Letunic and Bork 2019). A tree was also constructed by the Maximum Likelihood

(Huffman et al. 2002). The operon contains seven genes, and its expression is controlled by a two-component regulatory system encoded within the same operon (Rouch and Brown 1997). Proteins with similarity to PcoA have been detected in other gammaproteobacteria, CopA in *Pseudomonas syringae* (Cha and Cooksey 1991) and *Stenotrophomonas maltophilia* (Galai et al. 2011). Similar genes are present in *Pseudomonas* strains as part of a copper resistance operon, and the CopA and CopC enzymes were shown to exhibit activity for several organic substrates including lignin model compound (Granja-Travez and Bugg 2018). **A.3. Laccases in *Myxococcus xanthus*.** *Myxococcus xanthus*, a deltaproteobacterium with a complex life cycle, encodes three different laccases (CuoA, CuoB and CuoC) involved in copper resistance (Muñoz-Dorado et al. 2016). CuoB and CuoC cluster with an enzyme from the alphaproteobacterium *Pedomicrobium*, and CuoA is in a different branch (Fig. 1). CuoB is involved in an immediate response to copper addition to growth medium (Sanchez-Sutil et al. 2007), and the expression of *cuoB* was shown to be regulated by CorE, an extracytoplasmic sigma factor (ECF), whose activity is activated by Cu(II) and inactivated by Cu(I) (Marcos-Torres et al. 2016). The genes coding CuoA and CuoC are slowly copper-induced reaching a maximum at 24 h of growth. This maintenance response is regulated by CosSR, a two-component regulatory system, whose periplasmic domain is responsible for Cu sensing (Sánchez-Sutil et al. 2016). **A.4. Laccases in *Firmicutes*.** *Staphylococcus aureus* synthesises a laccase whose gene expression is induced by copper and whose deletion resulted in an increased copper sensitivity (Sitthisak et al. 2005). The genes coding for the *S. aureus* laccase have been detected in mobile genetic elements present in many clinical isolates and proposed to play a role in virulence (Zapotoczna et al. 2018). *Pediococcus* and *Lactobacillus* laccases that are very close in the phylogenetic tree to the *S. aureus* enzyme have been shown to oxidise and degrade biogenic amines (Callejon et al. 2016, 2017). **A.5. Other laccases involved in copper resistance.** The gene coding for CutO from *Rhodobacter capsulatus* is induced at high copper concentrations through a regulatory mechanism operating at the post-transcriptional level (Rademacher et al. 2012). Similarly the gene coding for a *Campylobacter jejuni* laccase is induced at high copper concentrations, and its deletion increases the bacterium sensitivity to copper (Hall et al. 2008).

(B) Proteins involved in spore pigmentation. This group comprises the most well-studied bacterial laccase, CotA, from *Bacillus subtilis* (Hullo et al. 2001; Martins et al. 2002). Most of the laccases identified in the *Bacillus* cluster are in the same group (Fig. 1). These enzymes are located at the outer coat of spores and have been implicated in the biosynthesis of a brown melanin-like pigment that protects spores from UV light and hydrogen peroxide. Closely related to those proteins is a laccase synthesised by the archaeum *Haloferax volcanii* that was

---

Fig. 1 (continued) (ML) method. Numbers at branches indicate bootstrap values higher than 80% for the NJ method and the bootstrap value obtained with the ML method in those branches. The enzymes in the same group are marked with the same colour and are annotated as explained in the text

suggested to be acquired through horizontal gene transfer from bacteria of the phylum *Firmicutes* (Uthandi et al. 2010). Some proteins of actinobacteria are found closely associated with the *Bacillus* CotA. For example, the phenoxazinone synthase (Phs) from *Streptomyces antibioticus* that catalyses aminophenol oxidation is required for the synthesis of secondary metabolites, e.g. the antibiotic actinomycin. It was additionally proposed that Phs could participate in spore pigment synthesis similar to CotA laccases (Jones 2000).

(C) Small laccases in *Streptomyces*. Some actinobacteria laccases from the order *Actinomycetales* are in a clearly differentiated branch. This group contains the secreted enzymes denominated "Small laccases" (SLACS, see below) where *Streptomyces coelicolor* SLAC is the most extensively characterised (Machczynski et al. 2004). These enzymes have been implicated in physiological roles related to morphogenesis, sporulation, pigmentation, ligninocellulose degradation, bacteria-bacteria interactions or antibiotic production (Arias et al. 2003; Eugenio et al. 2011; Majumdar et al. 2014).

(D) Laccases in plant-associated proteobacteria. The *Marinomonas mediterranea* laccase PpoA was one of the first bacterial laccases described (Solano et al. 1997). This protein is a membrane-associated protein showing an N-terminal His-rich region which could be related to its capacity to oxidise not only the substrates characteristic of laccases but also the monophenol L-tyrosine (Sanchez-Amat et al. 2001; Solano et al. 2001). The physiological role of PpoA is unknown; mutant strains lacking this protein are not affected in pigment synthesis, and no evidences for involvement in Cu resistance were found (Fernandez et al. 1999). The *Ralstonia solanacearum* laccase was proposed to participate in defence against phenolic compounds (Hernandez-Romero et al. 2005). This could be related to the association of the producing bacteria with plants: *M. mediterranea* was isolated from the microbiota of the marine plant *Posidonia oceanica* (Espinosa et al. 2010), *R. solanacearum* is a plant pathogen and *Rhizobium* is a symbiont of plants (Pawlik et al. 2016).

(E) Laccases in thermophilic bacteria and archaea. Several laccases from thermophilic and hyperthermophilic bacteria cluster in the same group (Miyazaki 2005; Fernandes et al. 2007; Kalyani et al. 2016; Navas et al. 2019). Notably, the *Thermus thermophilus* Tth laccase shows extreme stability at high temperatures with a half-life at 80 °C of around 14 h (Miyazaki 2005). The gene coding for McoA from *Aquifex aeolicus* that grows optimally at 89 °C is part of a putative copper resistance determinant, and the enzyme is hyperthermoactive and shown to exhibit higher efficiency for Cu(I) and Fe(II) metal ions than for aromatic substrates (Fernandes et al. 2007). This group also comprises the only hyperthermophilic archaeal-type laccase described to date, McoP from *Pyrobaculum aerophilum* (Fernandes et al. 2010).

(F and G) Other groups. There are other clusters revealed by the phylogenetic analyses for which a physiological role has not been yet proposed. For example, the small *Thermus thermophilus* HJA laccase shows a low molecular mass (Kim et al. 2015) and is not included in the same group that other enzymes from the same species which are in group E. Genes encoding proteins similar to those in group F can be detected in other bacteria, including some *E. coli* strains, but their physiological role remains unknown.

2 Structure, Mechanisms and Biochemical Properties

The coordination of the copper ions is largely conserved among laccases (Fig. 2). The T1 Cu is coordinated by two histidine nitrogen atoms and a cysteine sulphur, and it is characterised by an intense $S(\pi)\rightarrow Cu(d_{x^2-y^2})$ charge transfer absorption band at around 600 nm, $\varepsilon_{600nm} > 3000$/M/cm responsible for the intense blue colour of the enzymes. The T2 copper site, strategically positioned close to the T3 binuclear copper centre, is usually coordinated by two histidine residues and a water (or hydroxyl) molecule, while each T3 copper is coordinated by three histidines and a bridging ligand such as a hydroxyl moiety, displaying an absorption in the near-UV, with $\lambda_{max} = 30$ nm. The mononuclear T1 Cu site interacts with the trinuclear cluster T2/T3 through the highly conserved HCH motif, where the cysteine in the T1 binding Cu shuttles electrons to each of the two histidines that coordinate T3 copper ions. The routes of electron transfer, the oxidation states of the Cu centres and the mechanism of oxygen reduction in laccases were recently reviewed; please see (Jones and Solomon 2015). The oxidation of reducing substrates occurs at the T1-Cu site, while the reduction of O_2 occurs at the T2/T3 trinuclear cluster (Fig. 2) (Jones and Solomon 2015).

The first three-dimensional structures of prokaryotic laccases were reported for *E. coli* CueO (Roberts et al. 2002) and *Bacillus subtilis* CotA laccase (Enguita et al. 2003), followed by *Streptomyces coelicolor* SLAC (Skalova et al. 2009), *Pyrobaculum aerophilum* McoP (Sakuraba et al. 2011), *Campylobacter jejuni*

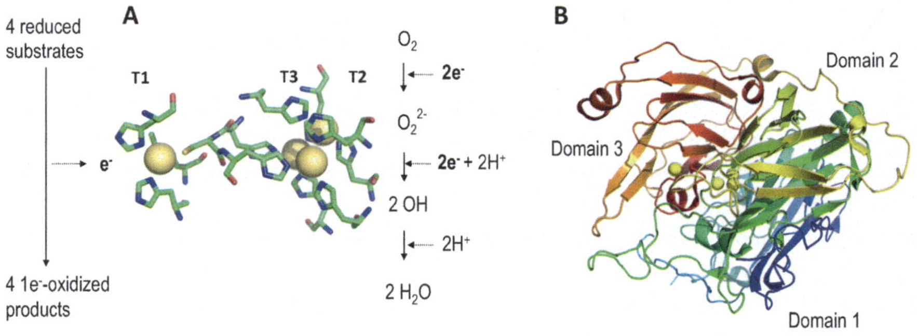

Fig. 2 Three-dimensional representation of copper centres arrangement (**a**) and the overall structure of the *Bacillus subtilis* CotA laccase (Enguita et al. 2003). In picture (**a**) the mononuclear T1 centre is on the left and the trinuclear centre is on the right. The distance between the T1 and T2/T3 centre is ≈13 Å. In picture (**b**) the cupredoxin domains are coloured differently (residues 1–173, domain 1: blue to green; residues 182–340, domain 2: green to yellow; residues 369–501, domain 3: yellow to red). The four copper atoms are shown as yellow spheres. Pictures drawn in PyMol software using the deposited structure PDB1w6L

McoC (Silva et al. 2012), *Streptomyces sviceus* Ssl1 (Gunne et al. 2014) and more recently *Ochrobactrum* sp. CueO (Granja-Travez et al. 2018). Structural information is also available for the Tht *Thermus thermophilus* HB27 laccase in the Protein Database Bank (Serrano-Posada et al. 2011). The overall bacterial laccase structural fold comprises the typical three cupredoxin-type domains, characterised by a Greek key β-barrel topology (Hakulinen and Rouvinen 2015) (Fig. 2). Laccases from *Streptomyces*, as well as *Amycolatopsis* sp. 75iv2, designed SLACs (*s*mall *lac*cases), lack the second of the three domains, and their crystal structures reveal a trimeric quaternary arrangement of the two-domain protein chains (Skalova et al. 2009; Komori et al. 2009; Gunne et al. 2014; Majumdar et al. 2014). Evolutionary theories proposed two-domain proteins as the intermediate step in the evolution between one-domain proteins with a cupredoxin fold and three-domain laccases (Nakamura and Go 2005; Komori et al. 2009).

In bacterial laccases the T1 Cu is weakly coordinated by an axial ligand, a methionine, in tetrahedral geometry. Fungal laccases have non-coordinating phenylalanine or leucine at this position, favouring a trigonal planar geometry for the site, which may contribute to the higher redox potential (~800 mV) observed in these enzymes as compared to bacterial ones (~500 mV) (Xu et al. 1999; Durao et al. 2006; Hong et al. 2011). However, variations in redox potential of the T1 centre observed among laccases cannot be assigned to this single structural feature but to a sum of factors including the nature of the second sphere residues influencing solvent accessibility, hydrogen bonding and dielectric anisotropy around the site (Durao et al. 2008b; Matera et al. 2008; Marshall et al. 2009).

In general, laccases show a broad binding pocket buried between domains 1 and 3 and close to the T1-Cu centre primarily stabilised through hydrophobic interactions (Hakulinen and Rouvinen 2015). In *B. subtilis* CotA laccase, the T1 Cu centre is sited at the bottom of the substrate binding region, relatively exposed to the solvent and interacting with the substrate molecules through the imidazole ring of one of its His ligands (Enguita et al. 2004) similar to what was observed in fungal structures complexed with phenolics or related substrates (Bertrand et al. 2002; Hakulinen et al. 2002; Matera et al. 2008). However, in the remainder of three-domain bacterial laccases whose structures have been reported, the T1 site is occluded by several secondary structure elements, with varied length, composition and structure (Fig. 3). The small substrate binding pocket of these enzymes seems to be in accordance with their higher specificity to small metal ions Cu(I) and Fe (II) (Singh et al. 2004; Fernandes et al. 2010; Silva et al. 2012), indicating that these laccases should behave predominantly as metallo-oxidases in their native microorganisms with a role in copper resistance. However, even at reduced efficiency, these enzymes are also able to oxidise phenolic or non-phenolic substrates, and it is plausible that they possess additional binding pockets to accommodate bulkier substrates (Liu et al. 2011; Bello et al. 2014).

One important structural distinctive feature of bacterial laccases as compared to those of fungal origin is the lack of negatively charged residues in the substrate binding pocket. Fungal laccases contain a conserved acidic residue (either an Asp or Glu) close to the T1 Cu centre that in their carboxylate form are able to hydrogen

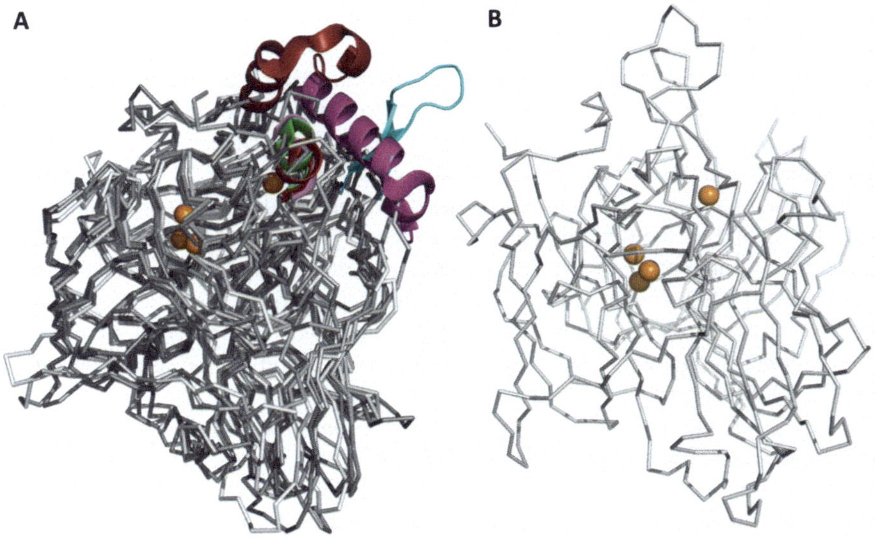

Fig. 3 (**a**) Ribbon superimposition highlighting the major secondary structure elements interfering with the access to the T1-Cu centre in *E. coli* CueO (magenta, PDB 1KV7), *C. jejuni* McoC (red, PDB 3ZX1), *P. aerophilum* McOP (green, PDB 3AW5) and *T. thermophilus* Tth (cyan, 2XU9). (**b**) *B. subtilis* CotA laccase (PDB 1w6L) shows a widely exposed T1-Cu centre. Cu ions are represented as dark orange spheres

bond with OH or NH_2 groups of substrates aiding in their deprotonation by stabilising the radicals formed during the catalytic reaction (Bertrand et al. 2002; Kallio et al. 2009; Madzak et al. 2006). In the case of bacterial laccases such as *B. subtilis* CotA laccase, in the absence of negatively charged residues close to the T1 Cu centre, the oxidation relies mainly on the substrates' structures, in particular in their protonation/deprotonation equilibrium, which is dependent on the pKa values of substrates (d'Acunzo and Galli 2003). Considering that oxidative reactions are more prone to occur after deprotonation of OH or NH_2 groups of phenolic and aromatic amine substrates, bacterial laccases show an optimal pH at the neutral to alkaline range in clear contrast to fungal laccases that show preference for the acidic range (Rosado et al. 2012).

3 Role of Copper and Thermostability Determinants

Copper incorporation in laccases is still a poorly understood process and remains an important issue of discussion in the current literature (Kosman 2017). Understanding this mechanism is important both at the biochemical and structural viewpoints but also from a biotechnological perspective since copper depletion is a limiting factor to achieve full catalytic activity and stability with consequences in the implementation

Box 1

CotA-laccase

Holo-CotA laccase	$N \leftrightarrow N_{no\ copper} \rightarrow U$	Durão et al., 2006
Apo-CotA laccase	$N \leftrightarrow I \leftrightarrow U$	Fernandes et al., 2012
Holo-CotA laccase	$N \leftrightarrow N_{no\ copper} \leftrightarrow I \leftrightarrow U \rightarrow \rightarrow Agg$	Fernandes et al., 2012

Box 1 Chemically induced unfolding pathways of CotA laccase where N, I and U are the native, intermediate and unfolded states, respectively, and Agg is an aggregated state

of cost-effective industrial applications. Bacterial laccases are typically heterologously overproduced in *E. coli* cells under aerobic conditions. However in these conditions, limiting concentrations of copper (10 μM) are present in the cytoplasm of *E. coli* due to the action of copper homeostasis mechanisms (Changela et al. 2003). In order to allow the production of fully copper-loaded recombinant laccases, *E. coli* cells require cultivation in copper-supplemented media under microaerobic growth conditions where significantly higher amounts of copper accumulate (Durao et al. 2008a).

The key role of copper to the stability and folding of CotA laccase. The CotA laccase from *B. subtilis* has been used as one model to dissect stability features and determinants of multi-copper oxidases. The long-term stability of CotA revealed that the enzyme denatures irreversibly according to a simple pathway of unfolding and deactivation (first-order process) with a half-life of inactivation of 50–172 min (at 80 °C) depending on the pH and copper content (Durao et al. 2008a; Fernandes et al. 2011; Brander et al. 2014). This value is larger than the half-life of *B. clausii* CotA (20 min) but smaller than the 350, 540 and 868 min measured for the thermophilic laccases from *T. terrenum*, *C. thermarum* and *T. thermophilus*, respectively (Miyazaki 2005; Brander et al. 2014, 2015; Ghatge et al. 2018). The thermodynamic stability of *B. subtilis* CotA was further characterised showing that at 25 °C the native state is very stable with a free energy change of 10 kcal/mol and a midpoint of 4.6 M guanidinium hydrochloride upon unfolding of the tertiary structure (Durao et al. 2006). Furthermore, it was observed that copper depletion from the T1 site precedes the unfolding of the secondary and tertiary structures similar to what was observed in other laccases of plant and fungal origin (Agostinelli et al. 1995; Koroleva et al. 2001) and is the key event in the inactivation of CotA laccase as shown in the top line of Box 1 (Durao et al. 2006). Copper depletion affects the k_{cat} values of the enzymes and the T1 Cu site redox potential: $E°_{T1}$ of the holoprotein is 525 mV as compared to 455 mV of the partially depleted enzyme (2.5 moles of copper per mole of protein) (Durao et al. 2006, 2008a).

The role of copper in CotA laccase was addressed in more detail to reveal a subtle balance between copper loading and enzyme folding. Incorporation of copper was shown to be a critical aspect in the fine-tuning of the enzyme folding in the cytoplasm of *E. coli*. The data obtained clearly indicated that Cu(I) is the most efficient Cu redox state to be incorporated during in vivo enzyme folding and its presence is key to achieve a fully copper-loaded and fully functional and stable enzyme (Durao et al. 2008a). The incorporation of copper in CotA occurs sequentially, with the T1 site being the first to be reconstituted, followed by the T2 and T3 centres (Durao et al. 2008a) similar to what was described in *E. coli* CueO (Galli et al. 2004). It is the T1 site that provides a template for the assembly of the native protein indicating the ancestral cupredoxin fold as the starting point for the assembly of a fully metallated protein (Kosman 2017). Structural tuning resulting from copper incorporation seems to be a feature of the cupredoxin fold. The similarity in the crystal structure of apo- and holo-form of *Pseudomonas aeruginosa* azurin (a 128-residue cupredoxin) actually obscures two occurrences: copper binding imparts stability to both thermal and mechanical stress, and the metal-binding region becomes less flexible at least in the holoprotein (Giannotti et al. 2015; Kosman 2017). Molecular dynamics simulations of the holo- and apo-forms of *T. thermophilus* HB27 laccase also revealed a different conformation for the linker connecting the β-strands 21 and 24 of the cupredoxin fold with possible implications in the process of electron transfer (Bello et al. 2012). The mutation G304K in CueO laccase that resulted in enhanced activity may well be a good example of structural tuning imparted by copper binding resulting in induced conformational changes in a methionine-rich helix and in a regulatory loop (Wang et al. 2018). Timely copper incorporation into the multi-copper oxidase ceruloplasmin was found crucial to prevent misfolding and assure successful biosynthesis in vivo (Sedlak and Wittung-Stafshede 2007). Deeper insight into the unfolding pathway and copper incorporation in CotA was gathered through acid-induced unfolding and double-jump stopped-flow experiments (Fernandes et al. 2012). In these double-jump experiments, CotA was first unfolded by a decrease in pH, and after fixed time intervals when the enzyme unfolds in a delay loop (20–1000 ms), refolding was promoted by mixing the unfolded enzyme preparation with a buffer at higher pH. If CotA unfolded according to a two-state process where only the native and the unfolded states accumulate, a simple refolding phase would be observed. On the contrary, if an intermediate accumulated between the native and the unfolded states, two refolding phases would be observed especially for longer delay times where the accumulation of the unfolded state becomes more significant. In other words, one refolding phase observed at short delay times will relate to the refolding of the intermediate, and the second phase evident at longer delay times would result from the refolding of the unfolded state. Apo-CotA displays two refolding phases at delay times over 200 ms indicating that a three-state process describes the unfolding of the enzyme. Even in the absence of copper, there is an intermediate in between the native and the unfolded only revealed through kinetic measurements as shown in Box 1. The apo-form of human ceruloplasmin also unfolds with at least one intermediate (Palm-Espling et al. 2012). Double-jump experiments for holo-CotA also revealed the presence of an intermediate in between the native and the unfolded state; however, longer unfolding times resulted in significant light scattering upon refolding in the second jump. This indicates that the presence of copper ions in the unfolded holo-protein prevents its refolding back to the native state due to protein aggregation as summarised in Box 1. In summary, copper has to be incorporated at later stages of in vitro folding (Fernandes et al. 2012), but this incorporation is important to adjust the structure to obtain a fully functional and stable enzyme (Durao et al. 2008a).

The contribution of a disulphide bond to the stability and copper incorporation into CotA laccase was investigated as these bonds are expected to improve protein stability mostly by reducing the conformational entropy of the unfolded state relative to the native state, thereby decreasing the entropy gain upon unfolding (Pace et al. 1998; Sawle and Ghosh 2011). This seems to be the reason for disulphide bond richness found in a subset of thermophilic proteins (Beeby et al. 2005). Surprisingly, the removal of the single intradomain disulphide bridge of the CotA laccase showed no significant effect neither on the thermodynamic and thermal stability nor on the catalytic activity (Fernandes et al. 2011). The three-dimensional structure solved to 2.25 Å resolution, the secondary structure evaluated by circular dichroism, the UV-Vis spectra and the electronic structure of the copper centres probed by electron paramagnetic resonance are very similar between the wild-type and the mutant protein with no disulphide bridge. A decrease in stability should be observed if the main effect of disulphide bonds was to decrease the entropy of the unfolded state increasing thus the free energy of the unfolded state relatively to that of the native state. The unfolding kinetics of CotA measured by stopped-flow, which report on the free energy change between the native and the transition state, are faster for the mutant protein. This indicates that the native state should be less stable in the absence of the disulphide bridge and this destabilisation has to be offset by any effect on the final state to explain why there is no effect on the equilibrium stability of the enzyme. Additionally, the removal of the disulphide bridge in CotA laccase increased the kinetics of copper release indicating conformational dynamics alterations upon the disulphide bond disruption. Conformational dynamics are known to be affected by copper binding such as in the case of the metallo-oxidase Fet3p and the copper chaperones Atox1 and CopZ (Sedlak and Wittung-Stafshede 2007; Palm-Espling et al. 2012) and the *Thermus thermophilus* HB27 laccase (Bello et al. 2012). The view that different proteins use disparate mechanisms for stabilisation has been widely noted (Petsko 2001). The disulphide bond solution prevalent in certain thermophiles (Beeby et al. 2005) was not utilised to increase stability but probably to tune copper binding as the key parameter for folding and stability of CotA.

The hyperthermophilic nature of the *A. aeolicus* McoA and *P. aerophilum* McoP. The thermal stability of the hyperthermostable metallo-oxidases from *A. aeolicus* (McoA) (Fernandes et al. 2009) and *P. aerophilum* (McoP) (Fernandes et al. 2010) was studied by DSC revealing a complex process characterised by three independent thermal transitions showing Tm values of 105, 110 and 114 °C for McoA and 97, 102 and 112 °C for McoP. Three independent thermal transitions were also observed for ascorbate oxidase (Savini et al. 1990), ceruloplasmin (Bonaccorsi di Patti et al. 1990) and CotA laccase (Durao et al. 2008a) and seem to correlate with the structural organisation of three cupredoxin-like domains of MCOs. The high thermal stability of both McoA and McoP is in agreement with a

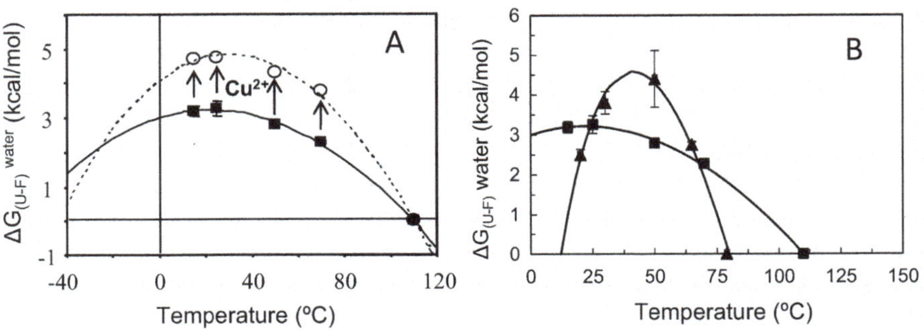

Fig. 4 (**a**) Temperature dependence of McoA stability built with an average Tm measured by DSC and chemical unfolding induced by guanidinium hydrochloride at different temperatures. Experimental data was fitted with the Gibbs-Helmholtz equation, and the dotted line is the fit using the same equation after adding 1.5 kcal/mol due to the stabilising effect of copper. Reprinted from BBA-Proteins Proteom 1794 (1), Fernandes AT, Martins LO, Melo EP, The hyperthermophilic nature of the metallo-oxidase from *Aquifex aeolicus*, 75–83, copyright © 2009, with permission from Elsevier. (**b**) Protein stability curves of McoA (squares) and CotA (triangles) at pH 3, fitted according to the Gibbs-Helmholtz equation

significant long-term stability; McoP displays a half-life time of 5.5 h at 80 °C (Fernandes et al. 2010), while McoA shows a more complex deactivation pathway due to protein aggregation, but the first-order deactivation step is characterised by a half-life time of 5.7 h at 80 °C (Fernandes et al. 2007). Despite the hyperthermophilic nature of both enzymes, the chemical-induced unfolding reveals a relatively low thermodynamic stability especially for McoA with a midpoint of 2.7 M guanidinium hydrochloride and a free energy change in buffer at 25 °C of 2.8 kcal/mol (Fernandes et al. 2009). Copper depletion from McoA precedes largely the unfolding of the tertiary structure and is the key event for inactivation in the presence of chemical denaturants as mentioned previously for CotA (Durão et al. 2006). The low equilibrium thermodynamic stability measured for McoA refers thus to the apo-form (Fernandes et al. 2009; Pozdnyakova et al. 2001). A fast unfolding rate constant for the apo-form compared to the holo-form was measured by stopped-flow, and this increase, reflecting the stabilising effect of copper on the native state, was quantified as 1.5 kcal/mol, which is very similar to the value of 1.4 kcal/mol imparted by copper on CotA (Durao et al. 2006). Copper loading increases the stability of McoA, and the dependence of stability on temperature becomes less flat when the effect of copper is considered (Fig. 4a). Interestingly, a flat dependence of the free energy change on temperature is one of the mechanisms that leads to enhanced thermal stability in proteins from thermophiles (Fitter and Heberle 2000; Robic et al. 2003). The comparison between the stability curves of McoA and CotA clearly shows that the flat dependence of stability on temperature observed for McoA allows the protein to remain folded over a wider range of temperatures as compared with CotA (Fig. 4b). For McoA, the flat dependence of stability on temperature imposed by a low heat capacity change upon unfolding (0.5 compared to 2.6 kcal/mol/K for CotA) results from the aggregation of the enzyme as proved by gel filtration chromatography and light scattering measurements. From a mechanistic point of view, the aggregation process confers residual structure to the final state resulting in a low heat capacity change and a flat dependence of stability on temperature. One hypothesis is that McoA suffers kinetic partitioning between aggregation and first-order deactivation which is an uncommon feature seldom observed for other enzymes (Baptista et al. 2003) that can explain the low heat capacity change. This hypothesis is in accordance with the low cooperativeness of the transition from the folded to further states under equilibrium conditions as evaluated by the thermodynamic parameter m (-1.0 ± 0.1 kcal/mol/M); both the heat capacity change and m value correlate to the amount of surface area exposed upon unfolding. This atypical behaviour of McoA explains the low thermodynamic stability observed upon chemical-induced destabilisation but simultaneously contributes to an increased stability over a wider range of temperatures and thus to the hyperthermophilic nature of McoA.

4 Application in Oxidative Bioprocesses

Oxidation of lignin-related phenolic and non-phenolic compounds using bacterial laccases. Bacterial laccases have been successfully tested in the oxidation of eucalyptus kraft pulp (Arias et al. 2003; Eugenio et al. 2011), wheat straw pulps (Zheng et al. 2012), kraft lignin (Huang et al. 2013), diverse hardwood and softwood lignocellulose samples (Sondhi et al. 2015; Singh et al. 2017), woody materials (Goacher et al. 2018; Navas et al. 2019), lignin samples and a broad range of low molecular weight lignin model compounds (Moya et al. 2011; Huang et al. 2013; Reiss et al. 2013; Majumdar et al. 2014; Hamalainen et al. 2018) confirming their potential in lignin degradation and valorisation.

The oxidation of lignin-related phenolic acids including sinapic, ferulic, caffeic and coumaric acids using bacterial laccases results in the formation of dimeric structures (Koschorreck et al. 2008; Ricklefs et al. 2014; Xie et al. 2015). The enzymatic oxidation of phenols is based on one-electron abstraction which generates resonance-stabilised phenoxy radicals. These may undergo a second enzyme-catalysed oxidation into quinonic structures, or react further in non-enzymatic reactions (hydration, disproportionation or homo- and heteromolecular coupling reactions giving rise to oligo- or polymeric products) (Kudanga et al. 2017; Romero-Guido et al. 2018). The oxidation of phenolic compounds is mostly affected by substituents in the *ortho* positions, and a preference in the oxidation of syringyl-type over the guaiacyl or hydroxyphenyl subunits has been observed. Phenolic and non-phenolic dimeric β-O-4 compounds, e.g. guaiacyl glycerol-β-guaicyl ether (GGE) and adlerol, representing more than 50% of lignin structure, are good models to study the Cα-Cβ bond cleavage and, were tested as substrates by several bacterial laccases, including SLACs from *S. coelicolor* A3(2), *S. lividans* TK24, *S. viridosporus* T7A and *Amycolatopsis* sp. 75iv2 (Majumdar et al. 2014), CopA from *P. putida* KT2440 and *P. fluorescens* Pf-5 (Granja-Travez and Bugg 2018),

CueO from *Ochrobactrum* sp. (Granja-Travez et al. 2018), CtLac from *C. thermarum* (Ghatge et al. 2018) and the *Thermus* sp. 2.9 laccase (Navas et al. 2019). Overall, the tested bacterial laccases catalyse the oxidative coupling of GGE units, but not the cleavage of the Cα-Cβ bond with the exception of *B. amyloliquefaciens* CotA that degraded GGE in a set of products identified as guaiacol, 4-vinylguaiacol, vanillin and other related compounds (Yang et al. 2018). The presence of these compounds indicates that the main degradation pathway of GGE follows the oxidation of Cα and the cleavage of β-O-4 and Cβ-Cγ bonds, highlighting the potential of the *B. amyloliquefaciens* CotA laccase for lignin degradation.

A mechanism of lignin-related phenolics oxidation and their role in mediating non-phenolics by bacterial laccases was described (Fig. 5) (Rosado et al. 2012). The oxidation rates for three syringyl-type phenolic models [syringaldehyde (SA), acetosyringone (AS) and methyl syringate (MS)] were measured showing that the enzymatic rates (SA $>$ AS $\gg$ MS) are in line with the electron-withdrawing capability of the *para*-substituting groups [aldehyde (SA) $>$ ketone (AS) $>$ ester (MS)] and concordant with the pKa values (and oxidation potential) of compounds (SA $<$ AS $<$ MS). The positive effect of electron-withdrawn substituents in the rates of enzymatic oxidation suggests that these improve the stabilisation of the phenolate anions. *Trametes versicolor* TvL laccase and CotA laccase oxidise the three phenolic substrates at maximal rates at pH 4 and pH 8–9, respectively. Considering that the pK_a of the compounds is between pH 7 and 9, the results indicate that the fungal laccase oxidises the substrates in the phenolic form and the bacterial laccase in the phenolate form, i.e. after deprotonation of the phenolic group. These phenolic compounds were further tested as laccase mediators in the oxidation of the non-phenolics, veratraldehyde, 4-methoxybenzyl alcohol and 3,4,5-trimethoxybenzyl alcohol. Interestingly, the mediator activity of the syringyl-type phenolic reflected a balance between reactivity and stability of the radicals formed as the measured conversion yields of non-phenolics were higher for MS $>$ AS $>$ SA, showing that the stability of radicals is improved by the presence of electron donor groups at the *para*-position. Moreover, maximal conversion yields (up to 80%) were achieved at large excess of mediators (10:1) implying an interplay of competitive routes (Fig. 5). Phenoxy radical intermediates are involved in (1) the oxidation of the non-phenolic monomers, forming hydroxybenzyl radicals that are converted to the correspondent aldehydes, (2) the conversion to 2,6-dimethoxy-*p*-benzoquinone and, finally (3) coupling reactions including radical recombination, cross-coupling and self-coupling, with or without release of substituent groups. This catalytic cycle points out the need to control the reaction conditions in order to guide the depolymerisation/re-polymerisation balance of lignin samples (Rosado et al. 2012).

Although laccases can play a major role in lignin depolymerisation, the achievement of a correct balance between the depolymerisation and re-polymerisation processes is a great challenge (Roth and Spiess 2015; Hamalainen et al. 2018). This balance depends on several reaction conditions including the substrate structures, the difference between the redox potentials of enzyme and substrate, pH of reaction, temperature and solvent. Recent advances highlighted the potential of

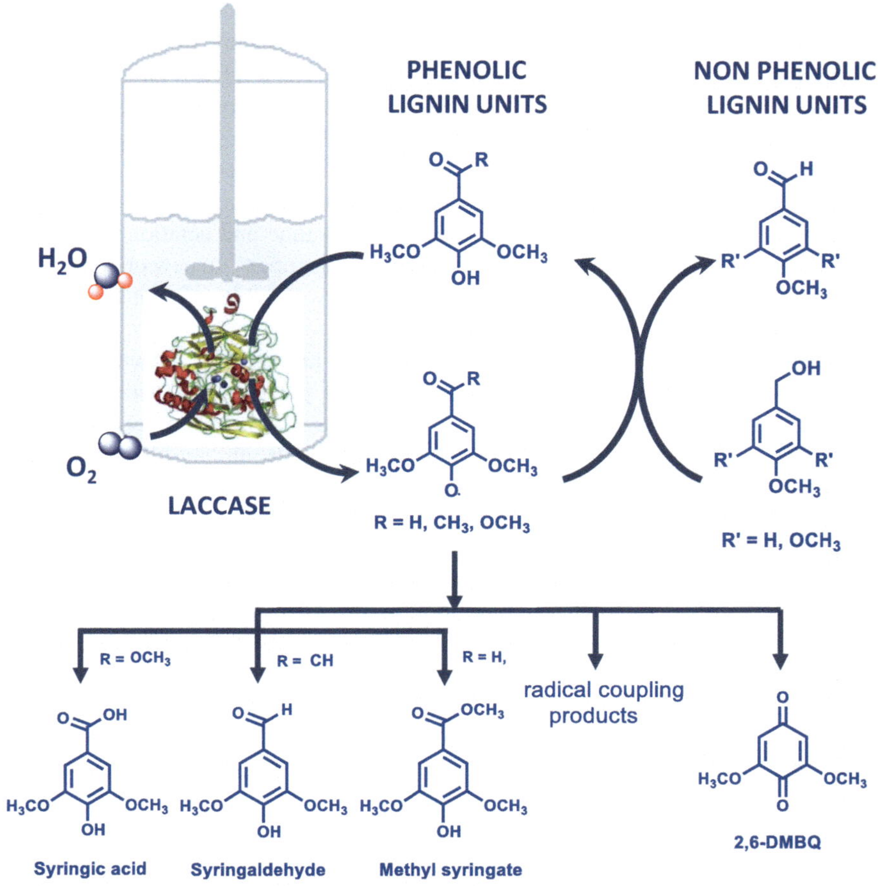

Fig. 5 The catalytic cycle of laccase-mediator systems showing the formation of intermediate radicals upon oxidation of the phenolic substrates and the routes leading to the formation of different oxidation products (adapted from Bioresource Technol 124, Rosado T, Bernardo P, Koci K, Coelho AV, Robalo MP, Martins LO, Methyl syringate: an efficient phenolic mediator for bacterial and fungal laccases, 371–378, copyright © 2012, with permission from Elsevier). The phenoxy radicals are not only involved in the oxidation of the non-phenolics but additionally are involved in the formation of 2,6-dimethoxy-*p*-benzoquinone (DMBQ) and other phenolic compounds and in coupling reactions, which can include radical recombination, cross-coupling and self-coupling proceeding with or without release of substituent groups

bacterial laccases as promising tools to tackle the lignin depolymerisation/repolymerisation bottleneck. MetZyme® LIGNO^{TM}, a laccase of bacterial origin, was successfully used for lignin depolymerisation at 50 °C and pH 10.5 where lignin is soluble in water (Hamalainen et al. 2018). After the enzymatic treatment, extensive lignin depolymerisation was observed accompanied by chemical activation via demethylation and benzylic oxidation as well as increased solubility in neutral and acidic pH. The reaction mechanism proposed for the laccase-mediated oxidation of lignin at alkaline pH is the oxidation at the Cα position of the lignin units (Fig. 6). This modification changes the electronic nature of the substituent leading to a drop in the *pKa* of the phenolic group from 9–11 to 7–8 and, for the phenolic units, is responsible for the increased lignin solubility at neutral to acidic pH. For the non-phenolic units, the electron-withdrawing nature of the *para*-substituent group promotes depolymerisation or demethylation reactions in the lignin polymer. Therefore, the authors demonstrate that depolymerisation of lignin prevailed over polymerisation pathways at strong alkaline and aeration conditions (Hamalainen et al. 2018). This enzymatic-based solution opens up new opportunities for lignin valorisation crucial to make economically feasible present and future lignocellulose biorefineries.

Oxidation of aromatic amines using bacterial laccases. The bacterial CotA laccase oxidises at the neutral to alkaline range of pH values a wide range of aromatic amines with different substitution patterns (*ortho*-phenylenediamines, substituted *para*-diphenyleneamines, *ortho*-amino-phenols, among others) that are precursors of dimeric and trimeric dyes (Sousa et al. 2013, 2016, 2019) as well as of substituted heterocyclic frameworks (phenazine, phenoxazinone, carbazole derivatives) (Sousa et al. 2014, 2015, 2018) (Fig. 7). Phenazine and phenoxazine cores are multifunctional and versatile building blocks widely distributed in a vast array of biologically active compounds, such as anti-tumour agents (Bolognese et al. 2002; Corona et al. 2009), antibiotics and antibacterial agents (McDonald et al. 1999; Borrero et al. 2014), agrochemicals (Starke et al. 2004), biosensors (Pauliukaite et al. 2010) and dyes and polymers (Yamamoto et al. 2003). Due to the importance and broad field of applications of the aforementioned molecules, the development of new enzymatic, greener methodologies is crucial as alternative (bio)synthetic routes for the formation of these aromatic frameworks.

Homocoupling reactions of *para*-substituted aromatic amines using CotA laccase result in trimers with diaminated quinone-diimine structures (Fig. 8a) (Sousa et al. 2013) at very good to excellent overall conversion yields. Additionally, the involvement of some of these amines as primary bases in cross-coupling reactions with different couplers mediated by CotA laccase (Sousa et al. 2016) resulted in the production of indo dyes, widely used in the permanent hair and leather colouration (Morel and Christie 2011). The substrate's reactivity is dependent on the nature and position of the substituents in the aniline ring (Sousa et al. 2013, 2016), and in general, laccase's activity is enhanced by the presence of electron-donating groups. Additionally, the efficiency of the CotA laccase enzymatic system is strictly dependent on the difference of the redox potential between the enzyme (0.55 V) (Durao et al. 2008a) and the substrates. The pH of reaction is a paramount parameter since it affects both the catalytic activity of laccase and the redox potentials of the substrates, i.e. their susceptibility for oxidation is also a critical operational parameter (Sousa et al. 2015). In the pathway proposed for the oxidation of substituted aromatic amines (Fig. 8a), the initial step is the abstraction of an electron from aromatic amines, followed by deprotonation of the primary intermediate (oxidation base) and formation of short-lived intermediates (aminium cation radical or neutral radical species) or the benzoquinonediimine intermediate (**A**). Therefore, starting from the

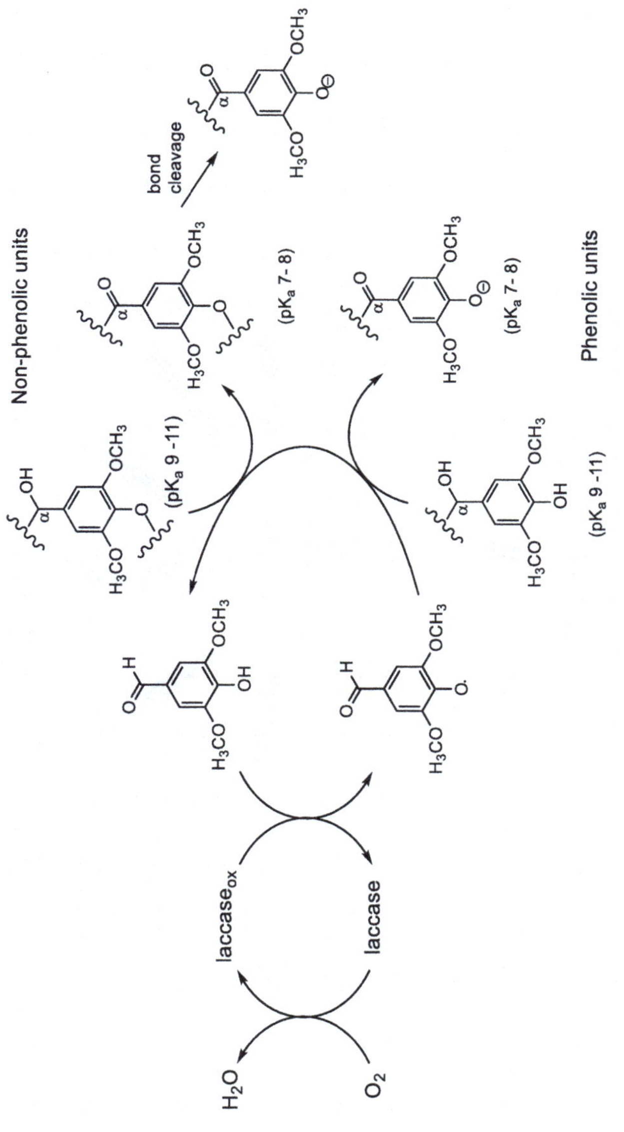

Fig. 6 Scheme of phenolic and non-phenolic lignin units oxidation using a MetZyme® LIGNO™ at alkaline conditions (pH 10.5). Adapted from Hamalainen et al. (2018; https://doi.org/10.3389/fbioe.2018.00020), an open-access article distributed under the terms of the Creative Commons Attribution License (CC BY; https://creativecommons.org/licenses/by/4.0/). Copyright © 2018 Hämäläinen, Grönroos, Suonpää, Heikkilä, Romein, Ihalainen, Malandra and Birikh

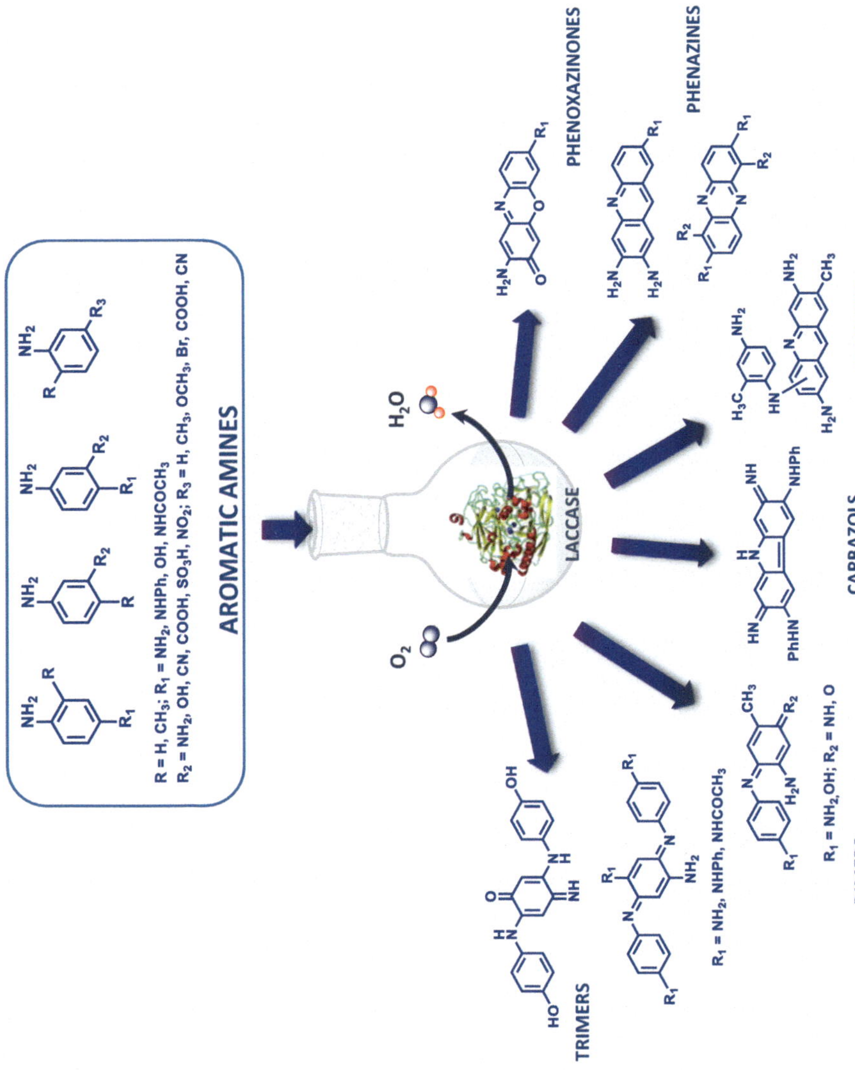

Fig. 7 Aromatic scaffolds obtained from oxidation of structurally different aromatic amines using CotA laccase

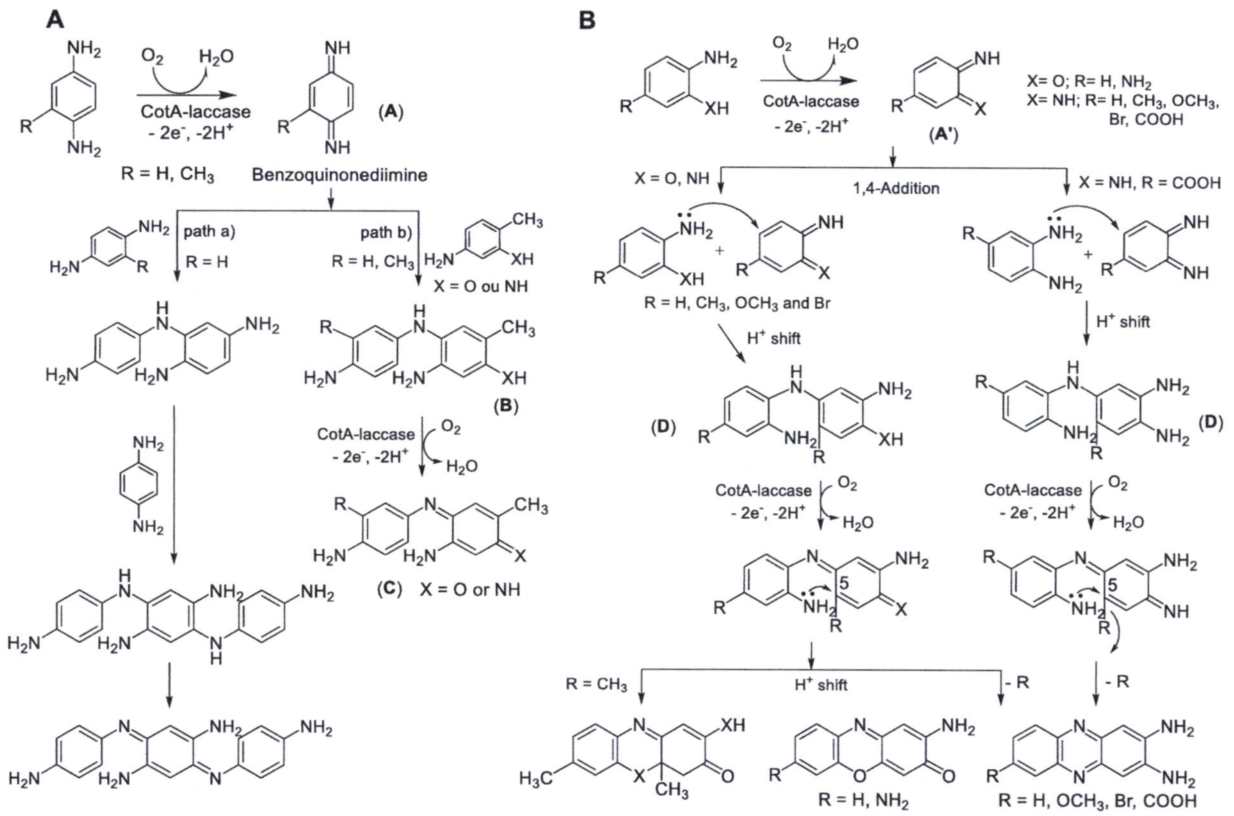

Fig. 8 Proposed pathways involved in the formation of symmetric (a) and asymmetric (b) phenazines and phenoxazines from substituted aromatic amines by CotA laccase. Adapted from Sousa AC, Oliveira MC, Martins LO, Robalo MP, Towards the rational biosynthesis of substituted phenazines and phenoxazinones

intermediate species, by sequential self-conjugation (**path a**) or cross-coupling with different couplers (**path b**), the reaction proceeds through N-C coupling in the ring activated positions. Path **a**) forms a homomolecular dimeric structure which is subsequently transformed in the final 1,4-substituted-2,5-benzoquinonediimine trimer, while path **b**) yields dinuclear leuco dyes (**B**) and results in the final indoaniline or aminoindamine dyes (**C**). According to the proposed pathway, the stability of the radicals, enhanced by the presence of electron donor substituents, seems to be of major importance for the laccase catalytic efficiency (Sousa et al. 2013, 2016).

The formation of different heterocyclic scaffolds, e.g. symmetric and asymmetric phenazines, phenoxazinones and carbazoles, by oxidation of structurally different aromatic substrates assisted by CotA laccase was also reported (Sousa et al. 2014, 2015, 2018) (see Fig. 8b). The efficiency of the reactions leading to the formation of heterocoupled dinuclear or trinuclear dyes was found strongly dependent on the presence of electron donor groups on the *meta*-substituted couplers. A mechanistic pathway for the synthesis of substituted symmetric and asymmetric heterocycles with phenazine and phenoxazine frameworks was proposed (Fig. 8) where an *ortho*-diamine or *ortho*-aminophenol is enzymatically oxidised through two successive one-electron oxidations, generating *ortho*-quinone-diimine or *ortho*-quinone-imine intermediates (**A**). Under the reaction conditions, these species suffer rapid nucleophilic addition by other substrate molecules in its most electrophilic carbon atom, followed by a proton shift, yielding the first coupling intermediate (**D**). These non-isolable products underwent a second oxidation, probably mediated by laccase, and an intramolecular Michael addition of an amino group (or phenol) to the C5 atom, with the displacement of an R group, leads to an aminophenazine or fully reduced aminophenoxazine, which is spontaneously oxidised in air to produce the final heterocycle products (Sousa et al. 2014; Bruyneel et al. 2012). For the *meta, para*-disubstituted aromatic amines, the first step is the in situ generation of a *para*-benzoquinonediimine intermediate (**A′**) in a similar way as described above. This intermediate further reacts with the nucleophilic amino group of another molecule at the *ortho*-position, adjacent to the R_1 group leading to the formation of dimeric structures. This second step, followed by a proton loss, yields the first coupling intermediate (**D**). These non-isolable products underwent a second oxidation, probably mediated by laccase, and an intramolecular Michael addition of an amino group (or phenol) to the C5 atom, with the displacement of an R group, leads to an aminophenazine or fully reduced aminophenoxazine, which is spontaneously oxidised in air to produce the final heterocycle products. The synthesis of azo dyes was also reported as minor products of laccase's aromatic amines' oxidation (Sousa et al. 2013, 2014, 2018). Recently, the enzymatic oxidation of aromatic amines was

---

Fig. 8 (continued) by laccases, Green Chem 16:4127–4136 (2018), and from Sousa AC, Oliveira MC, Martins LO, Robalo MP, A Sustainable synthesis of asymmetric phenazines and phenoxazinones mediated by CotA-Laccase, Adv Synth Catal 360:575–583 (2018), with permission from John Wiley & Sons. (copyright © 2018 Wiley-VCH Verlag GmbH&Co. KGaA, Weinheim)

directed to the formation of *ortho-* and *para-*substituted azo dyes in the presence of ABTS as mediator, and the correspondent mechanistic pathway was proposed (Sousa et al. 2019). The variety of different aromatic scaffolds obtained by this enzymatic approach clearly shows that laccases are promising tools for aromatic amines' oxidation, boosting new eco-friendly alternatives to the production of value-added aromatic compounds.

5 Concluding Remarks

The studies of laccases during the last years have revealed that they are a group of enzymes widely distributed in the bacterial domain. They apparently show different physiological roles, being the resistance to copper the most frequently described among those enzymes characterised. Bacterial laccases show a redox potential around 500 mV, 200–300 mV lower than fungal counterparts. However, their optimal pH for phenolics and amines, at the neutral to alkaline range, in contrast to the acidic character of fungal laccases, their optimal temperatures between 50 and 80 °C and their improved stability to temperature and denaturating agents, such as high concentrations of salt, detergents and organic solvents, allow their applications in a wide range of biotechnological fields. It is expected that in the next decade, the scope of applications expands due to the application of protein engineering approaches that improve their production yields and kinetic parameters for exquisite substrates enhancing the number of laccase-based industrial oxidation processes, in particular in the lignocellulose biorefinery field and wastes valorisation.

Acknowledgements This work was supported by the project grants PTDC/BBB-EBB/0122/2014, PTDC/BII-BBF/29564/2017, UID/Multi/04551/2013, UID/Multi/04326/2019 and UID/QUI/00100/2013 from FCT, Fundação para a Ciência e Tecnologia, Portugal, and Project LISBOA-01-0145-FEDER-007660 (Microbiologia Molecular, Estrutural e Celular), Research Unit GREEN-IT (Bioresources for Sustainability), UID/Multi/04551/2013, funded by FEDER funds through COMPETE2020—Programa Operacional Competitividade e Internacionalização (POCI) and FCT, Portugal, and by the European Union's Horizon H2020 B-Ligzymes RISE project (GA 824017).

References

Agostinelli E, Cervoni L, Giartosio A, Morpurgo L (1995) Stability of Japanese-lacquer-tree (Rhus vernicifera) laccase to thermal and chemical denaturation – comparison with ascorbate oxidase. Biochem J 306:697–702. https://doi.org/10.1042/bj3060697

Arias ME, Arenas M, Rodriguez J, Soliveri J, Ball AS, Hernandez M (2003) Kraft pulp biobleaching and mediated oxidation of a nonphenolic substrate by laccase from *Streptomyces cyaneus* CECT 3335. Appl Environ Microb 69(4):1953–1958. https://doi.org/10.1128/Aem.69.4.1953-1958.2003

Baptista RP, Chen LY, Paixao A, Cabral JM, Melo EP (2003) A novel pathway to enzyme deactivation: the cutinase model. Biotechnol Bioeng 82(7):851–857. https://doi.org/10.1002/bit.10641

Beeby M, O'Connor BD, Ryttersgaard C, Boutz DR, Perry LJ, Yeates TO (2005) The genomics of disulfide bonding and protein stabilization in thermophiles. PLoS Biol 3(9):e309. https://doi.org/10.1371/journal.pbio.0030309

Bello M, Valderrama B, Serrano-Posada H, Rudino-Pinera E (2012) Molecular dynamics of a thermostable multicopper oxidase from *Thermus thermophilus* HB27: structural differences between the apo and holo forms. PLoS One 7(7):e40700. https://doi.org/10.1371/journal.pone.0040700

Bello M, Correa-Basurto J, Rudino-Pinera E (2014) Simulation of the cavity-binding site of three bacterial multicopper oxidases upon complex stabilization: interactional profile and electron transference pathways. J Biomol Struct Dyn 32(8):1303–1317. https://doi.org/10.1080/07391102.2013.817954

Bertrand T, Jolivalt C, Briozzo P, Caminade E, Joly N, Madzak C, Mougin C (2002) Crystal structure of a four-copper laccase complexed with an arylamine: insights into substrate recognition and correlation with kinetics. Biochemistry 41(23):7325–7333. https://doi.org/10.1021/Bi0201318

Bolognese A, Correale G, Manfra M, Lavecchia A, Mazzoni O, Novellino E, Barone V, La Colla P, Loddo R (2002) Antitumor agents. 2. Synthesis, structure-activity relationships, and biological evaluation of substituted 5H-pyridophenoxazin-5-ones with potent antiproliferative activity. J Med Chem 45(24):5217–5223

Bonaccorsi di Patti MCB, Musci G, Giartosio A, Dalessio S, Calabrese L (1990) The multidomain structure of ceruloplasmin from calorimetric and limited proteolysis studies. J Biol Chem 265 (34):21016–21022

Borrero NV, Bai F, Perez C, Duong BQ, Rocca JR, Jin S, Huigens RW (2014) Phenazine antibiotic inspired discovery of potent bromophenazine antibacterial agents against Staphylococcus aureus and Staphylococcus epidermidis. Org Biomol Chem 12(6):881–886. https://doi.org/10.1039/c3ob42416b

Brander S, Mikkelsen JD, Kepp KP (2014) Characterization of an alkali- and halide-resistant laccase expressed in *E. coli*: CotA from *Bacillus clausii*. PLoS One 9(6):e99402. https://doi.org/10.1371/journal.pone.0099402

Brander S, Mikkelsen JD, Kepp KP (2015) TtMCO: a highly thermostable laccase-like multicopper oxidase from the thermophilic *Thermobaculum terrenum*. J Mol Catal B 112:59–65. https://doi.org/10.1016/j.molcatb.2014.12.002

Brown NL, Barrett SR, Camakaris J, Lee BT, Rouch DA (1995) Molecular genetics and transport analysis of the copper-resistance determinant (pco) from *Escherichia coli* plasmid pRJ1004. Mol Microbiol 17(6):1153–1166. https://doi.org/10.1111/j.1365-2958.1995.mmi_17061153.x

Bruyneel F, Dive G, Marchand-Brynaert J (2012) Non-symmetrically substituted phenoxazinones from laccase-mediated oxidative cross-coupling of aminophenols: an experimental and theoretical insight. Org Biomol Chem 10:1834–1846. https://doi.org/10.1039/C1OB05795B

Callejon S, Sendra R, Ferrer S, Pardo I (2016) Cloning and characterization of a new laccase from *Lactobacillus plantarum* J16 CECT 8944 catalyzing biogenic amines degradation. Appl Microbiol Biotechnol 100(7):3113–3124. https://doi.org/10.1007/s00253-015-7158-0

Callejon S, Sendra R, Ferrer S, Pardo I (2017) Recombinant laccase from *Pediococcus acidilactici* CECT 5930 with ability to degrade tyramine. PLoS One 12(10):e0186019. https://doi.org/10.1371/journal.pone.0186019

Cha J, Cooksey DA (1991) Copper resistance in *Pseudomonas syringae* mediated by periplasmic and outer membrane proteins. Proc Nat Acad Sci USA 88(20):8915–8919. https://doi.org/10.1073/pnas.88.20.8915

Chandra R, Chowdhary P (2015) Properties of bacterial laccases and their application in bioremediation of industrial wastes. Environ Sci Proc Impacts 17(2):326–342. https://doi.org/10.1039/c4em00627e

Changela A, Chen K, Xue Y, Holschen J, Outten CE, O'Halloran TV, Mondragon A (2003) Molecular basis of metal-ion selectivity and zeptomolar sensitivity by CueR. Science 301 (5638):1383–1387. https://doi.org/10.1126/science.1085950

Chauhan PS, Goradia B, Saxena A (2017) Bacterial laccase: recent update on production, properties and industrial applications. 3. Biotech 7(5):323. https://doi.org/10.1007/s13205-017-0955-7

Corona P, Carta A, Loriga M, Vitale G, Paglietti G (2009) Synthesis and in vitro antitumor activity of new quinoxaline derivatives. Eur J Med Chem 44(4):1579–1591. https://doi.org/10.1016/j.ejmech.2008.07.025

d'Acunzo F, Galli C (2003) First evidence of catalytic mediation by phenolic compounds in the laccase-induced oxidation of lignin models. Eur J Biochem 270(17):3634–3640. https://doi.org/10.1046/j.1432-1033.2003.03752.x

Durao P, Bento I, Fernandes AT, Melo EP, Lindley PF, Martins LO (2006) Perturbations of the T1 copper site in the CotA laccase from *Bacillus subtilis*: structural, biochemical, enzymatic and stability studies. J Biol Inorg Chem 11(4):514–526. https://doi.org/10.1007/s00775-006-0102-0

Durao P, Chen Z, Fernandes AT, Hildebrandt P, Murgida DH, Todorovic S, Pereira MM, Melo EP, Martins LO (2008a) Copper incorporation into recombinant CotA laccase from Bacillus subtilis: characterization of fully copper loaded enzymes. J Biol Inorg Chem 13(2):183–193. https://doi.org/10.1007/s00775-007-0312-0

Durao P, Chen Z, Silva CS, Soares CM, Pereira MM, Todorovic S, Hildebrandt P, Bento I, Lindley PF, Martins LO (2008b) Proximal mutations at the type 1 copper site of CotA laccase: spectroscopic, redox, kinetic and structural characterization of I494A and L386A mutants. Biochem J 412(2):339–346. https://doi.org/10.1042/BJ20080166

Enguita FJ, Martins LO, Henriques AO, Carrondo MA (2003) Crystal structure of a bacterial endospore coat component. A laccase with enhanced thermostability properties. J Biol Chem 278(21):19416–19425. https://doi.org/10.1074/jbc.M301251200

Enguita FJ, Marcal D, Martins LO, Grenha R, Henriques AO, Lindley PF, Carrondo MA (2004) Substrate and doxygen binding to the endospore coat laccase from *Bacillus subtilis*. J Biol Chem 279(22):23472–23476. https://doi.org/10.1074/jbc.M314000200

Espinosa E, Marco-Noales E, Gomez D, Lucas-Elio P, Ordax M, Garcias-Bonet N, Duarte CM, Sanchez-Amat A (2010) Taxonomic study of Marinomonas strains isolated from the seagrass *Posidonia oceanica*, with descriptions of *Marinomonas balearica* sp. nov. and *Marinomonas pollencensis* sp. nov. Int J Syst Evol Microbiol 60:93–98. https://doi.org/10.1099/ijs.0.008607-0

Eugenio ME, Hernandez M, Moya R, Martin-Sampedro R, Villar JC, Arias ME (2011) Evaluation of a new laccase produced by *Streptomyces Ipomoea* on biobleaching and ageing of kraft pulps. Bioresources 6(3):3231–3241. https://doi.org/10.15376/biores.6.3.3231-3241

Fernandes AT, Soares CM, Pereira MM, Huber R, Grass G, Martins LO (2007) A robust metallo-oxidase from the hyperthermophilic bacterium *Aquifex aeolicus*. FEBS J 274(11):2683–2694. https://doi.org/10.1111/j.1742-4658.2007.05803.x

Fernandes AT, Martins LO, Melo EP (2009) The hyperthermophilic nature of the metallo-oxidase from *Aquifex aeolicus*. BBA-Proteins Proteom 1794(1):75–83. https://doi.org/10.1016/j.bbapap.2008.09.006

Fernandes AT, Damas JM, Todorovic S, Huber R, Baratto MC, Pogni R, Soares CM, Martins LO (2010) The multicopper oxidase from the archaeon *Pyrobaculum aerophilum* shows nitrous oxide reductase activity. FEBS J 277(15):3176–3189. https://doi.org/10.1111/j.1742-4658.2010.07725.x

Fernandes AT, Pereira MM, Silva CS, Lindley PF, Bento I, Melo EP, Martins LO (2011) The removal of a disulfide bridge in CotA-laccase changes the slower motion dynamics involved in copper binding but has no effect on the thermodynamic stability. J Biol Inorg Chem 16 (4):641–651. https://doi.org/10.1007/s00775-011-0768-9

Fernandes AT, Lopes C, Martins LO, Melo EP (2012) Unfolding pathway of CotA-laccase and the role of copper on the prevention of refolding through aggregation of the unfolded state. Biochem Biophys Res Commun 422(3):442–446. https://doi.org/10.1016/j.bbrc.2012.05.011

Fernandez E, Sanchez-Amat A, Solano F (1999) Location and catalytic characteristics of a multipotent bacterial polyphenol oxidase. Pigment Cell Res 12(5):331–339. https://doi.org/10.1111/j.1600-0749.1999.tb00767.x

Fitter J, Heberle J (2000) Structural equilibrium fluctuations in mesophilic and thermophilic alpha-amylase. Biophys J 79(3):1629–1636. https://doi.org/10.1016/S0006-3495(00)76413-7

Galai S, Lucas-Elio P, Marzouki MN, Sanchez-Amat A (2011) Molecular cloning of a copper-dependent laccase from the dye-decolorizing strain *Stenotrophomonas maltophilia* AAP56. J Appl Microbiol 111(6):1394–1405. https://doi.org/10.1111/j.1365-2672.2011.05164.x

Galli I, Musci G, Bonaccorsi di Patti MC (2004) Sequential reconstitution of copper sites in the multicopper oxidase CueO. J Biol Inorg Chem 9(1):90–95. https://doi.org/10.1007/s00775-003-0501-4

Ghatge S, Yang Y, Song WY, Kim TY, Hur HG (2018) A novel laccase from thermoalkaliphilic bacterium Caldalkalibacillus thermarum strain TA2.A1 able to catalyze dimerization of a lignin model compound. Appl Microbiol Biotechnol 102(9):4075–4086. https://doi.org/10.1007/s00253-018-8898-4

Giannotti MI, Cabeza de Vaca I, Artes JM, Sanz F, Guallar V, Gorostiza P (2015) Direct measurement of the nanomechanical stability of a redox protein active site and its dependence upon metal binding. J Phys Chem B 119(36):12050–12058. https://doi.org/10.1021/acs.jpcb.5b06382

Goacher RE, Braham EJ, Michienzi CL, Flick RM, Yakunin AF, Master ER (2018) Direct analysis by time-of-flight secondary ion mass spectrometry reveals action of bacterial laccase-mediator systems on both hardwood and softwood samples. Physiol Plant 164(1):5–16. https://doi.org/10.1111/ppl.12688

Granja-Travez RS, Bugg TDH (2018) Characterization of multicopper oxidase CopA from *Pseudomonas putida* KT2440 and *Pseudomonas fluorescens* Pf-5: involvement in bacterial lignin oxidation. Arch Biochem Biophys 660:97–107. https://doi.org/10.1016/j.abb.2018.10.012

Granja-Travez RS, Wilkinson RC, Persinoti GF, Squina FM, Fulop V, Bugg TDH (2018) Structural and functional characterisation of multi-copper oxidase CueO from lignin-degrading bacterium *Ochrobactrum* sp. reveal its activity towards lignin model compounds and lignosulfonate. FEBS J 285(9):1684–1700. https://doi.org/10.1111/febs.14437

Grass G, Rensing C (2001) CueO is a multi-copper oxidase that confers copper tolerance in Escherichia coli. Biochem Biophys Res Commun 286(5):902–908. https://doi.org/10.1006/bbrc.2001.5474

Grass G, Thakali K, Klebba PE, Thieme D, Muller A, Wildner GF, Rensing C (2004) Linkage between catecholate siderophores and the multicopper oxidase CueO in *Escherichia coli*. J Bacteriol 186(17):5826–5833. https://doi.org/10.1128/Jb.186.17.5826-5833.2004

Guan ZB, Luo Q, Wang HR, Chen Y, Liao XR (2018) Bacterial laccases: promising biological green tools for industrial applications. Cell Mol Life Sci 75(19):3569–3592. https://doi.org/10.1007/s00018-018-2883-z

Gunne M, Hoppner A, Hagedoorn PL, Urlacher VB (2014) Structural and redox properties of the small laccase Ssl1 from *Streptomyces sviceus*. FEBS J 281(18):4307–4318. https://doi.org/10.1111/febs.12755

Hakulinen N, Rouvinen J (2015) Three-dimensional structures of laccases. Cell Mol Life Sci 72(5):857–868. https://doi.org/10.1007/s00018-014-1827-5

Hakulinen N, Kiiskinen LL, Kruus K, Saloheimo M, Paananen A, Koivula A, Rouvinen J (2002) Crystal structure of a laccase from *Melanocarpus albomyces* with an intact trinuclear copper site. Nat Struct Biol 9(8):601–605. https://doi.org/10.1038/Nsb823

Hall SJ, Hitchcock A, Butler CS, Kelly DJ (2008) A multicopper oxidase (Cj1516) and a CopA homologue (Cj1161) are major components of the copper homeostasis system of *Campylobacter jejuni*. J Bacteriol 190(24):8075–8085. https://doi.org/10.1128/Jb.00821-08

Hamalainen V, Gronroos T, Suonpaa A, Hekkila MW, Romein B, Ihalainen P, Malandra S, Birikh KR (2018) Enzymatic processes to unlock the lignin value. Front Bioeng Biotechnol 6(20). https://doi.org/10.3389/fbioe.2018.00020

Hernandez-Romero D, Solano F, Sanchez-Amat A (2005) Polyphenol oxidase activity expression in *Ralstonia solanacearum*. Appl Environ Microb 71(11):6808–6815. https://doi.org/10.1128/Aem.71.11.6808-6815.2005

Hong G, Ivnitski DM, Johnson GR, Atanassov P, Pachter R (2011) Design parameters for tuning the type 1 Cu multicopper oxidase redox potential: insight from a combination of first principles and empirical molecular dynamics simulations. J Am Chem Soc 133(13):4802–4809. https://doi.org/10.1021/ja105586q

Huang X-F, Santhanam N, Badri DV, Hunter WJ, Manter DK, Vivanco JM, Reardon KF (2013) Isolation and characterization of lignin-degrading bacteria from rainforest soils. Biotechnol Bioeng 110:1616–1626

Huffman DL, Huyett J, Outten FW, Doan PE, Finney LA, Hoffman BM, O'Halloran TV (2002) Spectroscopy of Cu(II)-PcoC and the multicopper oxidase function of PcoA, two essential components of *Escherichia coli* pco copper resistance operon. Biochemistry 41(31):10046–10055. https://doi.org/10.1021/bi0259960

Hullo MF, Moszer I, Danchin A, Martin-Verstraete I (2001) CotA of *Bacillus subtilis* is a copper-dependent laccase. J Bacteriol 183(18):5426–5430. https://doi.org/10.1128/JB.183.18.5426-5430.2001

Jones GH (2000) Actinomycin production persists in a strain of *Streptomyces antibioticus* lacking phenoxazinone synthase. Antimicrob Agents Chemother 44(5):1322–1327

Jones SM, Solomon EI (2015) Electron transfer and reaction mechanism of laccases. Cell Mol Life Sci 72(5):869–883. https://doi.org/10.1007/s00018-014-1826-6

Kallio JP, Auer S, Janis J, Andberg M, Kruus K, Rouvinen J, Koivula A, Hakulinen N (2009) Structure-function studies of a *Melanocarpus albomyces* laccase suggest a pathway for oxidation of phenolic compounds. J Mol Biol 392(4):895–909. https://doi.org/10.1016/j.jmb.2009.06.053

Kalyani D, Munk L, Mikkelsen JD, Meyer AS (2016) Molecular and biochemical characterization of a new thermostable bacterial laccase from *Meiothermus ruber* DSM 1279. RSC Adv 6:3910–3918. https://doi.org/10.1039/C5RA24374B

Kim HW, Lee SY, Park H, Jeon SJ (2015) Expression, refolding, and characterization of a small laccase from *Thermus thermophilus* HJ6. Prot Expr Purif 114:37–43. https://doi.org/10.1016/j.pep.2015.06.004

Komori H, Miyazaki K, Higuchi Y (2009) X-ray structure of a two-domain type laccase: a missing link in the evolution of multi-copper proteins. FEBS Lett 583(7):1189–1195. https://doi.org/10.1016/j.febslet.2009.03.008

Koroleva OV, Stepanova EV, Binukov VI, Timofeev VP, Pfeil W (2001) Temperature-induced changes in copper centers and protein conformation of two fungal laccases from *Coriolus hirsutus* and *Coriolus zonatus*. BBA-Protein Struct M 1547(2):397–407. https://doi.org/10.1016/S0167-4838(01)00209-6

Koschorreck K, Richter SM, Ene AB, Roduner E, Schmid RD, Urlacher VB (2008) Cloning and characterization of a new laccase from *Bacillus licheniformis* catalyzing dimerization of phenolic acids. Appl Microbiol Biotechnol 79(2):217–224. https://doi.org/10.1007/s00253-008-1417-2

Kosman DJ (2017) Multicopper oxidase assembly. In: Encyclopedia of inorganic and bioinorganic chemistry. Wiley. https://doi.org/10.1002/9781119951438.eibc2475

Kudanga T, Nemadziva B, Le Roes-Hill M (2017) Laccase catalysis for the synthesis of bioactive compounds. Appl Microbiol Biotechnol 101(1):13–33. https://doi.org/10.1007/s00253-016-7987-5

Kumar S, Stecher G, Li M, Knyaz C, Tamura K (2018) MEGA X: molecular evolutionary genetics analysis across computing platforms. Mol Biol Evol 35(6):1547–1549. https://doi.org/10.1093/molbev/msy096

Letunic I, Bork P (2019) Interactive tree of life (iTOL) v4: recent updates and new developments. Nucleic Acids Res. https://doi.org/10.1093/nar/gkz239

Liu C, Yang GY, Wu L, Tian GH, Zhang ZM, Feng Y (2011) Switch of substrate specificity of hyperthermophilic acylaminoacyl peptidase by combination of protein and solvent engineering. Protein Cell 2(6):497–506. https://doi.org/10.1007/s13238-011-1057-7

Machczynski MC, Vijgenboom E, Samyn B, Canters GW (2004) Characterization of SLAC: a small laccase from *Streptomyces coelicolor* with unprecedented activity. Prot Sci 13 (9):2388–2397. https://doi.org/10.1110/Ps.04759104

Madzak C, Mimmi MC, Caminade E, Brault A, Baumberger S, Briozzo P, Mougin C, Jolivalt C (2006) Shifting the optimal pH of activity for a laccase from the fungus *Trametes versicolor* by structure-based mutagenesis. Prot Eng Design Sel 19(2):77–84. https://doi.org/10.1093/protein/gzj004

Majumdar S, Lukk T, Solbiati JO, Bauer S, Nair SK, Cronan JE, Gerlt JA (2014) Roles of small laccases from *Streptomyces* in lignin degradation. Biochemistry 53:4047–4058. https://doi.org/10.1021/bi500285t

Marcos-Torres FJ, Pérez J, Gómez-Santos N, Moraleda-Muñoz A, Muñoz-Dorado J (2016) In depth analysis of the mechanism of action of metal-dependent sigma factors: characterization of CorE2 from *Myxococcus xanthus*. Nucleic Acids Res 44(12):5571–5584. https://doi.org/10.1093/nar/gkw150

Marshall NM, Garner DK, Wilson TD, Gao YG, Robinson H, Nilges MJ, Lu Y (2009) Rationally tuning the reduction potential of a single cupredoxin beyond the natural range. Nature 462 (7269):113–116. https://doi.org/10.1038/nature08551

Martins LO, Soares CM, Pereira MM, Teixeira M, Costa T, Jones GH, Henriques AO (2002) Molecular and biochemical characterization of a highly stable bacterial laccase that occurs as a structural component of the *Bacillus subtilis* endospore coat. J Biol Chem 277 (21):18849–18859. https://doi.org/10.1074/jbc.M200827200

Martins LO, Durao P, Brissos V, Lindley PF (2015) Laccases of prokaryotic origin: enzymes at the interface of protein science and protein technology. Cell Mol Life Sci 72(5):911–922. https://doi.org/10.1007/s00018-014-1822-x

Matera I, Gullotto A, Tilli S, Ferraroni M, Scozzafava A, Briganti F (2008) Crystal structure of the blue multicopper oxidase from the white-rot fungus *Trametes trogii* complexed with p-toluate. Inorg Chim Acta 361(14–15):4129–4137. https://doi.org/10.1016/j.ica.2008.03.091

Mayer AM, Staples RC (2002) Laccase: new functions for an old enzyme. Phytochem 60 (6):551–565. https://doi.org/10.1016/S0031-9422(02)00171-1

McDonald M, Wilkinson B, Van't Land CW, Mocek U, Lee S, Floss HG (1999) Biosynthesis of phenazine antibiotics in Streptomyces antibioticus: stereochemistry of methyl transfer from carbon-2 of acetate. J Am Chem Soc 121(24):5619–5624. https://doi.org/10.1021/ja991159i

Miyazaki K (2005) A hyperthermophilic laccase from *Thermus thermophilus* HB27. Extremophiles 9(6):415–425. https://doi.org/10.1007/s00792-005-0458-z

Morel OJX, Christie RM (2011) Current trends in the chemistry of permanent hair dyeing. Chem Rev 111(4):2537–2561. https://doi.org/10.1021/cr1000145

Moya R, Saastamoinen P, Hernandez M, Suurnakki A, Arias E, Mattinen ML (2011) Reactivity of bacterial and fungal laccases with lignin under alkaline conditions. Bioresour Technol 102 (21):10006–10012. https://doi.org/10.1016/j.biortech.2011.08.046

Muñoz-Dorado J, Marcos-Torres FJ, García-Bravo E, Moraleda-Muñoz A, Pérez J (2016) Myxobacteria: moving, killing, feeding, and surviving together. Front Microbiol 7:781. https://doi.org/10.3389/fmicb.2016.00781

Nakamura K, Go N (2005) Function and molecular evolution of multicopper blue proteins. Cell Mol Life Sci 62(18):2050–2066. https://doi.org/10.1007/s00018-004-5076-5

Navas LE, Martinez FD, Taverna ME, Fetherolf MM, Eltis LD, Nicolau V, Estenoz D, Campos E, Benintende GB, Berretta MF (2019) A thermostable laccase from *Thermus* sp. 2.9 and its potential for delignification of Eucalyptus biomass. AMB Express 9(1):24. https://doi.org/10.1186/s13568-019-0748-y

Outten FW, Outten CE, Hale J, O'Halloran TV (2000) Transcriptional activation of an *Escherichia coli* copper efflux regulon by the chromosomal MerR homologue, cueR. J Biol Chem 275 (40):31024–31029. https://doi.org/10.1074/jbc.M006508200

Outten FW, Huffman DL, Hale JA, O'Halloran TV (2001) The independent *cue* and *cus* systems confer copper tolerance during aerobic and anaerobic growth in *Escherichia coli*. J Biol Chem 276(33):30670–30677. https://doi.org/10.1074/jbc.M104122200

Pace CN, Hebert EJ, Shaw KL, Schell D, Both V, Krajcikova D, Sevcik J, Wilson KS, Dauter Z, Hartley RW, Grimsley GR (1998) Conformational stability and thermodynamics of folding of ribonucleases Sa, Sa2 and Sa3. J Mol Biol 279(1):271–286. https://doi.org/10.1006/jmbi.1998.1760

Palm-Espling ME, Niemiec MS, Wittung-Stafshede P (2012) Role of metal in folding and stability of copper proteins in vitro. BBA-Protein Struct M 1823(9):1594–1603. https://doi.org/10.1016/j.bbamcr.2012.01.013

Pauliukaite R, Ghica ME, Barsan MM, Brett CMA (2010) Phenazines and polyphenazines in electrochemical sensors and biosensors. Anal Lett 43(10–11):1588–1608. https://doi.org/10.1080/00032711003653791

Pawlik A, Wojcik M, Rulka K, Motyl-Gorzel K, Osinska-Jaroszuk M, Wielbo J, Marek-Kozaczuk-M, Skorupska A, Rogalski J, Janusz G (2016) Purification and characterization of laccase from Sinorhizobium meliloti and analysis of the lacc gene. Int J Biol Macromol 92:138–147. https://doi.org/10.1016/j.ijbiomac.2016.07.012

Pérez J, Muñoz-Dorado J, Moraleda-Muñoz A (2018) The complex global response to copper in the multicellular bacterium *Myxococcus xanthus*. Metallomics 10(7):876–886. https://doi.org/10.1039/c8mt00121a

Petsko GA (2001) Structural basis of thermostability in hyperthermophilic proteins, or "there's more than one way to skin a cat". Methods Enzymol 334:469–478

Pozdnyakova I, Guidry J, Wittung-Stafshede P (2001) Copper stabilizes azurin by decreasing the unfolding rate. Arch Biochem Biophys 390(1):146–148. https://doi.org/10.1006/abbi.2000.2369

Rademacher C, Masepohl B (2012) Copper-responsive gene regulation in bacteria. Microbiology 158(Pt 10):2451–2464. https://doi.org/10.1099/mic.0.058487-0

Rademacher C, Moser R, Lackmann JW, Klinkert B, Narberhaus F, Masepohl B (2012) Transcriptional and posttranscriptional events control copper-responsive expression of a *Rhodobacter capsulatus* multicopper oxidase. J Bacteriol 194(8):1849–1859. https://doi.org/10.1128/Jb.06274-11

Reiss R, Ihssen J, Richter M, Eichhorn E, Schilling B, Thony-Meyer L (2013) Laccase versus laccase-like multi-copper oxidase: a comparative study of similar enzymes with diverse substrate spectra. PLoS One 8(6):e65633. https://doi.org/10.1371/journal.pone.0065633

Ricklefs E, Winkler N, Koschorreck K, Urlacher VB (2014) Expanding the laccase-toolbox: a laccase from *Corynebacterium glutamicum* with phenol coupling and cuprous oxidase activity. J Biotechnol 191:46–53. https://doi.org/10.1016/j.jbiotec.2014.05.031

Roberts SA, Weichsel A, Grass G, Thakali K, Hazzard JT, Tollin G, Rensing C, Montfort WR (2002) Crystal structure and electron transfer kinetics of CueO, a multicopper oxidase required for copper homeostasis in *Escherichia coli*. Proc Natl Acad Sci USA 99(5):2766–2771. https://doi.org/10.1073/pnas.052710499

Robic S, Guzman-Casado M, Sanchez-Ruiz JM, Marqusee S (2003) Role of residual structure in the unfolded state of a thermophilic protein. Proc Natl Acad Sci USA 100(20):11345–11349. https://doi.org/10.1073/pnas.1635051100

Romero-Guido C, Baez A, Torres E (2018) Dioxygen activation by laccases: green chemistry for fine chemical synthesis. Catalysts 8:223. https://doi.org/10.3390/catal8060223

Rosado T, Bernardo P, Koci K, Coelho AV, Robalo MP, Martins LO (2012) Methyl syringate: an efficient phenolic mediator for bacterial and fungal laccases. Bioresource Technol 124:371–378. https://doi.org/10.1016/j.biortech.2012.08.023

Roth S, Spiess AC (2015) Laccases for biorefinery applications: a critical review on challenges and perspectives. Bioprocess Bios Eng 38(12):2285–2313. https://doi.org/10.1007/s00449-015-1475-7

Rouch DA, Brown NL (1997) Copper-inducible transcriptional regulation at two promoters in the Escherichia coli copper resistance determinant pco. Microbiology 143(Pt 4):1191–1202. https://doi.org/10.1099/00221287-143-4-1191

Sakuraba H, Koga K, Yoneda K, Kashima Y, Ohshima T (2011) Structure of a multicopper oxidase from the hyperthermophilic archaeon *Pyrobaculum aerophilum*. Acta Crystallogr Sect F-Struct Biol Crystall Commun 67:753–757. https://doi.org/10.1107/S1744309111018173

Sanchez-Amat A, Lucas-Elio P, Fernandez E, Garcia-Borron JC, Solano F (2001) Molecular cloning and functional characterization of a unique multipotent polyphenol oxidase from *Marinomonas mediterranea*. BBA-Protein Struct M 1547(1):104–116. https://doi.org/10.1016/S0167-4838(01)00174-1

Sanchez-Sutil MC, Gomez-Santos N, Moraleda-Munoz A, Martins LO, Perez J, Munoz-Dorado J (2007) Differential expression of the three multicopper oxidases from *Myxococcus xanthus*. J Bacteriol 189(13):4887–4898. https://doi.org/10.1128/JB.00309-07

Sánchez-Sutil MC, Marcos-Torres FJ, Pérez J, Ruiz-González M, García-Bravo E, Martínez-Cayuela M, Gómez-Santos N, Moraleda-Muñoz A, Muñoz-Dorado J (2016) Dissection of the sensor domain of the copper-responsive histidine kinase CorS from *Myxococcus xanthus*. Environ Microbiol Rep 8(3):363–370. https://doi.org/10.1111/1758-2229.12389

Savini I, Dalessio S, Giartosio A, Morpurgo L, Avigliano L (1990) The role of copper in the stability of ascorbate oxidase towards denaturing agents. Eur J Biochem 190(3):491–495. https://doi.org/10.1111/j.1432-1033.1990.tb15600.x

Sawle L, Ghosh K (2011) How do thermophilic proteins and proteomes withstand high temperature? Biophys J 101(1):217–227. https://doi.org/10.1016/j.bpj.2011.05.059

Sedlak E, Wittung-Stafshede P (2007) Discrete roles of copper ions in chemical unfolding of human ceruloplasmin. Biochemistry 46(33):9638–9644. https://doi.org/10.1021/bi700715e

Serrano-Posada H, Valderrama B, Stojanoff V, Rudino-Pinera E (2011) Thermostable multicopper oxidase from *Thermus thermophilus* HB27: crystallization and preliminary X-ray diffraction analysis of apo and holo forms. Acta Crystallogr Sect F 67(Pt 12):1595–1598. https://doi.org/10.1107/S174430911103805X

Silva CS, Durao P, Fillat A, Lindley PF, Martins LO, Bento I (2012) Crystal structure of the multicopper oxidase from the pathogenic bacterium *Campylobacter jejuni* CGUG11284: characterization of a metallo-oxidase. Metallomics 4(1):37–47. https://doi.org/10.1039/c1mt00156f

Singh SK, Grass G, Rensing C, Montfort WR (2004) Cuprous oxidase activity of CueO from Escherichia coli. J Bacteriol 186(22):7815–7817. https://doi.org/10.1128/Jb.186.22.7815-7817.2004

Singh R, Hu J, Regner MR, Round JW, Ralph J, Saddler JN, Eltis LD (2017) Enhanced delignification of steam-pretreated poplar by a bacterial laccase. Sci Rep 7:42121. https://doi.org/10.1038/srep42121

Sitthisak S, Howieson K, Amezola C, Jayaswal RK (2005) Characterization of a multicopper oxidase gene from *Staphylococcus aureus*. Appl Env Microbiol 71(9):5650–5653. https://doi.org/10.1128/AEM.71.9.5650-5653.2005

Skalova T, Dohnalek J, Ostergaard LH, Osteryaard PR, Kolenko P, Duskova J, Stepankova A, Hasek J (2009) The structure of the small laccase from *Streptomyces coelicolor* reveals a link between laccases and nitrite reductases. J Mol Biol 385(4):1165–1178. https://doi.org/10.1016/j.jmb.2008.11.024

Solano F, Garcia E, deEgea EP, SanchezAmat A (1997) Isolation and characterization of strain MMB-1 (CECT 4803), a novel melanogenic marine bacterium. Appl Environ Microb 63(9):3499–3506

Solano F, Lucas-Elio P, Lopez-Serrano D, Fernandez E, Sanchez-Amat A (2001) Dimethoxyphenol oxidase activity of different microbial blue multicopper proteins. FEMS Microbiol Lett 204(1):175–181. https://doi.org/10.1016/S0378-1097(01)00400-1

Sondhi S, Sharma P, George N, Chauhan PS, Puri N, Gupta N (2015) An extracellular thermo-alkali-stable laccase from *Bacillus tequilensis* SN4, with a potential to biobleach softwood pulp. 3. Biotechnology 5(2):175–185. https://doi.org/10.1007/s13205-014-0207-z

Sousa AC, Martins LO, Robalo MP (2013) Laccase-catalysed homocoupling of primary amines towards the biosynthesis of dyes. Adv Synth Catal 355:1857–1865. https://doi.org/10.1002/adsc.201300501

Sousa AC, Oliveira MC, Martins LO, Robalo MP (2014) Towards the rational biosynthesis of substituted phenazines and phenoxazinones by laccases. Green Chem 16:4127–4136. https://doi.org/10.1039/C4GC00901K

Sousa AC, Piedade MFMM, Martins LO, Robalo MP (2015) An enzymatic route for the synthesis of carbazole frameworks using bacterial CotA laccase. Green Chem 17:1429–1433. https://doi.org/10.1039/C4GC02511

Sousa AC, Piedade MFMM, Martins LO, Robalo MP (2016) Eco-friendly synthesis of indo dyes mediated by a bacterial laccase. Green Chem 18(22):6063–6070. https://doi.org/10.1039/c6gc02050j

Sousa AC, Oliveira MC, Martins LO, Robalo MP (2018) A sustainable synthesis of asymmetric phenazines and phenoxazinones mediated by CotA-Laccase. Adv Synth Catal 360:575–583. https://doi.org/10.1002/adsc.201701228

Sousa AC, Baptista SR, Martins LO, Robalo MP (2019) Synthesis of azobenzene dyes mediated by CotA laccase. Chem Asian J 14:187–193. https://doi.org/10.1002/asia.201801450

Starke I, Sarodnick G, Ovcharenko VV, Pihlaja K, Kleinpeter E (2004) Quinoxalines. Part 13: Synthesis and mass spectrometric study of aryloxymethylquinoxalines and benzo[b]furylquinoxalines. Tetrahedron 60:6063–6078

Tonin F, Melis R, Cordes A, Sanchez-Amat A, Pollegioni L, Rosini E (2016) Comparison of different microbial laccases as tools for industrial uses. New Biotechnol 33(3):387–398. https://doi.org/10.1016/j.nbt.2016.01.007

Uthandi S, Saad B, Humbard MA, Maupin-Furlow JA (2010) LccA, an archaeal laccase secreted as a highly stable glycoprotein into the extracellular medium by *Haloferax volcanii*. Appl Environ Microb 76(3):733–743. https://doi.org/10.1128/Aem.01757-09

Wang H, Liu X, Zhao J, Yue Q, Yan Y, Gao Z, Dong Y, Zhang Z, Fan Y, Tian J, Wu N, Gong Y (2018) Crystal structures of multicopper oxidase CueO G304K mutant: structural basis of the increased laccase activity. Sci Rep 8(1):14252. https://doi.org/10.1038/s41598-018-32446-7

Xie T, Liu Z, Liu Q, Wang G (2015) Structural insight into the oxidation of sinapic acid by CotA laccase. J Struct Biol 190(2):155–161. https://doi.org/10.1016/j.jsb.2015.03.005

Xu F, Palmer AE, Yaver DS, Berka RM, Gambetta GA, Brown SH, Solomon EI (1999) Targeted mutations in a *Trametes villosa* laccase – axial perturbations of the T1 copper. J Biol Chem 274 (18):12372–12375. https://doi.org/10.1074/jbc.274.18.12372

Yamamoto T, Lee BL, Kokubo H, Kishida H, Hirota K, Wakabayash T, Okamoto H (2003) Synthesis of a new thiophene/quinoxaline CT-type copolymer with high solubility and its basic optical properties. Macromol Rapid Commun 24:440–443. https://doi.org/10.1002/marc.200390067

Yang J, Gao MY, Li M, Li ZZ, Li H, Li HY (2018) *Bacillus amyloliquefaciens* CotA degradation of the lignin model compound guaiacylglycerol-beta-guaiacyl ether. Lett Appl Microbiol 67 (5):491–496. https://doi.org/10.1111/lam.13060

Zapotoczna M, Riboldi GP, Moustafa AM, Dickson E, Narechania A, Morrissey JA, Planet PJ, Holden MTG, Waldron KJ, Geoghegan JA (2018) Mobile-genetic-element-encoded hypertolerance to copper protects *Staphylococcus aureus* from killing by host phagocytes. Mbio 9(5). https://doi.org/10.1128/mBio.00550-18

Zheng ZQ, Li HZ, Li L, Shao WL (2012) Biobleaching of wheat straw pulp with recombinant laccase from the hyperthermophilic *Thermus thermophilus*. Biotechnol Lett 34(3):541–547. https://doi.org/10.1007/s10529-011-0796-0

Old Enzymes at the Forefront of Lignocellulosic Waste Valorization

Simona Giacobbe, Cinzia Pezzella, Giovanni Sannia, and Alessandra Piscitelli

Contents

Abstract Besides the major and widespread use of laccases in bioremediation of dyes and many other toxic chemical wastes, there is an increasing interest towards their exploitation in the valorization of lignocellulosic biomass in frame with the biorefinery concept. Biochemical conversion of lignocelluloses includes three major steps: pretreatment, enzymatic hydrolysis, and fermentation. Laccases find application in the pretreatment phase since they are effective both in removing/modifying the lignin polymer and in reducing the amount of phenolic compounds of already pretreated lignocellulosic materials. As a fact, pretreatment steps often release some biomass-derived products inhibiting both enzymes involved in saccharification and microorganisms used for the fermentation. Together with detoxification processes, laccases can be used for delignification of lignocellulosic biomasses thanks to their direct action on phenolic units of lignin polymer. Laccase-based pretreatments avoid sugars degradation, thanks to the mild conditions of reaction. Conversely, the hard conditions usually adopted for chemical-physical pretreatments, cause sugars degradation. Furthermore, laccase effect in delignification is improved by the use of mediator compounds which allow enzyme to oxidize both phenolic and

S. Giacobbe · G. Sannia · A. Piscitelli (✉)
Department of Chemical Sciences, University Federico II, Naples, Italy
e-mail: apiscite@unina.it

C. Pezzella
Department of Agricultural Sciences, University Federico II, Naples, Italy

D. Schlosser (ed.), *Laccases in Bioremediation and Waste Valorisation*, Microbiology Monographs 33, https://doi.org/10.1007/978-3-030-47906-0_3

non-phenolic component of lignin moieties, producing an extensive cleavage of covalent bonds in lignin. A survey of uses of laccase in lignocellulosic waste valorization is offered in this chapter with the aim to underline their potential. Indeed, applications of laccases in the valuable conversion of different wastes have been deeply described, focusing on available examples of laccase integration in biorefineries.

1 Introduction

The constant increase in the world energy consumption needs for effective alternatives to gradually replace the fossil fuels, as encouraged by the EU Bioeconomy Strategy. In this context, biomass and most particularly lignocellulosics can offer a concrete option for the production of chemicals and energy in future biorefineries, alternative to current oil refineries. Lignocellulosic raw materials, including forest residues, agricultural wastes, and agrofood wastes (AFW) (Fig. 1), are the most abundant source of organic material in the world: their production is estimated to be around 1.3 billion tons per year (Baruah et al. 2018).

Hence, lignocellulosic biomass represents a promising platform for the production of bio-based products, even if its potential is still untapped.

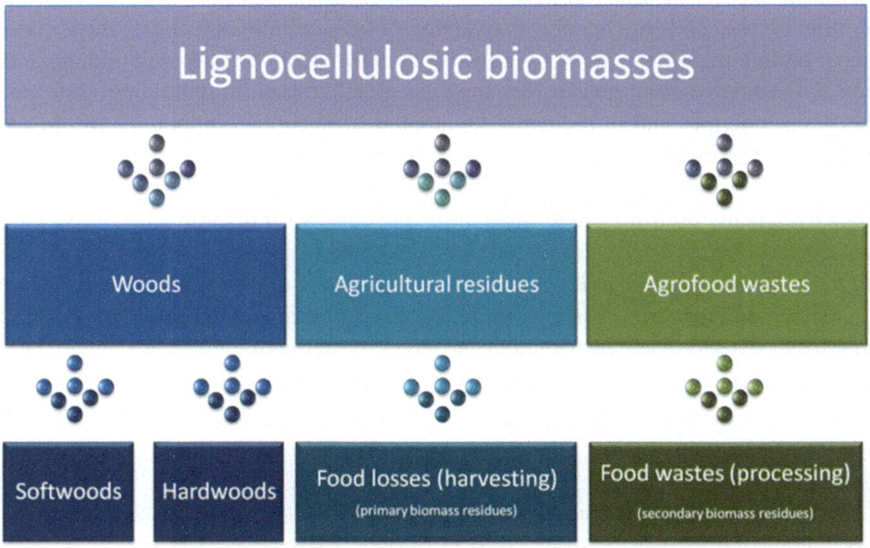

Fig. 1 Classification of lignocellulosic biomasses. Woods have been further subdivided in two groups, softwood and hardwood, based on their lignin and hemicellulose content

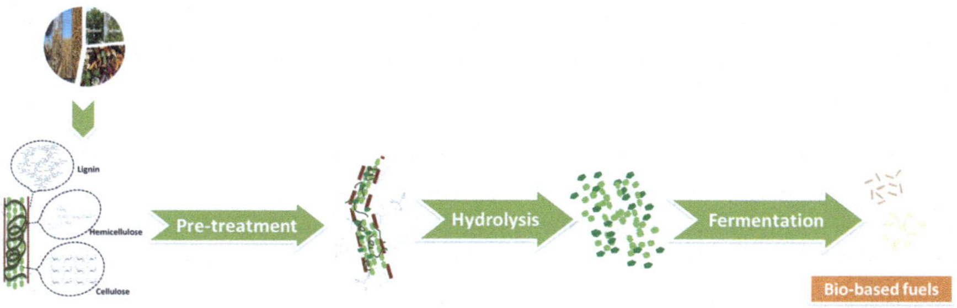

Fig. 2 Schematic representation of lignocellulosic conversion into added-value product showing the lignocelluloses composition along with the three main phases of the process

Albeit composition of lignocellulosic materials depends on many factors, such as plant species, age and growth conditions, and the eventual undergone process, generally these biomasses consist of 15–35% lignin, 40–50% cellulose, and 25–30% hemicelluloses (Pleissner et al. 2016). Very complex inter- and intramolecular network (covalent or noncovalent) exists between these components, rendering the utilization of these residues very challenging. Therefore, a pretreatment step is required to both destroy the layer of lignin that limits enzyme accessibility to cellulose and disassemble the crystalline cellulose structure.

The typical process for the valorization of lignocellulosic materials conventionally includes three primary steps (Fig. 2):

1. Pretreatment, aimed at disrupting the fibrous matrix and remove lignin
2. Hydrolysis, aimed at breaking down (hemi)cellulose and cellulose into monomers
3. Fermentation, aimed at transforming the obtained sugars into value-added chemicals, according to the metabolic abilities of the used yeasts

The pretreatment step is still a matter of investigation, since an effective method that combines environmental sustainability, sugar recovery, and cost-effectiveness is still missing (Fillat et al. 2017). A number of physical and chemical pretreatment methods have been tested; however they are often high-energy demanding and cause loss of carbohydrates and/or formation of by-products (furans, phenols, and weak acids) that may inhibit the following hydrolysis and/or fermentation steps. Therefore, biological approaches able to modify lignin and to reduce the amount of inhibitors represent interesting and greener alternatives to improve the efficiency of the bioconversion processes. These methods rely on the ability of some microorganisms to grow on ligninocellulosic substrates thanks to the secretion of ligninolytic enzymes, such as lignin peroxidases, manganese peroxidases, versatile peroxidase, and laccases (Faraco et al. 2009). On the other hand, the exploitation of ligninolytic enzymes over whole cells includes the possibility of using temperatures which are not necessarily optimal for the growth of the microorganisms. In this context, laccases have been widely investigated, and the obtained results recently reviewed (Kudanga et al. 2010; Roth and Spiess 2015; Fillat et al. 2017). Laccases have been proven able to contribute in different ways to the process of waste valorization. Indeed, they degrade the lignin, modify the structure of the lignocelluloses microfibre, and reduce the phenolic content of pretreated lignocellulosic materials, thus decreasing their inhibitory effect. Laccases have been used alone, or in mixture with other ligninolytic enzymes (Wang et al. 2013; Asgher et al. 2013; Ji et al. 2014). In this latter case, it is not easy to ascribe a defined role to laccase, since the different enzymes/molecules present in the mix can synergistically cooperate in lignin modification/degradation. Laccase delignification is often carried out in combination with mediators, easily oxidizable phenolic units that enable enzyme to indirectly oxidize large molecules and even non-phenolic substrates (Cañas and Camarero 2010). Nonetheless, the use of mediators also delivers various drawbacks, such as their high cost, the possible formation of toxic species, and the occurrence of grafting side reactions (Oliva-Taravilla et al. 2015a; Moreno et al. 2016b; Fillat et al. 2017). Natural mediators, such as lignin-derived soluble phenols (e.g. vanillin, acetosyringone, *p*-hydroxycinnamic acids), offer environmental and economic advantages compared to chemical mediators (Giacobbe et al. 2018). Nonetheless, the phenoxy radicals generated by laccase can be grafted onto pretreated material, with consequent increase of lignin content and decrease of sugar hydrolysis, depending on specific unpredictable combination among laccase, ligninolytic source, and mediator (Oliva-Taravilla et al. 2015a; Moreno et al. 2016b).

Very often the laccase treatment has been combined with other physical and chemical treatment methods, resulting in higher yields than the individual methods alone (see below).

2 Laccases in Biorefinery: Delignification and Detoxification of Lignocellulosic Biomasses

In this chapter, the laccase action in lignocellulosic valorization has been organized according to the different residues tested: softwood and hardwood forest residues, agricultural residues, and agrofood wastes (Fig. 1).

It is important to highlight that it is difficult to compare laccase efficiencies among the different reports. Indeed, not all phenolic compounds are susceptible to oxidation by laccase enzymes at the same extent, and lignin structure is highly dependent on biomass feedstock and/or on pretreatment conditions.

The aim of this chapter is to give a comprehensive picture of the potentiality of laccase enzymes in this interesting field.

2.1 Softwood and Hardwood Forest Residues

Among lignocellulosic biomasses, woody biomasses represent an interesting nonedible biomass for biofuels and biomaterials production. However their resistance to delignification hinders their sustainable transformation.

The lignin content of softwood (25–35% weight) is typically higher than that of hardwood (18–25% weight) (Azadi et al. 2013). Softwood's lignin is mainly composed by guaiacyl units (G), while hardwood contains also syringyl units (S). The G units reduce accessibility of hydrolytic enzymes; hence the first wood biomass is more difficult to be hydrolysed (Palonen and Viikari 2004; Jönsson et al. 2013). Hemicellulose in softwood is composed by galactoglucomannans (15–20%), and xylans represent only a small percentage (7–10%), while in hardwoods it is mainly composed by xylan (15–30%) (Palonen and Viikari 2004).

Several works demonstrated the ability of laccases and laccase mediator system (LMS) in delignification of woody biomasses, also in combination with other pretreatments. Laccases alone are not able to oxidize softwood biomasses; therefore these complex lignocellulosic substrates need to be pretreated with other methods. Two-step pretreatment of softwood has some attractive advantages, such as high ethanol yield and low consumption of enzymes (Galbe and Zacchi 2002). As summarized in Table 1, often LMS is required to improve delignification and saccharification yields from softwood since laccase is too large a molecule to penetrate into the fibre wall of native wood.

In 1999, Larsson's group (Larsson et al. 2003) demonstrated the ability of *T. versicolour* laccase to remove furans and phenolic compounds from steam-pretreated spruce with consequently improvement of ethanol production. Palonen and Viikari (2004) demonstrated that the lignin from pretreated spruce binds cellulase enzymes, but the action of the laccases, before the hydrolysis, improves the enzymatic hydrolysis of 13% by using laccase alone and by 21% by applying LMS. More recently, Moilanen et al. (2014) explored the action of different mediators to reduce hydrolytic enzymes adsorption with consequent increase in the enzymatic hydrolysis yields.

As regards hardwood biomasses, laccase action is more pronounced (Table 2). Kuila et al. (2011a, b) reported the ability of laccase alone to delignify unpretreated bamboo wood with consequent efficient saccharification. More recently, Kumar et al. (2017) set up a simultaneous pretreatment and saccharification process by utilizing laccases with cellulases in order to obtain sugars for biobutanol production from bamboo. Thanks to its rapid growth and high biomass-producing woody, eucalyptus is an interesting hardwood biomass for biofuel production and was successfully pretreated by laccase (Rico et al. 2014) obtaining a delignification of about 20%. The action of the laccase was enhanced by using methyl syringate as mediator (50% of delignification) with a consequent increment in saccharification yield (about 40%). Laccases were also applied on pretreated eucalypt wood with or without LMS. In all reported studies (Table 2), the effect of laccases was not only on delignification but also on detoxification of pretreated biomasses. Noteworthy are

Table 1 Application of laccases for detoxification and delignification of different pretreated softwoods

Pretreated softwood	Laccase	Mediator	Effect	Reference
Dilute acid steam-pretreated spruce	*Trametes versicolour*	–	Removal of furans and phenolic compounds by 93–95%. Ethanol yield similar to that obtained after detoxification with anion exchange chromatography	Larsson et al. (2003)
Steam-pretreated spruce *(Picea abies)*	*Cerrena unicolour PM170798*	–	The lignin removal directly correlated with increases (approximately 12%) in saccharification.	Moilanen et al. (2011)
Steam-treated softwood chips *(Picea abies)*	3 laccases preparations from *Mauginiella* sp., *Melanocarpus albomyces*, and *Trametes hirsuta*	N-hydroxy-N-phenylacetamide (NHA)	The highest conversion, 65% of polysaccharides, was obtained with the sequential use of the oxidative and hydrolytic enzymatic treatments	Palonen et al. (2003)
Acid steam-pretreated spruce	*T. hirsuta*	NHA	Lignin modification showing both modified hydrophobicity and surface charge. Enzymatic hydrolysis yield increased 1.61-fold compared to laccase alone	Palonen and Viikari (2004)
Acid steam-pretreated spruce	*T. hirsuta*	Acetosyringone, ABTS, HBT, and TEMPO	Reduction of hydrolytic enzymes adsorption except with HBT. Increment of enzymatic hydrolysis by 54% and 49% with ABTS and TEMPO, respectively. No positive effects with HBT	Moilanen et al. (2014)

the results reported by Gutiérrez and co-workers (Gutiérrez et al. 2012). The authors observed up to 48% lignin removal by using a combination of enzymatic pretreatment and alkaline extraction. The high delignification yield was linked to high glucose yield (61%) and ethanol production (over 4 g/L). To the best of our knowledge, only one paper reports the application of bacterial laccase to pretreated hardwood biomasses. Singh and co-workers (Singh et al. 2017) highlighted the biocatalytic potential of bacterial enzymes in delignification of steam-pretreated poplar. The small laccase (SLAC) from *Amycolatopsis* sp. *75iv3* allowed to reduce

Table 2 Application of laccases for detoxification and delignification of different hardwoods

Hardwood	Pretreatment	Mediator	Effect	Reference
Milled material from Thorny bamboo	*Pleurotus* sp.	–	Range of delignification between 84 and 89%. Better accessibility of hydrolytic enzymes and efficient saccharification of *Bambusa bambos*	Kuila et al. (2011a, b)
Bamboo	*Pleurotus djamor*	–	Simultaneous pretreatment and saccharification leading to 72.44% saccharification efficiency. Butanol content of 6.45 g/L	Kumar et al. (2017)
Eucalypt wood milled	Four cycles of *Myceliophthora thermophila* laccase and alkaline peroxide extraction	Methyl syringate	Up to 50% lignin removal directly correlated with increases (about 40%) in sugars yield. The pretreatment using laccase alone (without mediator) removed up to 20% of lignin from eucalypt wood	Rico et al. (2014)
Eucalyptus globules	Four cycles of *M. thermophila* laccase and alkaline peroxide extraction	–	Up to 20% of lignin loss after four cycles treatment with an increment of glucose production by 9%	Rico et al. (2015)
Steam-exploded eucalypt wood	Novozyme 51003	HBT	LMS treatment yielded an increase in delignification of 13.9% without affecting pulp properties. LMS increased the glucose yield from 24.7 to 27.1%, probably as a result of the detoxification of the substrate	Martin-Sampedro et al. (2011, 2012)
Milled materials from eucalypt wood	*T. villosa* laccase and alkaline extraction	HBT	Up to 48% lignin removal with high glucose yield (61%) and ethanol production (over 4 g/L). The treatments with laccase alone (without mediator) decreased the lignin content of only 5%	Gutiérrez et al. (2012)
Steam-pretreated poplar	Bacterial laccase from *Amycolatopsis* sp.		Reduction of molar mass lignin (about 50%). About 8% increment of glucose yield	Singh et al. (2017)

acid-soluble lignin of about 15% through oxidation of syringyl units of lignin, acting synergistically with commercial cellulases with a consequent increment of glucose production (about 8%).

2.2 Agricultural Residues

The broad term 'agricultural waste' is generally referred to any lignocellulosic residue produced by agrofood industries. Nevertheless, these residues can be further divided into two main categories depending if they are generated during the harvesting (primary biomass residues) and the processing (secondary biomass residues) of agricultural crops. Agricultural residues are produced during harvesting and are those considered in this paragraph. The residues most frequently tested are represented by wheat, and only few papers utilize different residues (corn, cotton, barley, rice, and pineapple).

When considering the laccase-mediated valorization of agricultural residues (Table 3), the most of examples rely on the exploitation of laccases as detoxification agent, rather than as delignifying one. Furthermore, even when laccases are used to delignify residues, they are very frequently used in combination with other pretreatment techniques and seldom alone (Hyeon et al. 2014; Davidi et al. 2016). Depending on laccases source, the treatments have been carried out at a wide range of pH, temperature, treatment time, and enzyme loading. Different phenol removal efficiencies have been described with different laccases, and reduced laccase effectiveness in detoxification has been reported at increasing viscosity of the medium (Moreno et al. 2013a). Moreno et al. tested the efficiency of laccase treatment before or after enzymatic hydrolysis, obtaining the highest phenol reduction (93–94%) when the laccase treatment was carried out before the enzymatic hydrolysis. Laccase action improved the fermentation performance of the thermotolerant yeast *Kluyveromyces marxianus* strain used, shortening its lag phase and enhancing the ethanol yields (Moreno et al. 2012). The ability of laccases in the detoxification of pretreated biomasses enabled working at high-substrate loading, resulting in the possibility to reduce freshwater and energy consumption without affecting the yields of the final product (Moreno et al. 2013b, c).

A unique example of laccase treatment with immobilized enzyme is present in the recent literature (Ludwig et al. 2013). The immobilized *T. versicolour* laccase efficiently removed phenolic compounds from a wheat straw organosolv fraction, with a consequence increase of the fermentation performance of the yeast *Pichia stipitis*. The immobilized laccase was efficiently reused, and an in situ product removal was achieved since insoluble products precipitated onto the carrier surface.

Few examples of residue valorization using bacterial laccases are present in the literature (Moreno et al. 2016a; De La Torre et al. 2017; Rocha-Martín et al. 2018), mainly due to the lower redox potential of these enzymes with respect to the fungal laccases. Moreno et al. (2016a) assayed the use of a novel bacterial laccase (MetZyme®) for enhancing the hydrolysability and fermentability of steam-exploded wheat straw. Albeit a low phenol reduction was measured, it was sufficient to improve the fermentation performance during a simultaneous saccharification and fermentation (SSF) process. De La Torre et al. (2017) compared the efficiencies of a commercial fungal laccase from *T. villosa* with a bacterial laccase from *S. ipomoeae* for delignification and detoxification of steam-exploded wheat straw. When the

Table 3 Application of laccases for detoxification and delignification of agricultural residues

Raw material	Laccase source	Role in processing	References
Steam-exploded wheat straw pretreated under soft conditions (water impregnation) or harsh conditions (impregnation with diluted acid)	*Coriolopsis rigida* and *T. villosa*	Removal of phenolic compounds	Jurado et al. (2009)
Corn stover and wheat straw	*Pycnoporus sanguineus*	Delignification	Lu et al. (2010)
Ensiled corn stover	*Trametes versicolour*	Delignification	Chen et al. (2012)
Liquid hot water pretreated wheat straw supernatants	Commercial enzyme from *T. versicolour*	Removal of phenolic compounds	Kolb et al. (2012)
Alkali-extracted corn straw	*T. hirsuta*	Delignification	Li et al. (2012)
Steam-exploded wheat straw slurries	*Pycnoporus cinnabarinus, T. villosa*	Removal of phenolic compounds	Moreno et al. (2012)
Hydrothermally processed rice straw	*Trametes* sp. Ha1	Removal of phenolic compounds	Nakanishi et al. (2012)
Steam-exploded wheat straw	*Sclerotium* sp.	Removal of phenolic compounds	Qiu and Chen (2012)
Corncob residue	*Trametes* sp. AH28-2		Zhang et al. (2012)
Steam-exploded wheat straw	*P. cinnabarinus*	Removal of phenolic compounds	Alvira et al. (2013)
Organosolv pretreated wheat straw	*T. versicolour*	Removal of phenolic compounds	Ludwig et al. (2013)
Steam-exploded wheat straw slurries	*P. cinnabarinus*	Removal of phenolic compounds	Moreno et al. (2013b)
Steam-exploded wheat straw slurries	*P. cinnabarinus*	Removal of phenolic compounds	Moreno et al. (2013c)
Steam-exploded wheat straw slurries	*P. cinnabarinus*	Removal of phenolic compounds	Moreno et al. (2013a)

(continued)

Table 3 (continued)

Raw material	Laccase source	Role in processing	References
Cotton gin trash	Commercial laccase mediator systems PrimaGreen® EcoFade LT100 from Genencor International (laccase from modified strains of *Cerrena unicolour* and the mediator 3,5-dimethoxy-4-hydroxybenzonitrile)	Removal of phenolic compounds	Plácido et al. (2013)
Air-dried corn stover	Novozymes; laccase, LiP and MnP and from *Coridus versicolour*; peroxidase from *Phanerochaete chrysosporium*	Delignification	Wang et al. (2013)
Acid-pretreated wheat straw	*T. versicolour*	Removal of phenolic compounds	Heap et al. (2014)
Barley straw	*Escherichia coli* (CUeO)	Delignification	Hyeon et al. (2014)
Cotton gin trash	Commercial laccase mediator systems PrimaGreen® EcoFade LT100 from Genencor International (laccase from modified strains of *Cerrena unicolour* and the mediator 3,5-dimethoxy-4-hydroxybenzonitrile)	Removal of phenolic compounds	Plácido and Capareda (2014)
Corn stover	Commercial laccase	Delignification	Schroyen et al. (2014)
Acidic steam-exploded corn stover	*Ganoderma lucidum*	Removal of phenolic compounds	Fang et al. (2015)
Dilute acid treated poplar straw and hot water treated rice straw	*T. versicolour*	Removal of phenolic compounds	Kapoor et al. (2015)
Steam-exploded wheat straw (water insoluble solids fraction)	*P. cinnabarinus*	Removal of phenolic compounds	Moreno et al. (2016b)
Steam-exploded wheat straw	*P. cinnabarinus*	Removal of phenolic compounds	Oliva-Taravilla et al. (2015a)
Steam-exploded wheat straw	Commercial laccase from *M. thermophila*	Removal of phenolic compounds	Oliva-Taravilla et al. (2015b)

Hemp, flax, corn stover, miscanthus, willow, ensilaged maize, and wheat straw	Commercial laccase	Delignification	Schroyen et al. (2015)
Ricinus communis, *Lantana camara*, *Saccharum officinarum* (top), *Saccharum spontaneum*, *Ananas comosus* (leaf wastes) and *Bambusa bambos*	*P. djamor*	Delignification	Avanthi and Banerjee (2016)
Wheat straw	*Thermobifida fusca*	Delignification	Davidi et al. (2016)
Steam-exploded wheat straw	MetZyme (commercial bacterial laccase)	Removal of phenolic compounds	Moreno et al. (2016a)
Steam-exploded wheat straw (water insoluble solids fraction)	*P. cinnabarinus*	Removal of phenolic compounds	Moreno et al. (2016b)
Steam-exploded wheat straw and water insoluble solids fraction	Commercial laccase from *M. thermophila*	Removal of phenolic compounds	Oliva-Taravilla et al. (2016)
Wheat straw	*P. cinnabarinus*	Delignification	Rencoret et al. (2016)
Steam-exploded wheat straw	*Streptomyces ipomoeae* (bacterial laccase) in comparison with *T. villosa* (commercial fungal laccase)	Removal of phenolic compounds	De La Torre et al. (2017)
Pretreated corn stover	*T. versicolour* (commercial from Sigma)	Removal of phenolic compounds	Liu et al. (2018)
Diluted acid/steam-exploded sugarcane straw and corn stover	Commercial laccases: bacterial (BLc) and fungal from *M. thermophila* (Novozymes)	Removal of phenolic compounds	Rocha-Martín et al. (2018)
Acid-pretreated and raw maize stovers	*Echinodontium taxodii*	Delignification and detoxification	Wang et al. (2017)
Pineapple leaf waste	*P. djamor*	Delignification	Banerjee et al. (2019)

reduction of lignin was considered, a slight reduction in lignin content and a consequent increase in both glucose and xylose production was detected with the bacterial laccase, while no effect was produced by the *T. villosa* laccase. Conversely, a higher efficiency in the decrease of the phenol content was measured using the fungal laccase with respect to the treatment performed with the *S. ipomoeae* laccase.

With the aim to get insights into lignin modifications, some authors carried out detailed analyses (Lu et al. 2010; Li et al. 2012; Qiu and Chen 2012; Wang et al. 2013; Rencoret et al. 2016; Banerjee et al. 2019). When the *Sclerotium* sp. laccase was employed in combination with steam explosion, the authors suggested that the laccase could oxidize the aromatic ring of lignin resulting in the formation of micropores which contributed to loosening the compact wrap of lignin-carbohydrate complex and consequently enhancing the enzymatic hydrolysis efficiency of cellulose (Qiu and Chen 2012). Rencoret et al. (2016) performed two-dimensional nuclear magnetic resonance (2D NMR) analysis of the whole wheat straw material pretreated with a *P. cinnabarinus* laccase in the presence of HBT as mediator. The authors highlighted the formation of Cα-oxidized lignin units during the enzymatic treatment, showing the selective action of laccase-mediator on the lignin moiety, with respect to the polysaccharide signals. Avanthi and Banerjee (2016) performed detailed analysis to verify the existence of a correlation between biomass composition and laccase-mediated lignin degradation. According to their results, G-type lignin has synergistic effect on laccase-mediated delignification, while the presence of S type lignin showed antagonistic relationship with laccase adsorption and delignification.

Contradictory effects due to laccase action have also been described due to the formation of laccase-derived compounds from phenols inhibiting cellulolytic enzymes and/or to the occurrence of grafting phenomena. For example, Oliva-Taravilla et al. (2015b) showed a decrease in the yield of the enzymatic hydrolysis, due to oligomeric products derived from the oxidative polymerization of vanillin and syringaldehyde by *M. thermophila* laccase. Interestingly, Rocha-Martin et al. (2018) highlighted the existence of a lower inhibiting effect of the enzymatic hydrolysis by bacterial laccase, when comparing the performance of commercial bacterial (*S. ipomoeae*) and fungal (*M. thermophila*) laccases on pretreated corn stover and sugarcane straw. As for grafting, lignin-derived phenols resulting from pretreatment of biomass can be oxidized by laccase to phenoxy radicals, which can undergo polymerization by radical coupling onto materials, resulting in an increment of the lignin content (Oliva-Taravilla et al. 2015a).

Particularly interesting are those works designed to obtain better processes integration through the simultaneous action of laccases and hydrolytic enzymes by genetic engineering (Zhang et al. 2012; Hyeon et al. 2014; Davidi et al. 2016) or by mixing extracellular products (Wang et al. 2017). Zhang et al. (2012) heterologously expressed a laccase from *Trametes* sp. AH28-2 in the cellulolytic fungus *Trichoderma reesei* and obtained an increased enzymatic saccharification of cellulosic materials by the laccase producing strains. The integration of laccase activity into cellulosome systems has been successfully obtained (Hyeon et al. 2014; Davidi et al. 2016). In particular, Davidi et al. (2016) incorporated the laccase from the aerobic bacterium *T. fusca* into a cellulase- and xylanase-containing cellulosome. The resulting complex allowed doubling the amount of reducing sugars released from untreated wheat straw compared with the same system lacking the laccase. Inspired by a similar idea, Wang and co-workers (Wang et al. 2017) built a composite enzymatic system mixing the extracellular products of the white-rot fungus *E. taxodii* with cellulolytic enzymes. A synergistic action between the extracellular products and cellulases was observed during the valorization of acid-pretreated and raw maize stovers.

2.3 Agrofood Wastes

The food supply chain is characterized by huge waste losses, whose volume and composition depend on the phase they are produced at (harvesting, processing, distribution, and consumption). Postharvest losses are very variable, being a function of the technology available in a country, the extent to which markets have developed for a specific crop, as well as of the crop perishability (Parfitt et al. 2010). Apart from harvesting, according to European Commission, up to 42% of food waste (FW) derives from households, 39% is produced by food manufacturing industry, 14% pertains to food service sector (catering, restaurants), and only the remaining 5% is a surplus of market chain. Despite the strong push towards the prevention of FW in all the stages of food life cycle, FW production is expected to reach about 126 Mt by 2020 (Mirabella et al. 2014). Besides representing an economic and environmental issue, most of waste generated from the food industry is lignocellulosic in nature, being potential substrate for the production of high-added value products (Ravindran and Jaiswal 2016).

Laccases were applied to valorization of different types of AFWs. Most of the reported examples refer to wastes derived from sugarcane biomass, only a few of them concern other sources (Tables 4, 5, and 6). In the frame of AFW valorization, laccases were exploited in two main processes: *(1)* delignification followed by saccharification (sequential process) and *(2)* concomitant delignification and saccharification (simultaneous process) (Tables 4 and 5).

In most of the described processes, laccases act directly on the biomass as the only delignificant agent (Table 4). Pretreatment conditions differ in operating parameters, such as, pH, temperature, enzyme loading, enzyme/biomass ratio, and incubation time and are strictly dependent on the biochemical properties of the tested laccases. Although fungal laccases are the catalysts of choice, the synergistic action of the whole ligninolytic enzymatic systems produced by selected fungi was effectively exploited for pretreatment of sugarcane bagasse (Asgher et al. 2013) and Jerusalem artichoke (Ji et al. 2014). In the latter case, the authors formulated a new enzymatic cocktail for simultaneous delignification and hydrolysis, based on a psychrophilic lignocellulosic enzymatic mixture from *C. cladosporioides* and a commercial xylanase. The wide thermal adaptability in mesophilic and low temperatures of the characterized enzymatic mixture provided consistency of hydrolysis

Table 4 Laccase-based processes for AFW pretreatment

AFW	Process	Laccase source	Reference
Sugarcane bagasse	Sequential	*Pleurotus ostreatus* ligninolytic mixture	Asgher et al. (2013)
Sugarcane biomass	Simultaneous	Bacterial laccase from *T. fusca*	Chen et al. (2013)
Bagasse and Jerusa-lem artchoke	Simultaneous	*Cladosporium cladosporioides* ligninolytic mixture	Ji et al. (2014)
Palm oil fruit residues	Sequential	Laccase 51003 from *M. thermophila*	Rashid Shah et al. (2016)
Sugarcane tops	Sequential	Laccase from *P. djamorr*	Sherpa et al. (2018)
Apple pomace Coffe silver skin Potato peels	Sequential	Native Laccases mix from *P. ostreatus* Recombinant *P. ostreatus* laccase	Giacobbe et al. (2018)
Brewer's spent grains	Sequential	Native laccases mix from *P. ostreatus* Recombinant *P. ostreatus* laccase	Giacobbe et al. (2019)

Table 5 Laccase-based combined processes for AFW pretreatment

AFW	Combined treatment	Process	Laccase source	Reference
Sugarcane bagasse	Steam explosion followed by laccase	Simultaneous	*Ganoderma lucidum* *T. versicolour*	Sitarz et al. (2013)
Palm oil fruit residues	Superheated steam (SHS) followed by laccase	Sequential	*T. versicolour*	Rizal et al. (2018)
Corncobs	Hydrodynamic cavi-tation with laccase	Delignification	*T. versicolour*	Thangavelu et al. (2018)
Agave bio-mass waste	Autoclaving followed by laccase	Sequential	*T. versicolour*	Sánchez-Ramírez et al. (2018)
Sugarcane bagasse	Laccase followed by microwave	Delignification and hemicellulose extraction	Novozyme laccase OMNO 7015	Nie (2019)
Sugarcane bagasse and straw	LMS followed by alkaline peroxide extraction	Sequential	*P. cinnabarinus*	Rencoret et al. (2017)
Rice straw	Dilute acid followed by laccase	Simultaneous	*T. versicolour*	Kumar et al. (2019)

and further fermentation at ambient temperatures (Ji et al. 2014). Bacterial laccases were applied only in limited cases (Chen et al. 2013). When the laccase from the moderately thermophilic *T. fusca* was added to the enzymatic cocktail for sugarcane bagasse hydrolysis, it significantly enhanced the saccharification yield. The enzyme was recombinantly produced in *E. coli*, thus supporting the cost-effectiveness of the

Table 6 Laccases promoted detoxification of pretreated AFWs

AFW	Pretreatment method	Laccase source	Reference
Sugarcane bagasse	Steam-explosion and cellulotytic enzyme hydrolysis	*T. versicolour*	Martín et al. (2002)
Sugarcane bagasse	Acid hydrolysis	*Cyathus stercoreus*	Chandel et al. (2007)
Rice straw	Acid hydrolysis	*C. perennis* Novozyme laccase 51003	Kalyani et al. (2012)
Apple pomace	Acid hydrolysis	*T. versicolour*	Parmar and Rupasinghe (2012)
Sugarcane bagasse	Acid hydrolysis	*Bacillus subtilis*	Hu et al. (2018)

process (Piscitelli et al. 2010). Statistical optimization of process parameters was applied to laccase-promoted delignification process (Rashid Shah et al. 2016; Sherpa et al. 2018). Delignification of sugarcane tops by *P. djamor* laccase was optimized through Response surface methodology and its effectiveness substantiated by the consequent improvement in saccharification yield (Sherpa et al. 2018). The depolymerization effect of laccase on the substrate was also confirmed by structural analyses. Interestingly, low-molecular compounds formed during delignification were proposed to act as natural laccase mediators, contributing to process economy and sustainability (Sherpa et al. 2018). Particularly interesting is the process proposed by Giacobbe et al. for the treatment of different kinds of AFWs (Giacobbe et al. 2018, 2019). The authors tested different combinations of *P. ostreatus* laccases with and without the addition of a natural phenolic mediator, designing a sequential protocol in which delignification and following saccharification occur in the same bioreactor without any washing step. The two enzymatic steps are separated only by a short thermal deactivation phase, in order to overcome the observed inhibition of cellulase activity caused by laccases. It is worth to note that significant differences emerged in the way of action of the tested enzymes. The native mix from *P. ostreatus* displayed a tendency to graft phenols, causing increase in the lignin content, conversely the main action of the recombinant POXA1b laccase was in lignin degradation (Giacobbe et al. 2018, 2019).

Examples of AFW pretreatment based on the combination of laccases together with chemical/physical methods are listed in Table 5. The processes are aimed at improving biomass delignification (Thangavelu et al. 2018), hemicellulose extraction (Nie 2019), or saccharification yield (Sitarz et al. 2013; Rencoret et al. 2017; Rizal et al. 2018; Sánchez-Ramírez et al. 2018). Thangavelu and co-workers coupled *T. versicolour* laccase to hydrodynamic cavitation and applied the combined method to the pretreatment of corncobs. Both biomass and biocatalyst were circulated continuously throughout the reactor, leading to high delignification with low energy requirement. Morphological studies confirmed the deconstruction of biomass without modification of the cellulose structure (Thangavelu et al. 2018). A two-step process in which laccase pretreatment was followed by microwave-assisted alkaline extraction of hemicellulose was applied to sugarcane bagasse. Besides reducing lignin content, laccase treatment structurally altered the bagasse, increasing its specific surface area and porosity, thus enhancing the following hemicellulose extraction (Nie 2019).

In most of the reported examples, the efficiency of laccase-combined pretreatments was evaluated in terms of improvement of following saccharification step. A survey of reported results indicates that the effect of the applied treatment is strongly influenced by the choice of proper laccase as well as by the lignin composition of the AFW of interest. For example, Sitarz et al. (2013) verified that the addition of the laccase-rich *G. lucidum* broth to a state-of-art cellulase preparation significantly increased the cellulolytic yield of steam-exploded sugarcane bagasse. In contrast, the spiking with *T. versicolour* laccase produced an inhibitory effect on the same cellulase mixture. The authors hypothesized that the higher K_{cat}/K_M or redox potential of *G. lucidum* laccase with respect to *T. versicolour* one caused a faster and more extensive solubilization of lignin degradation products, promoting cellulase adhesion. Alternatively, the enzymes may display different selectivity towards potential cellulase inhibitors derived from lignin hydrolysis (Sitarz et al. 2013). When the same *T. versicolour* laccase was combined to superheated steam (SHS) for the pretreatment of oil palm biomass, an enhancement of subsequent enzymatic hydrolysis was observed, as a consequence of the structural rearrangement occurring on lignin (Rizal et al. 2018). However, in this case, a sequential process was applied by introducing a washing step between the delignification and saccharification, in order to prevent any inhibitory effect on cellulases (Rizal et al. 2018). Rencoret et al. (2017) pointed attention on the lignin structural features that may influence the mechanism of laccase action. The LMS composed of the high-redox potential laccase from *P. cinnabarinus* and 1HBT as mediator was exploited to remove and/or modify the lignins in the sugarcane bagasse (S-rich lignin) and straw residues (G-rich lignin). The LMS pretreatment, followed by alkaline peroxide extraction, produced a rather similar lignin removal from both materials despite their different lignin composition (H/G/S ratio), suggesting that other features, i.e. the presence of *p*-coumarates acylating the γ-OH of the lignin side chains, may adversely affect laccase action (Rencoret et al. 2017). This scenario entirely changed when the treatment was carried out in the absence of mediator. In this case, the difference in lignin composition between the two materials resulted in a more pronounced lignin degradation in the straw with respect to bagasse (Rencoret et al. 2017). The overall economy of laccase-based process would benefit from enzyme immobilization on solid support, facilitating its recovery and reuse. However, when free and immobilized *T. versicolour* laccases were tested on previously autoclaved agave biomass waste, only the free enzyme led to raise the subsequent saccharification yield, while immobilized samples strongly impaired it. It was proposed that the formation of a complex between the nanocomposite, reaction products and biomass may hinder the interaction of cellulose with hydrolytic enzymes, preventing its hydrolysis (Sánchez-Ramírez et al. 2018). Conversely, in a more recent report, Kumar et al. (2019) succeeded in setting up an immobilized enzyme cocktail comprising cellulases and the *T. versicolour* laccase for the simultaneous detoxification and saccharification of pretreated rice straw. The reusability of the enzyme cocktail in the simultaneous process was verified up to eight cycles, conferring an economic advantage to the process. The effectiveness of the strategy was further confirmed by the improvement in yeast performance and enhanced ethanol production in subsequent fermentation step (Kumar et al. 2019).

Finally, several examples of laccase-promoted detoxification of chemically/physically hydrolysed biomasses were described (Table 6). Treatment of sugarcane bagasse hydrolysates with laccases from different sources was effective in removing most of phenolic compounds, resulting in improvement of yeast performances in subsequent ethanolic fermentation of hydrolysates (Martín et al. 2002; Chandel et al. 2007). Although laccase treatment did not eliminate other classes of inhibitors such as acetic acid, furfural, and 5-hydroxy-methyl-furfural, it brought about negligible loss in total sugars when compared with other chemical or physical detoxification methods (Chandel et al. 2007). As already described for delignification process (Sitarz et al. 2013; Giacobbe et al. 2018), laccases from different sources displayed also different performances in detoxification. For example, a newly isolated laccase from *Coltricia perennis* displayed a higher efficiency in removal of phenolic compounds in acid-pretreated rice straw, with respect to a commercial Novozyme laccase (Kalyani et al. 2012). On the other hand, the application of a bacterial laccase to detoxification of hydrolysed sugarcane bagasse did not increase following biohydrogen production, suggesting that detoxification provided by laccase action was not suitable for the thermophilic anaerobic bacteria used in the fermentation step (Hu et al. 2018).

3 Conclusions/Future Prospective

Laccase-based processes allow achieving ethanol yield, glucose conversion, and delignification and detoxification percentages similar or higher than conventional methodologies. Hence, in the current biorefinery concept, laccases constitute a powerful biotechnological tool towards the achievement of the EU Bioeconomy Strategy. The most important obstacle to laccases exploitation is the cost of both enzymes and redox mediators.

Now days, marked efforts are being made in order to achieve both cheap overproduction of laccases and their engineering aimed at gaining more robust and active enzymes (Piscitelli et al. 2011; Pezzella et al. 2017).

Moreover, the possibility to recover natural mediators and to produce lignocellulolytic enzymes from wastes themselves represents a step forward the process circularity.

References

Alvira P, Moreno AD, Ibarra D et al (2013) Improving the fermentation performance of *Saccharomyces cerevisiae* by laccase during ethanol production from steam-exploded wheat straw at high-substrate loadings. Biotechnol Prog 29:74–82. https://doi.org/10.1002/btpr.1666

Asgher M, Ahmad Z, Iqbal HMN (2013) Alkali and enzymatic delignification of sugarcane bagasse to expose cellulose polymers for saccharification and bio-ethanol production. Ind Crops Prod 44:488–495. https://doi.org/10.1016/j.indcrop.2012.10.005

Avanthi A, Banerjee R (2016) A strategic laccase mediated lignin degradation of lignocellulosic feedstocks for ethanol production. Ind Crops Prod 92:174–185. https://doi.org/10.1016/j.indcrop.2016.08.009

Azadi P, Inderwildi OR, Farnood R, King DA (2013) Liquid fuels, hydrogen and chemicals from lignin: a critical review. Renew Sustain Energy Rev 21:506–523. https://doi.org/10.1016/J.RSER.2012.12.022

Banerjee R, Chintagunta AD, Ray S (2019) Laccase mediated delignification of pineapple leaf waste: an ecofriendly sustainable attempt towards valorization. BMC Chem 13. https://doi.org/10.1186/s13065-019-0576-9

Baruah J, Nath BK, Sharma R, et al (2018) Recent trends in the pretreatment of lignocellulosic biomass for value-added products. Front Energy Res 6. https://doi.org/10.3389/fenrg.2018.00141

Cañas AI, Camarero S (2010) Laccases and their natural mediators: biotechnological tools for sustainable eco-friendly processes. Biotechnol Adv 28:694–705

Chandel AK, Kapoor RK, Singh A, Kuhad RC (2007) Detoxification of sugarcane bagasse hydrolysate improves ethanol production by *Candida shehatae* NCIM 3501. Bioresour Technol 98:1947–1950. https://doi.org/10.1016/j.biortech.2006.07.047

Chen Q, Marshall MN, Geib SM et al (2012) Effects of laccase on lignin depolymerization and enzymatic hydrolysis of ensiled corn stover. Bioresour Technol 117:186–192. https://doi.org/10.1016/j.biortech.2012.04.085

Chen CY, Hsieh ZS, Cheepudom J et al (2013) A 24.7-kDa copper-containing oxidase, secreted by *Thermobifida fusca*, significantly increasing the xylanase/cellulase-catalyzed hydrolysis of sugarcane bagasse. Appl Microbiol Biotechnol 97:8977–8986. https://doi.org/10.1007/s00253-013-4727-y

Davidi L, Moraïs S, Artzi L et al (2016) Toward combined delignification and saccharification of wheat straw by a laccase-containing designer cellulosome. Proc Natl Acad Sci 113:10854–10859. https://doi.org/10.1073/pnas.1608012113

De La Torre M, Martín-Sampedro R, Fillat Ú et al (2017) Comparison of the efficiency of bacterial and fungal laccases in delignification and detoxification of steam-pretreated lignocellulosic biomass for bioethanol production. J Ind Microbiol Biotechnol 44:1561–1573. https://doi.org/10.1007/s10295-017-1977-1

Fang Z, Liu X, Chen L et al (2015) Identification of a laccase Glac15 from *Ganoderma lucidum* 77002 and its application in bioethanol production David Wilson. Biotechnol Biofuels 8. https://doi.org/10.1186/s13068-015-0235-x

Faraco V, Pezzella C, Giardina P et al (2009) Decolourization of textile dyes by the white-rot fungi *Phanerochaete chrysosporium* and *Pleurotus ostreatus*. J Chem Technol Biotechnol 84. https://doi.org/10.1002/jctb.2055

Fillat Ú, Ibarra D, Eugenio M et al (2017) Laccases as a potential tool for the efficient conversion of lignocellulosic biomass: a review. Fermentation 3:17. https://doi.org/10.3390/fermentation3020017

Galbe M, Zacchi G (2002) A review of the production of ethanol from softwood. Appl Microbiol Biotechnol 59:618–628. https://doi.org/10.1007/s00253-002-1058-9

Giacobbe S, Pezzella C, Lettera V et al (2018) Laccase pretreatment for agrofood wastes valorization. Bioresour Technol 265:59–65. https://doi.org/10.1016/j.biortech.2018.05.108

Giacobbe S, Piscitelli A, Raganati F et al (2019) Butanol production from laccase-pretreated brewer's spent grain. Biotechnol Biofuels 12. https://doi.org/10.1186/s13068-019-1383-1
Gutiérrez A, Rencoret J, Cadena EM et al (2012) Demonstration of laccase-based removal of lignin from wood and non-wood plant feedstocks. Bioresour Technol 119:114–122. https://doi.org/10.1016/j.biortech.2012.05.112
Heap L, Green A, Brown D et al (2014) Role of laccase as an enzymatic pretreatment method to improve lignocellulosic saccharification. Catal Sci Technol 4:2251–2259. https://doi.org/10.1039/c4cy00046c
Hu BB, Li MY, Wang YT, Zhu MJ (2018) Enhanced biohydrogen production from dilute acid pretreated sugarcane bagasse by detoxification and fermentation strategy. Int J Hydrogen Energy 43:19366–19374. https://doi.org/10.1016/j.ijhydene.2018.08.164
Hyeon JE, You SK, Kang DH et al (2014) Enzymatic degradation of lignocellulosic biomass by continuous process using laccase and cellulases with the aid of scaffoldin for ethanol production. Process Biochem 49:1266–1273. https://doi.org/10.1016/j.procbio.2014.05.004
Ji L, Yang J, Fan H et al (2014) Synergy of crude enzyme cocktail from cold-adapted *Cladosporium cladosporioides* Ch2-2 with commercial xylanase achieving high sugars yield at low cost. Biotechnol Biofuels 7. https://doi.org/10.1186/s13068-014-0130-x
Jönsson LJ, Alriksson B, Nilvebrant N-O (2013) Bioconversion of lignocellulose: inhibitors and detoxification. Biotechnol Biofuels 6:16. https://doi.org/10.1186/1754-6834-6-16
Jurado M, Prieto A, Martínez-Alcalá Á et al (2009) Laccase detoxification of steam-exploded wheat straw for second generation bioethanol. Bioresour Technol 100:6378–6384. https://doi.org/10.1016/j.biortech.2009.07.049
Kalyani D, Dhiman SS, Kim H et al (2012) Characterization of a novel laccase from the isolated *Coltricia perennis* and its application to detoxification of biomass. Process Biochem 47:671–678. https://doi.org/10.1016/j.procbio.2012.01.013
Kapoor RK, Rajan K, Carrier DJ (2015) Applications of *Trametes versicolor* crude culture filtrates in detoxification of biomass pretreatment hydrolyzates. Bioresour Technol 189:99–106. https://doi.org/10.1016/j.biortech.2015.03.100
Kolb M, Sieber V, Amann M et al (2012) Removal of monomer delignification products by laccase from *Trametes versicolor*. Bioresour Technol 104:298–304. https://doi.org/10.1016/j.biortech.2011.11.080
Kudanga T, Prasetyo EN, Sipilä J et al (2010) Reactivity of long chain alkylamines to lignin moieties: implications on hydrophobicity of lignocellulose materials. J Biotechnol 149:81–87. https://doi.org/10.1016/j.jbiotec.2010.06.020
Kuila A, Mukhopadhyay M, Tuli DK, Banerjee R (2011a) Accessibility of enzymatically delignified *Bambusa bambos* for efficient hydrolysis at minimum cellulase loading: an optimization study. Enzyme Res 2011:1–8. https://doi.org/10.4061/2011/805795
Kuila A, Mukhopadhyay M, Tuli DK, Banerjee R (2011b) Production of ethanol from lignocellulosics: an enzymatic venture. EXCLI J 10:85–96
Kumar S, Gujjala LKS, Banerjee R (2017) Simultaneous pretreatment and saccharification of bamboo for biobutanol production. Ind Crops Prod 101:21–28. https://doi.org/10.1016/j.indcrop.2017.02.028
Kumar V, Patel SKS, Gupta RK et al (2019) Enhanced saccharification and fermentation of rice straw by reducing the concentration of phenolic compounds using an immobilized enzyme cocktail. Biotechnol J 1800468. https://doi.org/10.1002/biot.201800468
Larsson S, Reimann A, Nilvebrant N-O, Jönsson LJ (2003) Comparison of different methods for the detoxification of lignocellulose hydrolyzates of spruce. Appl Biochem Biotechnol 77:91–104. https://doi.org/10.1385/abab:77:1-3:91
Li J, Sun F, Li X et al (2012) Enhanced saccharification of corn straw pretreated by alkali combining crude ligninolytic enzymes. J Chem Technol Biotechnol 87:1687–1693. https://doi.org/10.1002/jctb.3818

Liu ZH, Xie S, Lin F, et al (2018) Combinatorial pretreatment and fermentation optimization enabled a record yield on lignin bioconversion. Biotechnol Biofuels 11. https://doi.org/10.1186/s13068-018-1021-3

Lu C, Wang H, Luo Y, Guo L (2010) An efficient system for pre-delignification of gramineous biofuel feedstock in vitro: application of a laccase from *Pycnoporus sanguineus* H275. Process Biochem 45:1141–1147. https://doi.org/10.1016/j.procbio.2010.04.010

Ludwig D, Amann M, Hirth T et al (2013) Development and optimization of single and combined detoxification processes to improve the fermentability of lignocellulose hydrolyzates. Bioresour Technol 133:455–461. https://doi.org/10.1016/j.biortech.2013.01.053

Martín C, Galbe M, Wahlbom CF et al (2002) Ethanol production from enzymatic hydrolysates of sugarcane bagasse using recombinant xylose-utilising *Saccharomyces cerevisiae*

Martin-Sampedro R, Capanema EA, Hoeger I et al (2011) Lignin changes after steam explosion and laccase-mediator treatment of eucalyptus wood chips. J Agric Food Chem 59:8761–8769. https://doi.org/10.1021/jf201605f

Martín-Sampedro R, Eugenio ME, García JC et al (2012) Steam explosion and enzymatic pre-treatments as an approach to improve the enzymatic hydrolysis of *Eucalyptus globulus*. Biomass Bioenergy 42:97–106. https://doi.org/10.1016/j.biombioe.2012.03.032

Mirabella N, Castellani V, Sala S (2014) Current options for the valorization of food manufacturing waste: a review. J Clean Prod 65:28–41

Moilanen U, Kellock M, Galkin S, Viikari L (2011) The laccase-catalyzed modification of lignin for enzymatic hydrolysis. Enzyme Microb Technol 49:492–498. https://doi.org/10.1016/j.enzmictec.2011.09.012

Moilanen U, Kellock M, Várnai A et al (2014) Mechanisms of laccase-mediator treatments improving the enzymatic hydrolysis of pre-treated spruce. Biotechnol Biofuels 7:1–13. https://doi.org/10.1186/s13068-014-0177-8

Moreno AD, Ibarra D, Fernández JL, Ballesteros M (2012) Different laccase detoxification strategies for ethanol production from lignocellulosic biomass by the thermotolerant yeast *Kluyveromyces marxianus* CECT 10875. Bioresour Technol 106:101–109. https://doi.org/10.1016/j.biortech.2011.11.108

Moreno AD, Ibarra D, Ballesteros I et al (2013a) Comparing cell viability and ethanol fermentation of the thermotolerant yeast *Kluyveromyces marxianus* and *Saccharomyces cerevisiae* on steam-exploded biomass treated with laccase. Bioresour Technol 135:239–245. https://doi.org/10.1016/j.biortech.2012.11.095

Moreno AD, Ibarra D, Ballesteros I et al (2013b) Ethanol from laccase-detoxified lignocellulose by the thermotolerant yeast *Kluyveromyces*-Effects of steam pretreatment conditions, process configurations and substrate loadings. Biochem Eng J 79:94–103. https://doi.org/10.1016/j.bej.2013.07.006

Moreno AD, Tomás-Pejó E, Ibarra D et al (2013c) *In situ* laccase treatment enhances the fermentability of steam-exploded wheat straw in SSCF processes at high dry matter consistencies. Bioresour Technol 143:337–343. https://doi.org/10.1016/j.biortech.2013.06.011

Moreno A, Ibarra D, Mialon A, Ballesteros M (2016a) A bacterial laccase for enhancing saccharification and ethanol fermentation of steam-pretreated biomass. Fermentation 2:11. https://doi.org/10.3390/fermentation2020011

Moreno AD, Ibarra D, Alvira P et al (2016b) Exploring laccase and mediators behavior during saccharification and fermentation of steam-exploded wheat straw for bioethanol production. J Chem Technol Biotechnol 91:1816–1825. https://doi.org/10.1002/jctb.4774

Nakanishi A, Bae JG, Fukai K et al (2012) Effect of pretreatment of hydrothermally processed rice straw with laccase-displaying yeast on ethanol fermentation. Appl Microbiol Biotechnol 94:939–948. https://doi.org/10.1007/s00253-012-3876-8

Nie S (2019) Laccase pretreatment for enhancing microwave-assisted alkaline extraction of hemicellulose from bagasse. National Natural Science Foundation of China 31760192. The degradation of hexenuronic acid and its effect on AOX formation during xylanase aided chlorine dioxide bleaching of pulp. View project Guangxi Natural Science Foundation of China

(2018GXNSFDA281050). Fundamental study on conversion of lignocellulosic biomass into high-value materials. View project

Oliva-Taravilla A, Moreno AD, Demuez M et al (2015a) Unraveling the effects of laccase treatment on enzymatic hydrolysis of steam-exploded wheat straw. Bioresour Technol 175:209–215. https://doi.org/10.1016/j.biortech.2014.10.086

Oliva-Taravilla A, Tomás-Pejó E, Demuez M et al (2015b) Inhibition of cellulose enzymatic hydrolysis by laccase-derived compounds from phenols. Biotechnol Prog 31:700–706. https://doi.org/10.1002/btpr.2068

Oliva-Taravilla A, Tomás-Pejó E, Demuez M et al (2016) Phenols and lignin: key players in reducing enzymatic hydrolysis yields of steam-pretreated biomass in presence of laccase. J Biotechnol 218:94–101. https://doi.org/10.1016/j.jbiotec.2015.11.004

Palonen H, Viikari L (2004) Role of oxidative enzymatic treatments on enzymatic hydrolysis of softwood. Biotechnol Bioeng 86:550–557. https://doi.org/10.1002/bit.20135

Palonen H, Saloheimo M, Viikari L, Kruus K (2003) Purification, characterization and sequence analysis of a laccase from the ascomycete *Mauginiella* sp. Enzyme Microb Technol 33:854–862. https://doi.org/10.1016/S0141-0229(03)00247-3

Parfitt J, Barthel M, MacNaughton S (2010) Food waste within food supply chains: quantification and potential for change to 2050. Philos Trans R Soc B Biol Sci 365:3065–3081

Parmar I, Rupasinghe HPV (2012) Optimization of dilute acid-based pretreatment and application of laccase on apple pomace. Bioresour Technol 124:433–439. https://doi.org/10.1016/j.biortech.2012.07.030

Pezzella C, Giacobelli VG, Lettera V et al (2017) A step forward in laccase exploitation: recombinant production and evaluation of techno-economic feasibility of the process. J Biotechnol 259. https://doi.org/10.1016/j.jbiotec.2017.07.022

Piscitelli A, Pezzella C, Giardina P et al (2010) Heterologous laccase production. 1:252–262

Piscitelli A, Del Vecchio C, Faraco V et al (2011) Fungal laccases: versatile tools for lignocellulose transformation. Comptes Rendus – Biol 334:789–794. https://doi.org/10.1016/j.crvi.2011.06.007

Plácido J, Capareda S (2014) Analysis of alkali ultrasonication pretreatment in bioethanol production from cotton gin trash using FT-IR spectroscopy and principal component analysis. Bioresour Bioprocess 1. https://doi.org/10.1186/s40643-014-0023-7

Plácido J, Imam T, Capareda S (2013) Evaluation of ligninolytic enzymes, ultrasonication and liquid hot water as pretreatments for bioethanol production from cotton gin trash. Bioresour Technol 139:203–208. https://doi.org/10.1016/j.biortech.2013.04.012

Pleissner D, Qi Q, Gao C et al (2016) Valorization of organic residues for the production of added value chemicals: a contribution to the bio-based economy. Biochem Eng J 116:3–16

Qiu W, Chen H (2012) Enhanced the enzymatic hydrolysis efficiency of wheat straw after combined steam explosion and laccase pretreatment. Bioresour Technol 118:8–12. https://doi.org/10.1016/j.biortech.2012.05.033

Rashid Shah S, Chinonso Ishmael U, Vejayan Palliah J et al (2016) Optimization of the enzymatic saccharification process of empty fruit bunch pretreated with laccase enzyme. BioResources 11

Ravindran R, Jaiswal AK (2016) A comprehensive review on pre-treatment strategy for lignocellulosic food industry waste: challenges and opportunities. Bioresour Technol 199:92–102

Rencoret J, Pereira A, del Río JC et al (2016) Laccase-mediator pretreatment of wheat straw degrades lignin and improves saccharification. Bioenergy Res 9:917–930. https://doi.org/10.1007/s12155-016-9745-z

Rencoret J, Pereira A, Del Río JC et al (2017) Delignification and saccharification enhancement of sugarcane byproducts by a laccase-based pretreatment. ACS Sustain Chem Eng 5:7145–7154. https://doi.org/10.1021/acssuschemeng.7b01332

Rico A, Rencoret J, del Río JC et al (2014) Pretreatment with laccase and a phenolic mediator degrades lignin and enhances saccharification of Eucalyptus feedstock

Rico A, Rencoret J, del Río JC et al (2015) In-depth 2D NMR study of lignin modification during pretreatment of eucalyptus wood with laccase and mediators. Bioenergy Res 8:211–230. https://doi.org/10.1007/s12155-014-9505-x

Rizal NFAA, Ibrahim MF, Zakaria MR et al (2018) Combination of superheated steam with laccase pretreatment together with size reduction to enhance enzymatic hydrolysis of oil palm biomass. Molecules 23. https://doi.org/10.3390/molecules23040811

Rocha-Martín J, Martínez-Bernal C, Zamorano LS et al (2018) Inhibition of enzymatic hydrolysis of pretreated corn stover and sugar cane straw by laccases. Process Biochem 67:88–91. https://doi.org/10.1016/j.procbio.2018.01.021

Roth S, Spiess AC (2015) Laccases for biorefinery applications: a critical review on challenges and perspectives. Bioprocess Biosyst Eng 38:2285–2313

Sánchez-Ramírez J, Martínez-Hernández JL, López-Campos RG et al (2018) Laccase validation as pretreatment of agave waste prior to saccharification: free and immobilized in superparamagnetic nanoparticles enzyme preparations. Waste Biomass Valor 9:223–234. https://doi.org/10.1007/s12649-016-9774-z

Schroyen M, Vervaeren H, Van Hulle SWH, Raes K (2014) Impact of enzymatic pretreatment on corn stover degradation and biogas production. Bioresour Technol 173:59–66. https://doi.org/10.1016/j.biortech.2014.09.030

Schroyen M, Vervaeren H, Vandepitte H et al (2015) Effect of enzymatic pretreatment of various lignocellulosic substrates on production of phenolic compounds and biomethane potential. Bioresour Technol 192:696–702. https://doi.org/10.1016/j.biortech.2015.06.051

Sherpa KC, Ghangrekar MM, Banerjee R (2018) A green and sustainable approach on statistical optimization of laccase mediated delignification of sugarcane tops for enhanced saccharification. J Environ Manage 217:700–709. https://doi.org/10.1016/j.jenvman.2018.04.008

Singh R, Hu J, Regner MR et al (2017) Enhanced delignification of steam-pretreated poplar by a bacterial laccase. Sci Rep 7. https://doi.org/10.1038/srep42121

Sitarz AK, Mikkelsen JD, Højrup P, Meyer AS (2013) Identification of a laccase from *Ganoderma lucidum* CBS 229.93 having potential for enhancing cellulase catalyzed lignocellulose degradation. Enzyme Microb Technol 53:378–385. https://doi.org/10.1016/j.enzmictec.2013.08.003

Thangavelu K, Desikan R, Taran OP, Uthandi S (2018) Delignification of corncob via combined hydrodynamic cavitation and enzymatic pretreatment: process optimization by response surface methodology. Biotechnol Biofuels 11. https://doi.org/10.1186/s13068-018-1204-y

Wang FQ, Xie H, Chen W et al (2013) Biological pretreatment of corn stover with ligninolytic enzyme for high efficient enzymatic hydrolysis. Bioresour Technol 144:572–578. https://doi.org/10.1016/j.biortech.2013.07.012

Wang Y, Shao Y, Zou X et al (2017) Synergistic action between extracellular products from white-rot fungus and cellulase significantly improves enzymatic hydrolysis. Bioengineered:1–8. https://doi.org/10.1080/21655979.2017.1308991

Zhang J, Qu Y, Xiao P et al (2012) Improved biomass saccharification by *Trichoderma reesei* through heterologous expression of lacA gene from *Trametes* sp. AH28-2. J Biosci Bioeng 113:697–703. https://doi.org/10.1016/j.jbiosc.2012.01.016

Laccases in the Context of Potentially Cooperating Enzymes

Kamahldin Haghbeen and Dietmar Schlosser

Contents

Abstract Laccases have attracted scientists and practitioners already since a long time, due to their catalytic versatility, robustness, and potential usability for a wide spectrum of diverse applications. In lignocellulose decaying microorganisms, laccases and further enzymes frequently form suites of biocatalysts cooperating in decay type-specific concerted actions. Such possible natural enzymatic interactions have largely

The original version of this chapter was revised. A correction to this chapter can be found at https://doi.org/10.1007/978-3-030-47906-0_10

K. Haghbeen
Department of Plant Biotechnology, National Institute of Genetic Engineering and Biotechnology (NIGEB), Tehran, Iran
e-mail: Kamahl@nigeb.ac.ir

D. Schlosser (✉)
Department of Environmental Microbiology, Helmholtz Centre for Environmental Research – UFZ, Leipzig, Germany
e-mail: dietmar.schlosser@ufz.de

D. Schlosser (ed.), *Laccases in Bioremediation and Waste Valorisation*, Microbiology Monographs 33, https://doi.org/10.1007/978-3-030-47906-0_4

inspired biotechnological concepts aiming at the use of multi-enzyme cocktails or enzyme cascades for diverse applications such as wastewater treatment, biorefinery approaches targeting the production of biofuels and platform chemicals, as well as pharmaceutical and industrial biotransformations. This chapter illustrates both proven and potential (albeit still rather hypothetical) cooperations between laccases and other microbial enzymes. The thereby addressed types of cooperation involve (1) enzymes being structurally and catalytically similar to laccases such as tyrosinases, which may complement laccase reactivity; (2) enzymes (or subsequent reactions) fueled with substrates (or reactants) generated during laccase reactions, as exemplified for lignin-modifying manganese peroxidases and β-etherases; and (3) enzymes potentially producing substrates for laccases, e.g., unspecific peroxygenases.

1 Introduction

Laccases have attracted scientists and practitioners since decades, due to their catalytic versatility, robustness, and potential usability for a wide spectrum of applications (Giardina et al. 2010; Pezzella et al. 2015; Rodgers et al. 2010). These enzymes have been considered as biocatalysts for environmental biotechnology and biorefinery applications; in the textile, pharmaceutical, food, and cosmetic industries; as well as for polymer synthesis and grafting (Majeau et al. 2010; Pezzella et al. 2015; Roth and Spiess 2015). In fungi, laccases usually co-exist with numerous other enzymes active on certain polysaccharide and lignin components of lignocelluloses. Together these enzymes form cooperating suites of biocatalysts contributing to lignocellulose decomposition in decay type-specific concerted actions (Bissaro et al. 2018; Sutzl et al. 2018). Such natural enzymatic interactions and growing knowledge regarding the underlying biochemical mechanisms and ecological principles have led to biotechnological concepts targeting the use of multi-enzyme cocktails or enzyme cascades for diverse applications such as wastewater treatment, biorefinery approaches aiming at the valorization of lignocellulosic biomass for the production of biofuels and platform chemicals, and pharmaceutical or industrial biotransformations (Ba and Vinoth Kumar 2017; Martinez et al. 2017; Pollegioni et al. 2015; Roth and Spiess 2015).

This chapter attempts to illustrate both proven and potential cooperation between laccases and other enzymes of microorganisms, which may not necessarily be operative in nature but could potentially support the development of future multi-enzyme biocatalysts. Such cooperation could involve structurally and catalytically similar enzymes such as tyrosinases, which are frequently produced along with laccases, and may complement laccase reactivity and thus widen the scope of enzyme applicability (Ba and Vinoth Kumar 2017; Morin et al. 2012) as outlined in Sect. 2 (Fig. 1). Going beyond the focus of chapters "Fungal Laccases and Their Potential in Bioremediation Applications" and "Bacterial Laccases: Some Recent Advances and Applications", which describe fundamental structural characteristics and reaction mechanisms of fungal and bacterial laccases, Sect. 2 aims to provide a

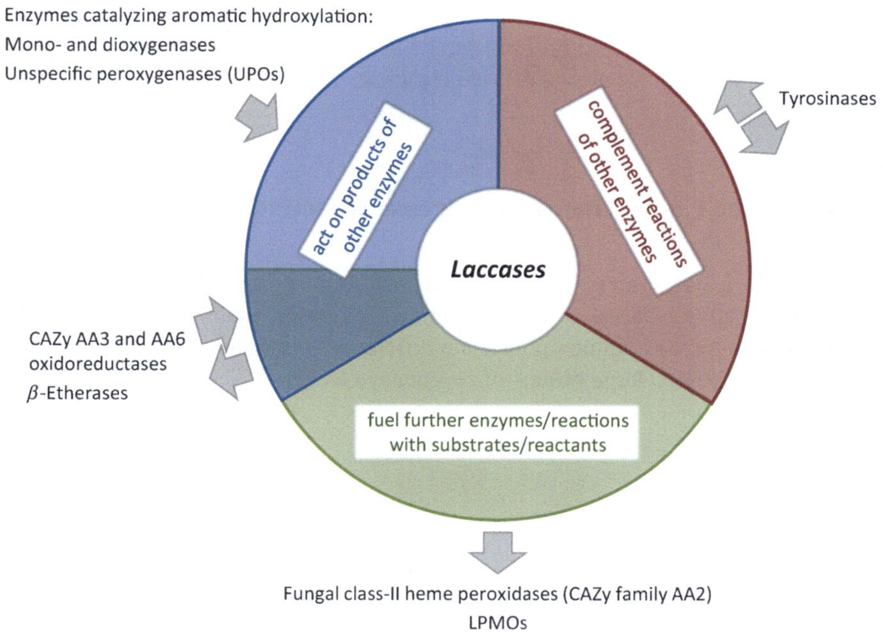

Fig. 1 Types of potential cooperation between laccases and other enzymes addressed in this chapter. Examples for potentially cooperating enzymes are also shown. It is worthwhile to note that in some cases laccases may supply their cooperation partners with substrates as well as receive substrates from them. For further explanations, please refer to Sect. 1

detailed and comparative reasoning for the selectivity and substrate specificity of laccases and tyrosinases and to explain important factors influencing the activity and reaction rates of these enzymes. Such characteristics would have to be considered as a starting point for successful design of multi-enzyme biocatalysts.

The following Sects. 3 and 4 are intended to extend the scope of potential biocatalytic cooperations to enzymes being structurally and catalytically more distant from laccases. In order to provide a broad overview of related possible interactions and their potential exploitation, Sects. 3 and 4 abstain from in-depth structure-function considerations and instead generally focus on reactions enabling the sequential action of different enzymes for an intended purpose. A potential type of cooperation in this context, explained in more detail in Sect. 3, refers to supply of other enzymes or subsequent reactions with substrates or reactants generated during laccase reactions (Fig. 1). Finally, laccases may utilize products formed by other enzymes as substrates (Fig. 1) as described in Sect. 4. It has to be noted that particular cooperation types, e.g., those potentially involving laccase and CAZy AA3 or AA6 oxidoreductases during quinone redox cycling, combine interaction types of Sects. 3 and 4 since in this case laccases would produce substrates (i.e., quinones) for quinone reductases and at the same time receive substrates (i.e., diphenols such as hydroquinones) from these reductases (Fig. 1). Similarly, *β*-

etherases may be supplied by laccases with substrates, but they could also produce potential laccase substrates (Fig. 1; Sects. 3.3 and 4.2). While primarily focusing on fungal enzymes, we have also addressed bacterial ones where this deemed obvious.

2 Enzymes Complementing Laccase Reactivity: Tyrosinase

To lay down a background for evaluation, possibilities, and challenges ahead of cooperation between laccases with polyphenol oxidases, mainly tyrosinases, this section overviews the common points and differences between the mechanisms and activities of these two large groups of cuproenzymes.

2.1 Laccase

In the following, major catalytic characteristics of laccases are explained in more detail together with the respective underlying structural determinants of these enzymes and their substrates.

2.1.1 Reaction and Mechanism

To reduce molecular oxygen to water, laccase (EC 1.10.3.2; benzenediol/oxygen oxidoreductase) accepts electron-donating substrates (Thurston 1994). During a catalytic cycle of laccase, four electrons are detached from four molecules of the reducing substrate through a *one-electron* mechanism. Then the electrons, one by one, are delivered to oxygen to produce two molecules of water. The oxidized substrate, as a result of losing one electron to laccase, is a free radical which inevitably pursues stability via further reactions (Thurston 1994).

$$O_2 + 4(\textbf{reduced substrate}) \xrightarrow{\textbf{Laccase}} 2H_2O + 4(\textbf{oxidised substrate})^{\bullet} \dashrightarrow \text{Further reactions}$$

The catalytic activity of laccase is orchestrated by a delicate layout of four copper ions in a conical space in its active site where a single copper ion (T1) is at the top of the cone and a trinuclear copper cluster (T2/T3) sits at the bottom (Bertrand et al. 2002). A bridge of His–Cys–His, about 12 Å, connects T1 to T2/T3 center. Laccase is activated when molecular oxygen sits on T2/T3 site, but electron capture from the reducing substrate happens at T1 site (Agrawal et al. 2018). As a result, the coordination atmosphere of this copper ion is very important to both catalytic activity and substrate selectivity of laccase.

2.1.2 Redox Potential of T1

Experimental evidence indicates that the electron transfer from the reducing substrate to T1 follows an outer-sphere mechanism. This means that the substrate does not need to be in the ligating distance of T1 and pose in a special direction to form a coordination bond. Therefore, because of the potential difference between the substrate and T1, electron transfer happens through the space once the substrate enters the substrate channel[1] of laccase (Xu et al. 1996). Taking this fact into account, laccases are categorized based on the redox potential (E^0 versus standard hydrogen electrode) of T1 to low (E^0 close to 430 mV), medium (430–710 mV), and high (E^0 about 780 mV) potential oxidases (Yang et al. 2017).

Sensibly, there must be a correlation between the T1 redox potential and its surrounding amino acids. As shown, laccases of *Trametes hirsuta*, *Trametes versicolor*, and *Trametes villosa* with the same amino acid composition around T1 had similar E^0, while laccases with more changes in that composition had lower potentials (Shleev et al. 2004). Nonetheless, this trend was not followed in all the examined cases. In a study on several fungal laccases, Xu et al. (1996) showed that, despite the large variation in their E^0, the blue copper sites of these enzymes had similar EPR and spectroscopic features indicating that the ligating residues of T1 were in similar structures. Accordingly, they concluded that the different E^0(s) could not be merely attributed to the ligating amino acids of T1. To verify this conclusion, Xu et al. (1998) changed some key surrounding residues of T1 of *Rhizoctonia solani* ($E^0 = 0.73$ V) and *Myceliophthora thermophila* ($E^0 = 0.48$ V) laccases by applying single and triplet site-directed mutations. Interestingly, neither the single nor the triplet mutants obtained significantly different E^0 as compared with those of the wild types. These results indicate that, in addition to the amino acid composition, the folding of the protein structure, which influences both the exact distances and the geometry of the surrounding residues with respect to T1, has important impact on the redox potential of T1 of a laccase.

2.1.3 Substrate Selectivity

Electronic Effect and Activity

The difference between the redox potentials (ΔE^0) works as the main driving force for the reaction of a laccase with a reducing substrate. An electron-donating substituent (EDS) on a reducing substrate increases the charge density of the whole molecule and decreases its redox potential (Roy et al. 2009). As a result, the chance of electron transfer from the substrate to laccase is increased. No doubt that an electron-withdrawing substituent (EWS) affects the chance of this event inversely.

---

[1]O_2 binds to the T2/T3 center. Therefore, the channel for molecular oxygen differs from the binding site of the reducing substrate, which approaches the T1 center.

Xu (1996) reported k_{cat} values of *Polyporus pinsitus* laccase (320, 2400, and 3600 min^{-1}) for 4-nitro guaiacol, guaiacol, and 4-methyl guaiacol, respectively. This data shows the effects of EWS and EDS on the laccase reaction. In parallel with this result, from a study on several derivatives of cinnamic acid, benzoic acid, and catechol, Glazunova et al. (2018) concluded that all the benzoic acid derivatives, in which the EWS ($–CO_2H$) was directly attached to the aromatic ring, had lower rates of reactions in comparison with those of cinnamic acid derivatives. However, in these studies, it was hard to separate the electronic effect of a substituent from the hydrophobic effect of the molecule because of different sizes of the examined substrates. For instance, Glazunova et al. (2018) observed that hydroxyl group had more positive effect on the rates of benzoic acids reactions than a methoxy group, whereas the opposite trend was true for cinnamic acid derivatives.

To examine the electronic effect of a substituent, Moshtaghioun et al. (2011) studied oxidation of some synthetic diazo derivatives of guaiacol (**S1**, Fig. 2) by the extracellular *Neurospora crassa* laccase. Since the structural skeleton of the examined substrates were the same and the enzymatic reactions were monitored directly through the consumption of the substrates, from the decrease in the substrate optical density at its λ_{max} and not the formation of products, the changes in the kinetics could be solely ascribed to the effect of the substituents. Results of this study clearly showed the instant rates of the reactions (−dA/min) had the trend of $–OCH_3 > –CH_3 \approx –H > –SO_2NH_2 > –Cl$, while the $V_{max(s)}$ of these substrates were in the same range (Moshtaghioun et al. 2011). This result could be explained based on the radical nature of laccase reaction as the electron demand of organic radicals varies dramatically with the structure and induction effect of a substituent. Therefore, the effect of EWS or EDS on stabilization of a radical species (the product, intermediate, or transition state) is different from what is seen for an ion. This results in different Hammett constants for EWS and EDS on a radical (Wang et al. 2010).

Redox Potential and T1 Cavity

The outer-sphere mechanism explains low selectivity of a laccase for oxidation of various molecules with little structural similarities such as ABTS, 2,6-dimethoxyphenol (Thurston 1994), and even $K_4[Fe(CN)_6]$ (Shleev et al. 2004). Nonetheless, the share of folding of the protein structure of a laccase which can play key role in architecting the shape and the size of the site where T1 is located should not be underestimated. The overlaid structures of *T. versicolor* laccase (LacIIIb) and *Coprinus cinereus* laccase (Lac-Cc) reveal that the T1 cavity of Lac-Cc has a wider opening (Bertrand et al. 2002). Similarly, by a comparative study on the crystal structures of an ascomycete (*Melanocarpus albomyces*) laccase with a basidiomycete (*C. cinereus*) laccase, Hakulinen et al. (2002) demonstrated that *M. albomyces* had a narrower T1 cavity in comparison with that of *C. cinereus* laccase.

The discrepancies between the T1 sites of various laccases are sensed in kinetics studies. A good example of these studies was reported by Xu (1996). He showed

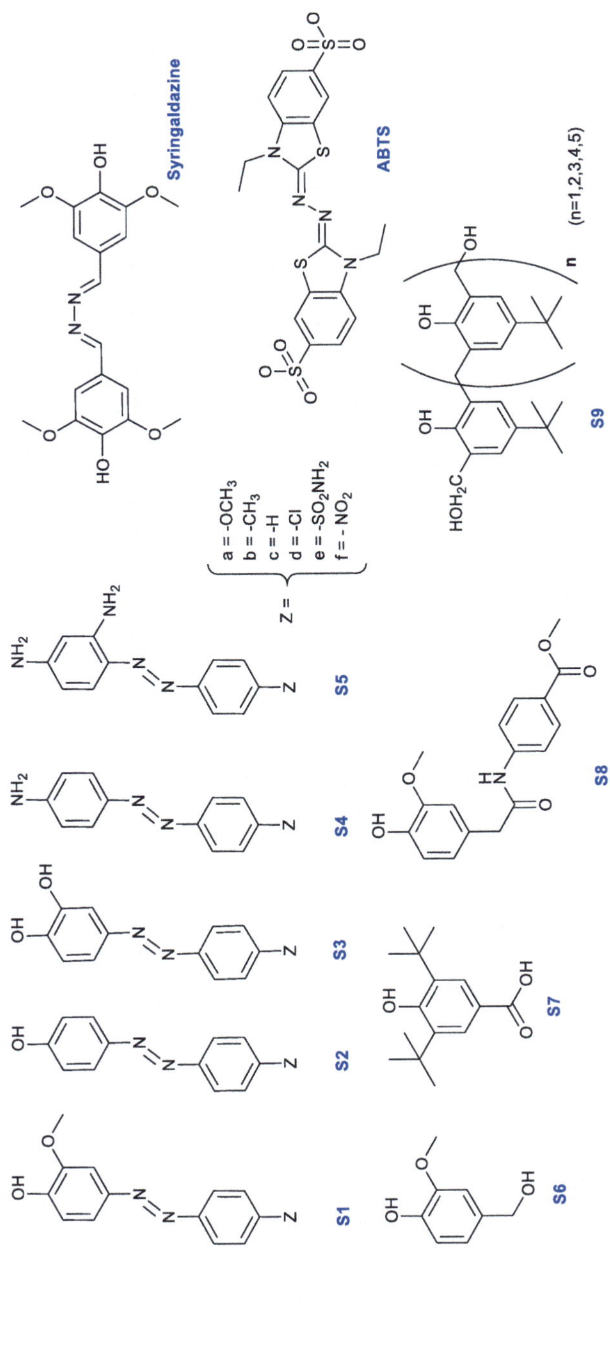

Fig. 2 Structures of the chemicals mentioned in Sect. 2

different recombinant laccases [(*P. pinsitus* laccase, $E^0 \approx 0.8$ V); (*R. solanilaccase* laccase, $E^0 \approx 0.8$ V); (*M. thermophila*, $E^0 \approx 0.5$ V); (*Scytalidium thermophilum*, $E^0 \approx 0.5$ V)] obtained distinctly different kinetic parameters, while the structure(s) of the selected substrate(s) and inhibitor(s)[2] were the same. For instance, despite their similar E^0, the K_m values of *M. thermophila* laccase and *S. thermophilum* laccase for **S6** (Fig. 2) were 0.92 and 0.13 mM with k_{cat} values of 23 and 4.9 min^{-1}, respectively. In the same manner, the IC_{50} of NaCl for *M. thermophila* laccase was 600 mM, while it was only 0.4 for *S. thermophilum* laccase (Xu 1996).

Size and Steric Hindrance

A part of Xu's work also elucidated the steric effect of different derivatives of phenolic compounds on the laccase activity. He found that the effect of a bulky substituent at *para* position on the laccase activity was insignificant. In contrast, a large substituent at *ortho* position of the phenolic ring could considerably decrease the laccase activity. For instance, *P. pinsitus* laccase showed almost no activity in the presence of **S7** (Fig. 2), but its activity in the presence of **S6** ($K_m = 0.1$ mM, $k_{cat} = 2200$ min^{-1}) and **S8** ($K_m = 0.27$ mM, $k_{cat} = 2100$ min^{-1}) was almost unaffected despite the large difference in the sizes of the *para* substituents of these substrates (Xu 1996). Using structure **S9** (Fig. 2), d'Acunzo et al. (2002) also showed that *P. pinsitus* laccase activity was hindered in the presence of derivatives with $n > 2$ even when a co-solvent was used to increase the solubility of these compounds. Another good example was reported by Glazunova et al. (2018). They found that the specific activity of *T. hirsuta*, *Coriolopsis caperata*, *Antrodiella faginea*, and *Steccherinum murashkinskyi* laccases in the presence of various coumaric acids followed the order of *para* > *ortho* > *meta*. For example, *A. faginea* laccase showed specific activities of 289 and 49 (μM μg^{-1} min^{-1}) with *para*- and *ortho*-coumaric acids, respectively, but it failed to show any activity with *meta*-coumaric acid. Interestingly, laccases of *A. faginea* and *S. murashkinskyi* did not show any activity on 2,5-xylidine which has only a methyl group at *meta* position. In a similar way, the large difference between the average specific activities of laccase with catechol (294,524 μM μg^{-1} min^{-1}) and orcinol (1374 μM μg^{-1} min^{-1}) can be reasoned (Glazunova et al. 2018).

2.1.4 pH and Activity

In the interaction between an enzyme and its substrate, pH of the medium works as a pivotal determinant. From the substrate side, depending on its K_a, pH controls the equilibrium between the molecular and the ionic forms. In laccase reaction with a phenolic substrate, ionization of the substrate increases as the pH rises. Surely, the

[2]It is known that small ions such as Cl^- and OH^- inhibit O_2 binding at T2/T3 site.

resulting phenoxide has different solubility and higher tendency toward reducing T1 copper. Kallio et al. (2011) showed that E^0 of 2,6-dimethoxy phenol, syringic acid, and methyl syringate decreased from 0.53, 0.57, and 0.69 to 0.40, 0.51, and 0.65 V, respectively, by rising the pH from 4.5 to 6.

From an enzymatic point of view, although changes in pH modify the hydrogen bonds lattice of an enzyme as well as its surface charge and the hydrophilicity of its microenvironments, it has been shown that E^0 of laccase is not affected (Xu 1997). Nonetheless, the pH changes can be injurious or beneficial to the activity of a laccase. The K_m value of *T. hirsuta* laccase, a basidiomycete laccase with $E^0 = 0.78$ V, for syringic acid decreased from 35 to 17 μM by increasing pH from 4.5 to 6. But the V_{max} of the reaction was also dropped from 8.3 to 2 (dA min^{-1} $nmol^{-1}$). In contrast, this increase in pH increased both the K_m value of recombinant *M. albomyces* laccase, an ascomycete laccase with $E^0 = 0.48$ V, for syringic acid from 122 to 132 μM and the V_{max} from 3.5 to 12.1 (dA min^{-1} $nmol^{-1}$) (Kallio et al. 2011).

The optimal pH for most of fungal laccases is below 6 (Agrawal et al. 2018). At pH > 7, hydroxide inhibits laccase as it competes with O_2 for binding to T2/T3 center. Nonetheless, there are some reports about laccases with optimal pH of activity above 7 (Agrawal et al. 2018). The correlation between some residues in T1 cavity and the optimal pH of laccase activity has been demonstrated via site-directed mutations, but the results seem to be premature to be conclusive (Rodgers et al. 2010).

Considering the facts that the electron transfer step is the bottleneck of a laccase reaction and deprotonation of a phenolic substrate enhances the chance of the electron transfer, it has been a matter of question where and how deprotonation happens. Some researchers speculated that the deprotonation of phenolic substrates happens by the help of an aspartate or glutamate residue close to T1 site (Kallio et al. 2009). However, this assumption has faced with some challenges. Recently, Ferraroni et al. (2017) have reported the crystal structure of *Aspergillus niger* laccase McoG ($E^0 = 0.453$ V) in which a histidine (His253) was instead of the common aspartate or glutamate. Through comparative kinetics study of McoG and its variants (H253D, H253A, and H253N), they showed the substitution of His253 with alanine, aspartate, or asparagine residue neither changed the E^0 of the enzyme nor improved the catalytic activity. Unexpectedly, all the variants had lower affinity for the examined phenolic substrates at lower reaction rates as compared with that of McoG. Likewise, all the variants showed lower affinity for the non-phenolic substrates, ABTS and *N,N*-dimethyl-*p*-phenylenediamine (Ferraroni et al. 2017). Despite all these valuable findings about laccases, substrate selectivity of a laccase is still a matter of question as it can be different from one species to another one. No doubt that modeling and simulation studies can be of great help (Pardo et al. 2016), but the existing results from both practical and computational experiments indicate that more works have to be done to enable us to explain molecular basis of the action of a laccase.

2.2 *Tyrosinase*

The following Sect. 2.2 aims to summarize tyrosinase selectivity and substrate specificity and to explain important factors influencing activity and reaction rates of tyrosinases.

2.2.1 Reactions of Tyrosinase

Nature has engineered tyrosinases for production of *ortho*-quinone(s) (oQn) from certain monophenols and *ortho*-diphenols (odP) in the presence of molecular oxygen (Haghbeen and Tan 2003). Due to the high reactivity, oQn readily takes part in further chemical reactions with its surrounding molecules as soon as it is released from the active site of a tyrosinase. The reactions of an oQn usually proceed to produce macromolecules known as melanins. Consequently, the resulting macromolecules owe their compositions to the existing molecules in the reaction mixture (Plonka and Grabacka 2006).

Tyrosinase is a bi-functional enzyme as it can start its activity with either a monophenol or an odP substrate. In the case of a monophenol, tyrosinase first hydroxylizes it at *ortho* position to form an odP compound. This is a monooxygenase activity known as cresolase (Fig. 3). Interestingly, the resulting odP is not released from the active site, but it is further oxidized to an oQn. Then, the oQn is released. When an odP is the substrate, tyrosinase oxidizes it directly to an oQn. This is an oxidase activity, known as catecholase, because both oxygens of O_2 are reduced to water molecules. Therefore, the stoichiometries of cresolase and catecholase reactions are different (Matoba et al. 2018).

A sub-family of tyrosinases, usually found in plants, is called catechol oxidase. They are structurally very similar to tyrosinases but unable to do *ortho*-hydroxylation. As a result, they show only catecholase activity (Solem et al. 2016). Comparing the reactions of tyrosinase with laccase reaction leads to an interesting point. The products of both enzymes have something in common from a reactivity point of view. Laccase products are unstable radicals that easily take part in further *radical reactions*, and tyrosinase products are very reactive molecules that easily participate in further *nucleophilic addition reactions* (Gandía-Herrero and García-Carmona 2013).

2.2.2 Mechanism of Tyrosinase Reaction

Similar to laccase, tyrosinase is activated by molecular oxygen. A binuclear copper center (type 3) is responsible for hosting O_2 in the active site of tyrosinase. Based on the oxidation state of the coppers and combination with oxygen, tyrosinase is named met-, deoxy-, and oxy-tyrosinase. In met-tyrosinase, both copper ions are Cu^{2+}, while they are both Cu^{+} in deoxy-tyrosinase. Met-tyrosinase does not show any

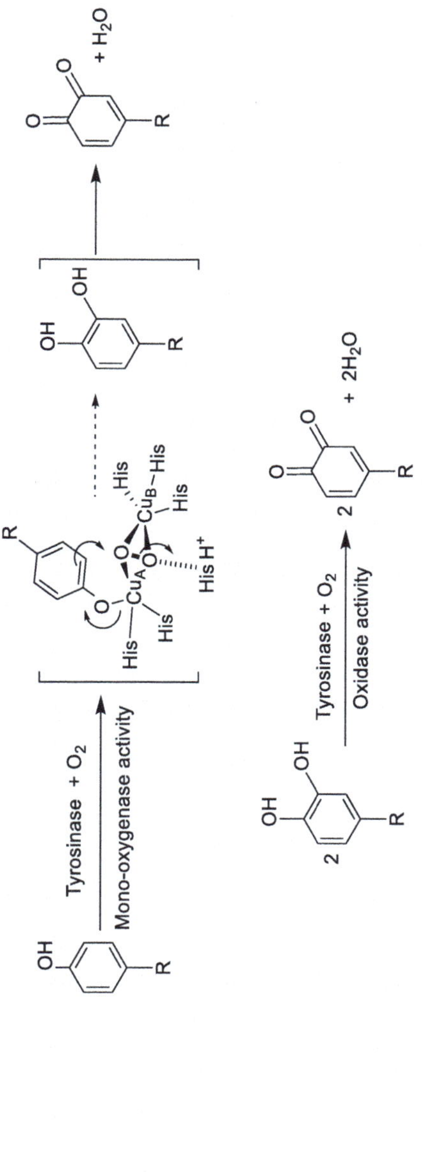

Fig. 3 Cresolase and catecholase reactions of tyrosinase

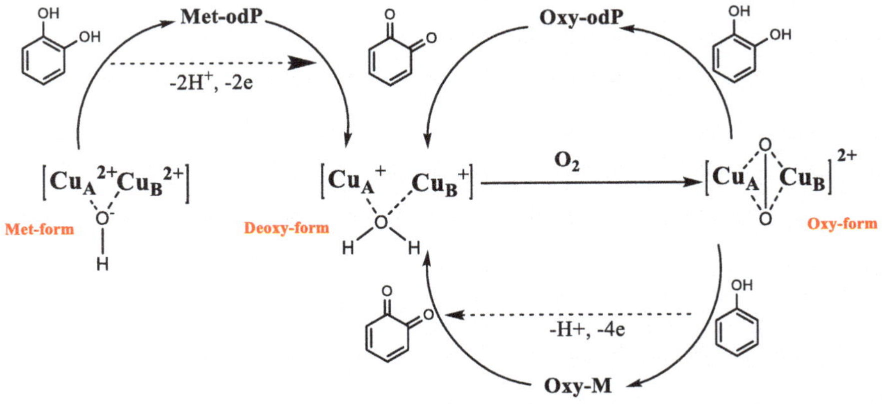

Fig. 4 Oxidation state of T3 copper center during the catalytic cycles of a tyrosinase

affinity for O_2 nor does it react with a monophenol (García-Molina et al. 2005). But, an odP is able to reduce Cu^{2+} to Cu^{+} because of its lower redox potentials as compared to E^0 of monophenols (Haghbeen and Tan 1998). Once met-tyrosinase is reduced to deoxy-tyrosinase, it reacts with O_2 and forms oxy-tyrosinase which is a peroxo-dicopper (Cu_2O_2) intermediate (Yoon et al. 2009). Oxy-tyrosinase can not only hydroxylize a monophenol but it also oxidizes an odP to an oQn (Fig. 4).

A key difference between tyrosinase and laccase is the fact that T1 in laccase involves in an outer-sphere electron transfer but tyrosinase follows an inner-sphere mechanism in its both cresolase and catecholase reactions. This means that the electron transfer during the reactions of tyrosinase happens through the bonds between oxy-tyrosinase and the substrate. Therefore, the reducing substrate must be in a ligating distance, about 2 Å, with T3 coppers. This is in contrast with an outer-sphere electron transfer that could happen from a distance longer than 15 Å (Gray and Ellis 1994). Consequently, for tyrosinase reaction, the reducing substrate has to fit in the pocket of tyrosinase active site so that it can form a coordination bond to one of the T3 coppers (Hassani et al. 2018). This constraint, in turn, narrows the spectrum of tyrosinase substrates.

2.2.3 Electronic Effects of a Substrate

Compared to laccase, another distinctive feature of tyrosinase mechanism correlates to its monooxygenase activity in which the C–H bond at *ortho* position of the phenolic substrate is cleaved through an electrophilic attack of peroxo-dicopper intermediate (Fig. 3) (Borojerdi et al. 2004). Given this fact, it is anticipated that an EDS increases the rate of cresolase reaction. This is seen in both the instant rates (dC/min = 1.64 and 0.6 μM min^{-1}) and $V_{max(s)}$ (6.7 and 1.9 μM min^{-1}) of **S2a** and **S2c** (Fig. 2) cresolase reactions with *Agaricus bisporus* tyrosinase (AbT) (Haghbeen and Tan 2003). In a similar way, an EDS facilitates catecholase reaction of tyrosinase. The instant rates of catecholase reactions of **S3b** and **S3e** were 1.15 and 0.34 μM min^{-1}, respectively. However, their $V_{max(s)}$ were almost the same because binding of the substrate to the enzyme is the bottleneck of a catecholase reaction and not the electron transfer step (Haghbeen and Tan 2003).

The binding step is important in both reactions of tyrosinase, and deprotonation of the substrate can be a key event for its binding to T3 copper center (Matoba et al. 2018). Hence, the electronic effect of a substituent can be determinant at this stage because it influences the K_a value of the substrate. An EDS increases the chance of electron transfer from a reducing substrate to oxy-tyrosinase, but, at the same time, it decreases K_a of the substrate. As a result, the substrates with lower K_a(s) have less inclination for deprotonation. This two-sided effect of a substituent is clearly seen in tyrosinase reactions and reveals the importance of the binding step in each reaction of this enzyme.

The EDS (–OCH_3) increased the instant rate of cresolase reaction of AbT with **S2a** because it increased the charge density of the whole substrate. But it decreased its affinity for AbT (K_m of 80.2 and 51 μM for **S2a** and **S2c**, respectively) because the EDS lowered the K_a of **S2a** as compared with that of **S2c** (Shamsipur et al. 2008). In case of catecholase activity, a similar phenomenon was not observed in the reactions of **S3b** and **S3e** though. The EDS (–CH_3) not only increased the rate of the reaction; it also increased the affinity of the substrate for the enzyme (K_m of 35 and 231 μM for **S3b** and **S3e**, respectively) (Haghbeen and Tan 2003). These results indicate that deprotonation is more important for cresolase substrates than the catecholase substrates. This conclusion was supported with further evidence (Solem et al. 2016).

2.2.4 Substrates of Different Families

Sulfur-containing compounds are not considered as good substrates for polyphenol oxidases because they are able either to form stable bonds with copper ions or reduce cupric ion to cuprous and disconnect the catalytic cycle of a cuproenzyme (Wood and Schallreuter 1991). Saboury et al. (2006) showed that benzenethiol was not oxidized by AbT. However, it could inhibit competitively both activities of tyrosinase. Xu (1996) studied reactions of *P. pinsitus* and *T. villosa* laccases with 1,2-benzenedithiol and *ortho*-methoxy benzenethiol. He proposed that the high affinity of these compounds was responsible for their anomalous kinetic parameters as compared with those of the phenolic counterparts.

To differentiate between a laccase and a tyrosinase, Flurkey et al. (1995) suggested tolidine as a test substrate because laccase could oxidize the examined arylamines, while tyrosinase had failed to do so. In contrast, there were also some reports about slow oxidation of aromatic amines by tyrosinase (Rescigno et al. 2002). To shed light on this subject, Mirazizi et al. (2018) examined oxidation of some diazo dyes of phenol, catechol, guaiacol, monoamine (aniline), and *meta*-diamine by AbT, *N. crassa* laccase (NcL), and horseradish peroxidase (hPOX).

AbT had no effect on diazo derivatives of guaiacol, aniline, and *meta*-diamines. But it could oxidize phenolic and catecholic dyes. However, AbT reactions with those carrying strong EWS such as **S2f** and **S3f** (Fig. 2) were very slow. Despite its high redox potential, NcL ($E^0 = 780$ mV; Piontek et al. 2002) also failed to oxidize aniline and *meta*-diamine derivatives (**S4** and **S5**, Fig. 2), but it oxidized guaiacolic and catecholic dyes excluding those with EWS such as **S3f** and **S1f** (Mirazizi et al. 2018). The inability of AbT in oxidation of arylamines could be ascribed to the low K_a of these compounds, which works as a barrier in the deprotonation step. But NcL failure to oxidize diazo arylamines and phenols could be related to the high redox potential of these compounds. The oxidation potentials of these compounds follow the order of aniline (1.059 V) > phenol (0.6 V) > guaiacol (0.52 V) > catechol (0.42 V) (Haghbeen and Tan 1998; Bhuvaneshwari and Elango 2006).

The interesting results belonged to hPOX as it could oxidize almost all of these compounds with exception of **S1d** and **S1f**, which were carrying strong EWS (Mirazizi et al. 2018). The other notable result of this work related to the comparative rates of the reactions. Since these dyes had comparable extinction coefficients, the values of dA/min could be used as a reliable index for the instant rate of substrate consumption during the corresponding enzymatic reaction. For all the examined substances, the rate of oxidation by hPOX was markedly higher than those of AbT and NcL. As an example, all the three enzymes could oxidize **S3b** (Fig. 2), but hPOX reaction was 3.4- and 60-fold faster than the rates of AbT and NcL, respectively (Mirazizi et al. 2018).

2.2.5 Stereospecificity and Steric Hindrance

There is no report for any kind of tyrosinase with absolute stereospecificity. However, most of tyrosinases show preference of L-enantiomers. Harrison et al. (1967) showed higher activity of AbT in the presence of L(−)-DOPA over D(+)-DOPA, L(+)-adrenaline over D(−)-adrenaline, and L(+)-noradrenaline over D(−)-noradrenaline. These results were in harmony with previous results of mammalian tyrosinases including human tyrosinase (Yasunobu et al. 1959). From a comparative study on tyrosinases of apple, potato, the white rot fungus *Pycnoporus sanguineus*, the filamentous fungus *Trichoderma reesei*, and AbT, Selinheimo et al. (2007) also showed the enzymes preference for L-stereo isomers. However, for some tyrosinases, it was very low. For instance, potato tyrosinase showed no preference for L-tyrosine over D-tyrosine (Selinheimo et al. 2007).

Studies on tyrosinase selectivity for *ortho*-, *meta*-, and *para*-substituted phenols have disclosed high selectivity of this enzyme for *para*-substituted substrates. Examination of oxidation of substituted phenols by the complexes which could mimic *ortho*-hydroxylation of tyrosinase revealed that *ortho*-substituted phenols were practically resistant, while the *meta*-substituted counterparts were oxidized but at slower pace as compared with their *para*-substituted ones (Chioccara et al. 1995). Similar results obtained from studying reactions of *Streptomyces antibioticus* tyrosinase with fluorophenols. The kinetic parameters for *para*- and *meta*-fluorophenols were $K_m = 2.9$ and 6.9 mM and $k_{cat} = 10.5$ and 3 S^{-1}, respectively. But, there was no reaction between *ortho*-fluorophenol and the applied tyrosinase (Battaini et al. 2002). Inactivity of tyrosinase in the presence of guaiacolic compounds is ascribed to the steric hindrance of methoxy group at the *ortho* position (Selinheimo et al. 2007), while this type of compounds are easily oxidized by laccases (Xu 1996).

2.2.6 Pocket of Active Site

Another contrasting difference between laccases and tyrosinases originates from the apparent roles of the pockets of their active sites. In tyrosinases from various sources, the evident variations among the constructing residues of the pockets affect mainly the performance of the enzymes. This usually yields different kinetics parameters for two isozymes of tyrosinase in the presence of the same substrate (Kanteev et al. 2015). In contrast, the changes in the constructing residues of T1 cavity may result in the advent of unprecedented physicochemical features (Agrawal et al. 2018). From a substrate point of view, this means that the alterations in T1 cavity not only may modify the performance; it may also change the spectrum of laccase substrates. This characteristic feature of laccases has fueled recent researches for discovery of new ones (Baldrian 2006; Yang et al. 2017).

As illustrated in Fig. 3, cresolase reaction advances through a process of bond breaking and bond formation, which is initiated by the electrophilic attack of copper superoxide to the aromatic ring of the phenolic substrate, whereas catecholase reaction is completed by an inner-sphere electron transfer. This is why catecholase k_{cat} of a tyrosinase is usually much higher than its cresolase[3] k_{cat} (Perez-Gilabert and García Carmona 2000). As mentioned, catechol oxidases are extreme examples as they show only catecholase, while their active sites are quite similar to tyrosinases. A great deal of research has been conducted to explain different ratios of catecholase to cresolase among various tyrosinases and inability of catechol oxidases for running cresolase. Based on the similarities between hemocyanins,[4] tyrosinases, and catechol oxidases, some researchers suggested that the phenylalanine, above Cu_A in the active site of catechol oxidase, works as a blocker and hampers coordination of a phenolic substrate to Cu_A. This assumption lost its credibility by advances in the structural studies on various tyrosinases. *Juglans regia* tyrosinase, which has the same phenylalanine above CuA, shows monooxygenase activity (Pretzler and Rompel 2018).

The optimal pH for most of tyrosinases is close or below 7. As a result, deprotonation of a phenolic substrate is a crucial step in its binding to the copper center of tyrosinase. Historically, scientists thought that the bridging hydroxide between two copper ions of a type 3 cuproenzyme was able to mediate in

[3]The ratio of catecholase/cresolase differs among different tyrosinase isoenzymes.

[4]Hemocyanins are oxygen carriers in invertebrates. They are also cuproproteins of type 3, but their active sites are not accessible to molecules larger than O_2 (Solomon et al. 2014).

deprotonation of the phenolic substrate. However, in more recent works, Kanteev et al. (2015) showed the key role of a water molecule, which is conserved in the structure of *Bacillus megaterium* tyrosinase by two conserved amino acids of Glu and Asp, in deprotonation of the phenolic substrate. Based on this evidence, the inability of catechol oxidase for doing cresolase was correlated to the lack of that Asp in its structure (Solem et al. 2016). This theory also seemed incompatible with the fact that a threonine sits instead of that asparagine in aurone synthase which shows both cresolase and catecholase reactions (Pretzler and Rompel 2018).

2.3 Co-occurrence and Concomitant Use of Laccase with Tyrosinase

Tyrosinase is a ubiquitous enzyme, and laccase exists in plants, fungi, and, in a limited range, some other microorganisms. Association and distinctive separation of these enzymes have been the subject of numerous researches. Results of such studies have shed light on physiological functions as well as their biotechnological potentials (Ba and Vinoth Kumar 2017). However, it seems that despite considerable progress, this field of research is young and new discoveries encourage more deep research in this field (Martinez et al. 2017).

Tyrosinases, often as intracellular enzymes, are responsible for melanin biosynthesis. Nonetheless, by the help of reducing compounds such as ascorbic acid or proper nucleophiles, nature has used tyrosinases in biosynthesis of natural compounds such as betalains, aurones, and robust macromolecules (Gandía-Herrero et al. 2005). There is also emerging evidence that tyrosinase may play role in biosynthesis of neurotransmitters (Eisenhofer et al. 2003). On the other hand, laccases, as extracellular enzymes, are responsible for biodegradation of lignins and other molecules, which have impact on the quality of soil, water, and dialogues at biotic-biotic and biotic-abiotic interfaces (Ba and Vinoth Kumar 2017). By the help of other enzymes, intracellular laccases mediate in biosynthesis of lignins and lignin precursors (Yang et al. 2017).

Due to biotechnological potentials, both tyrosinase and laccase have been at the center of attention for similar applications (Ba and Vinoth Kumar 2017). Yet, there are few reports about simultaneous use of these enzymes especially for valorization or any kind of treatment of wastes. Hence, the lack of practical information hinders bright judgment about the impact of the progress of a tyrosinase reaction on a laccase reaction and vice versa. Mechanistically, these enzymes are different, but it does not mean that they are destructive of each other. It is assumed that at the beginning of a simultaneous use of laccase and tyrosinase, an oQn resists against a radical due to the electron deficiency, but by the progress of the reaction, the resistance would lessen, and the oQn(s) may work in benefit of the research purposes.

Increasing the range of the targets, growing the length of the chain reactions, and boosting the efficacy of the radical products could be among the primary aims of the simultaneous use of laccase with a tyrosinase. Freire et al. (2002) have already demonstrated cooperative effect of concomitant immobilization of laccase and tyrosinase in a biosensor for detecting phenols with various structures. Accordingly, it is anticipated that simultaneous use of tyrosinase and a laccase with optimal pH of activity about 6 would increase the efficacy of bioremediation of polluted waters.

In addition to the general goals of enzyme biotechnology including broadening pH tolerance, hosting various substrates, and increasing functional and thermal stability of enzymes, planning for cooperating enzymes via selecting or designing mission-oriented substrates seems to be promising for both engineering new materials and boosting efficiencies of enzymes in the desired applications, a target which has already been approached in cross-linking proteins and enzymes by tyrosinase and laccase (Isaschar-Ovdat and Fishman 2018).

3 Enzymes and Reactions Fueled by Laccases

Section 3 addresses the supply of other enzymes or subsequent reactions with substrates or reactants generated during laccase reactions. In the following, important groups of enzymes and the corresponding modes of action potentially involved in the aforementioned type of cooperation with laccases are introduced.

3.1 Fungal Class II Heme Peroxidases (CAZy Family AA2) and Heme-Thiolate Peroxidases/Peroxygenases

Many fungi concomitantly produce laccases and H_2O_2-depending oxidoreductases (i.e., peroxidases and peroxygenases) of various types. In white rot basidiomycetes, laccases are frequently secreted along with lignin-modifying peroxidases. The heme-containing extracellular enzymes lignin peroxidase (LiP; EC 1.11.1.14), manganese peroxidase (MnP; EC 1.11.1.13), and versatile peroxidase (VP; EC 1.11.1.16) belong to class II peroxidases of the peroxidase-catalase superfamily. Owing to their comparatively high redox potentials (ca. 1.4–1.5 V for LiP, 1.0–1.2 V for MnP, and 1.4–1.5 V for VP) and extraordinarily broad ranges of organic substrates, these peroxidases contribute to fungal degradation of both lignin constituents of lignocellulose and many organic environmental pollutants (Harms et al. 2011; Hofrichter et al. 2010; Kues 2015; Martinez et al. 2017). They are listed in the auxiliary activity family 2 (AA2) of the Carbohydrate Active Enzymes CAZy database (http://www.cazy.org/AA2.html/; Levasseur et al. 2013; Lombard et al. 2014), together with other peroxidases briefly described in the following.

Dye-decolorizing peroxidases (DyPs; EC 1.11.1.19; CAZy AA2 family) are members of the peroxidase-chlorite dismutase (or CDE; i.e., including chlorite dismutase, DyP, and *Efe*B protein from *Escherichia coli*) superfamily of heme peroxidases. They are known from basidiomycetes and were also detected in mycetozoa, bacteria, and archaea. Fungal DyPs (redox potential ca. 1.2–1.5 V) are extracellular enzymes, which are phylogenetically distant from the fungal high and low redox potential class II peroxidases mentioned before. However, their biological functions still remain to be firmly established. DyPs oxidize some recalcitrant organic compounds only poorly attacked by other peroxidases, such as certain synthetic dyes and mono-nitrophenols. They have also been reported to oxidize Mn^{2+} to Mn^{3+} as MnP and VP do (http://peroxibase.toulouse.inra.fr/; Fawal et al. 2013; Harms et al. 2011; Hofrichter et al. 2010; Kues 2015; Martinez et al. 2017).

Unspecific peroxygenases (UPOs; EC 1.11.2.1; not listed in the CAZy AA2 family) are heme-containing extracellular enzymes first discovered in and characterized from basidiomycetes. They are seemingly widespread in the entire fungal kingdom (*Eumycota*) and seen in certain oomycetes too. UPOs belong to the heme-thiolate peroxidase (HTP) superfamily. They are able to transfer an oxygen atom from H_2O_2 to a very broad range of different organic substrates, e.g., aromatics, heterocycles, and linear and cyclic hydrocarbons. Oxyfunctionalization reactions catalyzed by UPOs include selective hydroxylation, epoxidation, and heteroatom (S, N) oxygenation. The related product patterns frequently resemble those of intracellular cytochrome P450 monooxygenase systems (Bormann et al. 2015; Hofrichter et al. 2015; Martinez et al. 2017).

Peroxidases and peroxygenases have widely been considered as biocatalysts for degradation and detoxification purposes in various bioremediation and waste treatment applications. As versatile biocatalysts, they may also be applied in diverse industrial organic syntheses as well as for the upgrading and valorization of plant biomass (wastes) within lignocellulose biorefineries (Bormann et al. 2015; Karich et al. 2017; Martinez et al. 2017). Although secreted oxidoreductases possess the common advantage of being generally more stable than intracellular membrane-bound or cytosolic enzymes, the practical applicability of corresponding peroxidase and peroxygenase representatives is hampered by their need for H_2O_2 (Bormann et al. 2015; Kues 2015; Martinez et al. 2017). Class II peroxidases like LiP and MnP are dependent on H_2O_2 or organic peroxides and may become partly inactivated by excess H_2O_2 in a reversible manner (Hofrichter et al. 2010). Similarly, UPOs are readily inactivated at elevated concentrations of H_2O_2, which must be taken into account for correct peroxide dosage to ensure proper enzyme functioning (Bormann et al. 2015; Karich et al. 2016). Various ways of external H_2O_2 addition or the alternative use of organic hydroperoxides are of only limited practical applicability for performing peroxidase or peroxygenase reactions beyond laboratory scale. Instead, the simultaneous in situ generation of H_2O_2 from molecular oxygen by addition of suitable enzymes, e.g., glucose oxidase, has been considered as a way out (Bormann et al. 2015; Kues 2015). However, technical problems like the formation of pH-changing gluconic acid, a high viscosity of concentrated glucose solutions, considerable by-product formation, and ethical concerns related to the use of edible glucose challenge the in situ H_2O_2 generation by glucose oxidase at large scale (Bormann et al. 2015).

Laccases have also been reported as a cause of H_2O_2 production. The laccase-catalyzed oxidation of diphenols can initiate the subsequent reductive activation of O_2, yielding reactive oxygen species (ROS) such as superoxide anion radical and H_2O_2 that may participate in various abiotic and enzymatic follow-up reactions. ROS formation via the redox cycling of hydroquinones as driven by the interplay of laccase and other enzymes is described in Sect. 3.2.

Another type of laccase-initiated ROS formation is based on the oxidative cleavage of certain organic dicarbonic acids such as oxalate, malonate, and glyoxylate by Mn^{3+} (Hofrichter et al. 1998; Schlosser and Hofer 2002; Urzua et al. 1998b). For oxalate, the following reactions have been described (Urzua et al. 1998a, b):

$$Mn^{3+} + COOH - COOH \rightarrow Mn^{2+} + CO_2^{\cdot -} + CO_2 + 2H^+ \quad (1)$$

$$CO_2^{\cdot -} + O_2 \rightarrow O_2^{\cdot -} + CO_2 \quad (2)$$

$$O_2^{\cdot -} + Mn^{2+} + 2H^+ \rightarrow Mn^{3+} + H_2O_2 \quad (3)$$

$$O_2^{\cdot -} + H^+ \rightarrow ½H_2O_2 + ½O_2 \quad (4)$$

Abiotic decomposition of malonate can proceed via the following reactions (Hofrichter et al. 1998):

$$Mn^{3+} + COOH - CH_2 - COOH \rightarrow Mn^{2+} + COOH - CH_2^{\cdot} + CO_2 + H^+ \quad (5)$$

$$COOH - CH_2^{\cdot} + O_2 \rightarrow COOH - CH_2OO^{\cdot} \quad (6)$$

$$COOH - CH_2OO^{\cdot} + O_2 \rightarrow O_2^{\cdot -} + COOH - COOH + H^+ \quad (7)$$

$$COOH - CH_2OO^{\cdot} + Mn^{2+} + H^+ \rightarrow Mn^{3+} + COOH - CH_2OOH \quad (8)$$

Superoxide and oxalate derived from Eq. (7) subsequently can contribute to Eqs. (1)–(4).

Certain laccases can oxidize Mn^{2+} to soluble Mn^{3+} complexes in the presence of suitable organic (e.g., oxalate, malonate, tartrate) and inorganic Mn chelators (e.g., pyrophosphate), albeit at very low rates undercutting those commonly observed with fungal MnPs by several orders of magnitude (Gorbacheva et al. 2009; Hofer and Schlosser 1999; Schlosser and Hofer 2002). As typical for laccase reactions, the electron transfer from Mn^{2+} to type 1 copper (T1) of laccase is thought to be the rate-limiting step of substrate oxidation, which is generally thermodynamically driven by the redox potential difference between type 1 copper and the substrate (Gorbacheva et al. 2009; Schlosser and Hofer 2002; Xu et al. 2000). In line with this mechanism, Mn^{2+} oxidation could only be demonstrated for basidiomycete laccases (which frequently represent high redox potential enzymes; Gorbacheva et al. 2009; Hofer and Schlosser 1999; Rodgers et al. 2010; Schlosser and Hofer 2002), whereas low redox potential laccases such as plant *Rhus vernicifera* laccase were found to be inactive on Mn^{2+} (Frasconi et al. 2010; Gorbacheva et al. 2009). Despite being

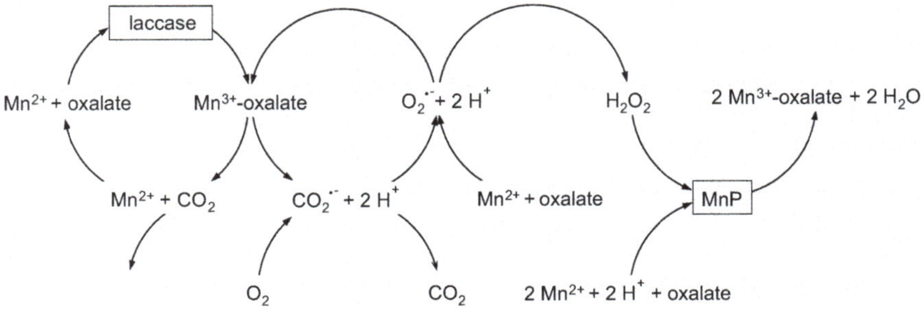

Fig. 5 H_2O_2 production resulting from the laccase-catalyzed Mn^{2+} oxidation in the presence of oxalate and its influence on MnP. Due to the possible involvement of complexes with various unknown ligand ratios, the stoichiometry has not been balanced for oxalate. Similar, albeit weaker cooperative effects between laccase and MnP with respect to Mn^{3+} formation were also observed in presence of malonate instead of oxalate. Reprinted with permission of the publisher from Schlosser and Hofer (2002), Laccase-catalysed oxidation of Mn^{2+} in the presence of natural Mn^{3+} chelators as a novel source of extracellular H_2O_2 production and its impact on manganese peroxidase. Appl Environ Microbiol 68(7):3514–3521, doi: https://doi.org/10.1128/AEM.68.7.3514-3521.2002. Copyright © 2002, American Society for Microbiology

extremely slow, the formation of Mn^{3+} complexes by basidiomycete laccases is sufficient to produce enough ROS such as superoxide anion radical and H_2O_2 to cause a significant impact on MnP reactions at physiologically relevant conditions of white rot basidiomycetes (Schlosser and Hofer 2002). In the presence of Mn^{2+} and appropriate organic chelators (e.g., oxalate and malonate), laccase produces Mn^{3+} complexes. The latter initialize a set of follow-up reactions finally resulting in H_2O_2 formation, which initiates or supports MnP reactions (Fig. 5). This type of laccase-MnP cooperation could be relevant for biodegradation of both lignin and xenobiotics (Schlosser and Hofer 2002) and potentially may be exploited for the in situ H_2O_2 generation in technical processes.

3.2 *Enzymes of the CAZy AA3 and AA6 Families*

The auxiliary activity family 3 (AA3) of the CAZy database (http://www.cazy.org/AA3.html/) harbors various members of the glucose-methanol-choline (GMC) family of oxidoreductases, which are flavoproteins containing a flavin-adenine dinucleotide (FAD)-binding domain (Bissaro et al. 2018; CAZypedia 2019; Kracher et al. 2016; Sutzl et al. 2018). Briefly, AA3 enzymes oxidize various carbohydrates or alcohols, which is accompanied by the reduction of FAD to $FADH_2$ (reductive half-reaction). Re-oxidation of $FADH_2$ (oxidative half-reaction) proceeds along with reduction of either O_2 to H_2O_2, (complexed) metal ions (e.g., Fe^{3+}) to reduced metal ions, or various quinones and phenoxy radicals to their corresponding hydroquinone counterparts (for a recent review, see Sutzl et al. 2018). CAZy family AA3 enzymes are widespread in yeasts and filamentous asco- and basidiomycetes. Due to

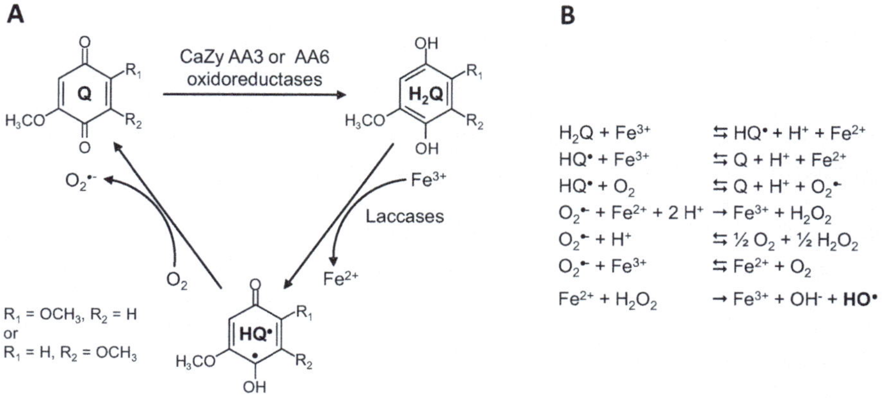

Fig. 6 (**a**) Possible interplay of laccases with AA3 and AA6 oxidoreductases during quinone redox cycling. Fungal or plant (lignin)-derived diphenols such as differently *o*-methoxy-substituted hydroquinones (**H_2Q**) are oxidized by laccases and Fe^{3+} (or only Fe^{3+} when laccase is absent), yielding semiquinone radicals (**HQ**·) which are then further oxidized to quinones (**Q**). The latter are reduced back to hydroquinones by CAZy AA3 or AA6 oxidoreductases. The figure was compiled based on information from the following references: An et al. (2015), Aranda et al. (2010), Gomez-Toribio et al. (2009a, b), Guillen et al. (1997, 2000), Krueger et al. (2016, 2015), Marco-Urrea et al. (2010), Sutzl et al. (2018), Vasiliadou et al. (2019), and Wei et al. (2010). Please also refer to Sect. 3.2 for further explanations. (**b**) Reactions accompanying quinone redox cycling, which lead to the formation of reactive oxygen species (ROS) including hydroxyl radical (HO·) (based on Kramer et al. 2004)

the aforementioned catalytic properties, members of different AA3 sub-families have been shown to fuel crystalline cellulose-attacking lytic polysaccharide monooxygenases (LPMOs; see also Sect. 3.4) with electrons either directly or via the reductive regeneration of fungal and plant-derived phenols (mostly benzenediols) in a quinone redox cycling mechanism (Kracher et al. 2016; Martinez 2016; Martinez et al. 2017). They are further considered as suppliers of H_2O_2 for lignin-modifying peroxidases (in white rot basidiomycetes) and extracellular Fenton-type reactions (in brown rot fungi) (Bissaro et al. 2018; CAZypedia 2019; Sutzl et al. 2018) (Fig. 6).

The possible interplay of certain CAZy AA3 family members with laccases could relate to quinone redox cycling aiming at production of highly reactive hydroxyl radicals (·OH) via extracellular Fenton-type reactions in both brown and white rot basidiomycetes (Bissaro et al. 2018; Harms et al. 2011; Sutzl et al. 2018) (Fig. 6). CAZy AA3 enzymes such as glucose dehydrogenase (GDH; EC 1.1.5.9), pyranose dehydrogenase (PDH; EC 1.1.99.29; not typical for brown rot fungi), pyranose oxidase (POx; EC 1.1.3.10; also not typical for brown rot fungi), and aryl-alcohol quinone oxidoreductases (AAQO), the latter showing high sequence identity to aryl-alcohol oxidase (AAO; EC 1.1.3.7) of white rot basidiomycetes, are known to reduce various (often methoxylated or methylated) quinones and phenoxy radicals efficiently (Bissaro et al. 2018; CAZypedia 2019; Kracher et al. 2016; Mathieu et al. 2016; Sutzl et al. 2018).

The CAZy family AA6 (http://www.cazy.org/AA6.html) contains 1,4-benzoquinone reductases (EC. 1.6.5.6), which contribute to the biodegradation of aromatic compounds and the protection of fungal cells from reactive quinone compounds (Levasseur et al. 2013). Quinone reductases are cell-bound enzymes found in diverse fungal groups. They are particularly prominent in white rot and brown rot basidiomycetes but not limited to wood- or litter-decaying basidiomycetes (Floudas et al. 2012; Jensen et al. 2002; Krueger et al. 2016; Lee et al. 2007). Alternatively to AA3 family enzymes, fungal AA6 benzoquinone reductases can contribute to quinone redox cycling leading to hydroxyl radical production (Bissaro et al. 2018; Harms et al. 2011; Sutzl et al. 2018) (Fig. 6).

Brown rot basidiomycetes are phylogenetically closely related to white rot fungi (Floudas et al. 2012), but have largely lost lignin-modifying enzymes (especially peroxidases). Instead, they employ Fenton chemistry to attack lignocellulose (Krueger et al. 2016; Martinez et al. 2009; Yelle et al. 2008). In brown rot fungi which do not express laccases or other phenol-oxidizing enzymes, secreted fungal diphenols such as 2,5-dimethoxyhydroquinone and 4,5-dimethoxycatechol can reduce Fe^{3+} to Fe^{2+} while becoming oxidized (Kerem et al. 1999; Krueger et al. 2016, 2015; Suzuki et al. 2006). Semiquinone radicals hereby produced from hydroquinones/diphenols upon one-electron oxidation steps reduce molecular oxygen, thus becoming further oxidized to quinones, while superoxide anion radicals are formed. These, in turn, can dismutate to H_2O_2, which reacts with Fe^{2+} in a Fenton-type reaction to yield hydroxyl radicals. Enzymes of the CAZy AA6 or AA3 family reduce benzoquinones back to diphenols/hydroquinones to complete the redox cycle. Alternatively to diphenols, other phenol derivatives such as phenolates and peptides produced by both brown and white rot basidiomycetes were also demonstrated to act as low molecular weight Fe^{3+} reductants (Arantes et al. 2011; Bissaro et al. 2018).

A few reports describe the expression of laccase and the occurrence of corresponding genes in certain brown rot fungi (An et al. 2015; Bissaro et al. 2018; Wei et al. 2009). Laccases may alternatively oxidize diphenols under conditions where abiotic oxidation of the latter is impeded by high concentrations of the strong Fe^{3+} chelator oxalate, which is produced by many brown rot basidiomycetes and secreted concomitantly with fungal diphenols (An et al. 2015; Wei et al. 2010). Laccases rapidly oxidize diphenols/hydroquinones to the corresponding semiquinones and thus can replace or boost abiotic diphenol oxidation, hereby becoming a part of the quinone redox cycling machinery described before.

Laccases were also demonstrated to be involved in quinone redox cycling in white rot fungi, where they are especially prominent (Aranda et al. 2010; Gomez-Toribio et al. 2009a, b; Guillen et al. 1997, 2000; Vasiliadou et al. 2019). Hydroquinone production by white rot fungi is less evident than that by brown rot basidiomycetes. Therefore, exogenously added quinones were often employed to demonstrate redox cycling effects in the corresponding studies (Aranda et al. 2010; Gomez-Toribio et al. 2009b; Krueger et al. 2016; Marco-Urrea et al. 2010).

Nevertheless, under natural conditions, lignin-derived diphenols (e.g., hydroquinones) stemming from *p*-hydroxyphenyl, guaiacyl, and syringyl units of lignin may act as redox cycling agents (Gomez-Toribio et al. 2009b; Sutzl et al. 2018). In white rot fungi, the redox cycling-based formation of H_2O_2 and hydroxyl radicals was found to be enhanced when Mn^{2+} was also present, due to high reactivity of the latter with superoxide anion radical (see also Eq. 3) (Gomez-Toribio et al. 2009a). Also in this fungal group, laccase may contribute to redox cycling in conjunction with secreted lignin-modifying peroxidases or also be replaced by them, owing to the easy oxidability of diphenols by all of such enzymes (Gomez-Toribio et al. 2009b; Marco-Urrea et al. 2010). Laccases may thus drive or support the oxidative part of quinone redox cycling, thereby cooperating with quinone reductases in the production of ROS such as H_2O_2 and highly reactive hydroxyl radicals as degradative agents. Beyond lignocellulose components, such extracellularly produced oxidants have been demonstrated to attack many aromatic and aliphatic pollutants such as benzene–toluene–ethylbenzene–xylene (BTEX) hydrocarbons, many pharmaceuticals, synthetic dyes, and polymers (Aranda et al. 2010; Gomez-Toribio et al. 2009a; Harms et al. 2011; Krueger et al. 2015, 2016; Marco-Urrea et al. 2010; Vasiliadou et al. 2019). Being widespread in the fungal kingdom (Giardina et al. 2010; Rodgers et al. 2010; Theerachat et al. 2019), laccases may potentially also contribute to quinone redox cycling in fungal groups beyond terrestrial lignocellulose decaying basidiomycetes (Krueger et al. 2016).

Cellobiose dehydrogenase (CDH; EC 1.11.99.18) of the CAZy sub-family AA3_1 is widespread among fungi and more common in white rot than in brown rot basidiomycetes. It oxidizes cellodextrins, mannodextrins, and lactoses while reducing a wide range of different two- and one-electron acceptors including Fe^{3+}, molecular oxygen, various quinones, and organic radicals (Larrondo et al. 2005; Sutzl et al. 2018; Zamocky et al. 2006). For a long time, CDH has widely been considered as a direct Fe^{3+} reductant, thus contributing to fungal Fenton-type reactions (Bissaro et al. 2018; Sutzl et al. 2018) (Fig. 5). A substantial increase in decolorization of recalcitrant synthetic dyes by laccase in the presence of CDH was attributed to the initial conversion of non-phenolic aromatic structures to phenolic ones by hydroxyl radicals, thus rendering molecules more susceptible to laccase attack (Ciullini et al. 2008; Enayatzamir et al. 2009). More recent reports indicate that CDHs directly fuel LPMOs with electrons as a major natural function (Bissaro et al. 2018; Kracher et al. 2016; Martinez et al. 2017).

3.3 *β*-Etherases

Lignin is a major renewable resource and could serve as feedstock for the future production of aromatic fine chemicals and building blocks for diverse applications, provided that a controlled and selective breakdown of the bonds linking the aromatic moieties of lignin could be achieved (Husarcikova et al. 2018; Marinovic et al. 2018; Pollegioni et al. 2015). Intracellular bacterial and fungal *β*-etherases selectively catalyze the reductive cleavage of *β-O*-4 aryl ether bonds, which form the most abundant bond type and account for 45–60% of the total linkages present in lignin (Husarcikova et al. 2018; Marinovic et al. 2018). *β*-Etherases belong to the glutathione-*S*-transferase (GST; EC 2.5.1.18) superfamily of proteins and have been functionally characterized from bacteria (*Sphingobium* and *Novosphingobium* species) and white rot basidiomycetes (*Dichomitus squalens*) (Husarcikova et al. 2018; Marinovic et al. 2018). Putative *β*-etherase encoding genes have widely been detected in bacterial and fungal genomes (Marinovic et al. 2018; Voss et al. 2020). Cleavage of *β-O*-4 aryl ether bonds of lignin by GSH-dependent *β*-etherases results in production of chiral glutathione adducts, which are then converted to oxidized glutathione (GSSG) and mono-aromatic building blocks of lignin by glutathione lyases (Husarcikova et al. 2018; Pollegioni et al. 2015). The known intracellular *β*-etherases from bacteria and fungi require a carbonyl function at the C*α* position adjacent to the *β*-carbon of *β-O*-4 bonds for activity, whereas hydroxyl groups at C*α* positions are present in natural lignin (Husarcikova et al. 2018; Marinovic et al. 2018). In *Sphingobium*, several NAD^+-dependent alcohol dehydrogenases first oxidize C*α* hydroxyl groups to the corresponding keto functions, thereby enabling the subsequent cleavage of *β-O*-4 aryl ether bond by *β*-etherases (Becker and Wittmann 2019; Husarcikova et al. 2018; Marinovic et al. 2018).

Possible practical applications of intracellular *β*-etherases necessitate the regeneration of the essential cofactors GSH and NAD^+ (Husarcikova and Schallmey 2019; Husarcikova et al. 2018; Roth and Spiess 2015). Picart et al. (2017) used a laccase-redox mediator system (LMS) instead of a C*α*-acting alcohol dehydrogenase in order to produce Cα carbonyl groups required for *β*-etherase activity, finally resulting in the successful depolymerization of natural lignin upon subsequent treatment with *β*-etherases and a glutathione lyase. This research demonstrated that complex cofactor-dependent steps involved in natural *β-O*-4 aryl ether cleavage pathways can be replaced by a less complicated and more user-friendly biocatalytic system. Nonetheless, remaining problems related to different pH optima of the involved enzymes and the partial repolymerization of lignin building blocks still need to be solved (Husarcikova et al. 2018; Picart et al. 2017; Roth and Spiess 2015).

In contrast to the intracellular *β*-etherases described above, an extracellular *β*-etherase from an ascomycete of the genus *Chaetomium* sp. has been reported to cleave a phenolic *β-O*-4 lignin model dimer containing a hydroxyl group at the C*α* position neighboring the *β*-carbon of the *β-O*-4 bond in a GSH-independent manner (Otsuka et al. 2003; Pollegioni et al. 2015).

3.4 Lytic Polysaccharide Monooxygenases

Fungal and bacterial LPMOs (CAZy families AA9, AA10, AA11, AA13, AA14, and AA15) oxidatively cleave the existing polysaccharide chains in cellulose, starch, and chitin. These enzymes depend on the supply of electrons for their catalytic action (Bissaro et al. 2018; Harris et al. 2010; Martinez et al. 2017; Sutzl et al. 2018;

Vaaje-Kolstad et al. 2010). Soluble lignin fragments arising from the redox mediator-assisted laccase attack on lignin, which may still contain phenolic groups, have been demonstrated to activate LPMOs via donation of electrons (Bissaro et al. 2018; Brenelli et al. 2018). However, the competition of laccases and LPMOs for oxygen and their different oxygen consumption rates have to be considered with regard to potential co-applications (Brenelli et al. 2018). Apart from laccases, other polyphenol oxidases such as atypical tyrosinases differing from the common fungal ones are also seemingly widespread in asco- and basidiomycetes along with LPMOs (Frommhagen et al. 2017). Such enzymes were reported to boost cellulose degradation by LPMO via initial *ortho*-hydroxylation of lignin-derived methoxylated monophenols to the corresponding 1,2-benzenediols, which can easily serve as electron donors (Frommhagen et al. 2017).

4 Enzymes and Reactions Producing Potential Laccase Substrates

Laccases may also utilize products formed by other enzymes as substrates. Important groups of enzymes and reactions potentially leading to the production of laccase substrates are presented in Sect. 4.

4.1 Enzymes and Reactions Catalyzing Aromatic Hydroxylation

Methoxy-substituted phenols, *o*- and *p*-diphenols, aminophenols, benzenethiols, polyphenols, and polyamines are typical substrates of laccases. However, these enzymes are not able to oxidize non-phenolic lignin and other high redox potential aromatic compounds directly (Arantes et al. 2011; Harms et al. 2011; Hofrichter et al. 2010; Munk et al. 2015). Biochemical reactions introducing hydroxyl groups to aromatic compounds, which are not (or only poorly) directly oxidized by laccases, could thus potentially produce (more easily oxidable) laccase substrates. However, direct evidence for such interactions is rather rare (Arantes et al. 2011; Harms et al. 2011). While the prominence of laccases in fungi is evident since a long time, bacterial lignin-modifying enzymes such as laccases and DyPs were revealed only in recent years (Bugg et al. 2011a; de Gonzalo et al. 2016; Granja-Travez et al. 2018).

Certain intracellular mono- and dioxygenases, which are involved in the fungal and bacterial metabolism of both natural and xenobiotic compounds, are able to hydroxylate aromatic rings. Such reactions facilitate further oxidation of the resulting products by various phenol oxidases including laccases and thus can qualify recalcitrant chemical structures for subsequent laccase attack. Certain fungal members of the cytochrome P450 monooxygenase superfamily are especially prominent in wood and plant litter decay basidiomycetes and catalyze hydroxylations at aromatic carbon atoms of plant constituents (lignin, plant defense chemicals) and many xenobiotic chemicals. Such reactions promote the detoxification and initiate the degradation of the aforementioned compounds (Harms et al. 2011; Hussain et al. 2020; Kues 2015; Morel et al. 2013; Syed et al. 2014). Fungal non-heme mixed-function oxidases such as phenol 2-monooxygenases (EC 1.14.13.7) produce *o*-diphenols (catechols) from phenols. Such enzymes have been implicated in fungal catabolism of chlorophenols and lignin-related *p*-hydroxyphenyl units (Harms et al. 2011; Lubbers et al. 2019). Similarly, mainly bacterial di- but also monooxygenases hydroxylate both natural (plant-derived) and xenobiotic compounds (Becker and Wittmann 2019; Chang and Zylstra 2008; de Gonzalo et al. 2016; Fuentes et al. 2014; Lubbers et al. 2019; Peng et al. 2010; Wang et al. 2017). For instance, certain bacterial dioxygenases produce 2,3-dihydroxybiphenyl structures from biphenyl and less chlorinated biphenyls (Bugg et al. 2011a, b).

Selective hydroxylation of non-activated C–H bonds in aromatic structures (oxyfunctionalization) by UPOs which can attack non-phenolic lignin structures as well as many different environmental pollutants (see Sect. 3.1) (Bormann et al. 2015; Hofrichter and Ullrich 2014; Karich et al. 2017; Kinne et al. 2011; Martinez et al. 2017) could also potentially make recalcitrant compounds (more) susceptible to subsequent laccase oxidation and possibly open up new routes for enzyme cascade applications, e.g., in the treatment of contaminated waste streams. UPOs would be advantageous over intracellular aromatic ring-hydroxylating mono- and dioxygenases with respect to enzyme stability and avoidance of reduced cofactors such as NAD(P)H.

Aromatic hydroxylation may further arise from the oxidative attack to natural (lignin) and xenobiotic aromatic structures by hydroxyl radicals, which are produced during extracellular Fenton-type reactions operative in both brown and white rot basidiomycetes and employed as degradative tools by these organisms (Bissaro et al. 2018; Bugg et al. 2011b; Harms et al. 2011; Sutzl et al. 2018). Hydroxyl radical-generating processes and involved enzymes are presented in more detail in Sect. 3.2.

4.2 *Enzymes Releasing Potential Laccase Substrates from Parent Compounds*

Potential laccase substrates could also be produced during biochemical cleavage of more complex parent compounds. For instance, guaiacol (*o*-methoxy phenol), which is a known laccase substrate, was demonstrated to be released during the cleavage of both phenolic and non-phenolic *β*-*O*-4 aryl ether dimeric lignin model compounds by bacterial and fungal *β*-etherases (see Sect. 3.3 for further details concerning these enzymes) (Husarcikova et al. 2018; Marinovic et al. 2018; Otsuka et al. 2003). The reductive enzymatic cleavage of *β*-*O*-4 aryl ether bonds in lignin by *β*-etherases hence yields phenolic products representing potential laccase substrates. Lignostilbene dioxygenase is a non-heme iron-dependent enzyme involved in the bacterial metabolism of diarylpropane lignin model compounds. It oxidatively cleaves the dimeric intermediate lignostilbene to yield two molecules of vanillin (Bugg et al. 2011b), another *o*-methoxy-substituted phenolic laccase substrate. Fungal lignin-modifying peroxidases catalyze the oxidative cleavage of $C\alpha$-$C\beta$ bonds in β-*O*-4 aryl ether and diarylpropane units found in lignin. In dependence on the respective parent compound structure(s), these enzymes can also release *o*-methoxy-substituted phenolic and diphenolic monolignin products (Becker and Wittmann 2019; Bugg et al. 2011b; Mäkelä et al. 2015; Pollegioni et al. 2015). However, due to the radical-based oxidation mechanisms employed by lignin-modifying peroxidases, such monolignin products usually occur in mixture and are easily prone to phenoxy radical coupling and polymerization processes (Becker and Wittmann 2019; Pollegioni et al. 2015), thus questioning a substantial practical role as laccase substrates.

Various bacterial and fungal aryl-*O*-demethylases can remove methyl moieties from methoxy-substituted aromatic compounds such as vanillate-, syringate-, or guaiacol-related lignin building blocks, thereby generating *o*-methoxy-substituted phenolic or catechol-like compounds potentially serving as laccase substrates (Abdelaziz et al. 2016; Becker and Wittmann 2019; Lubbers et al. 2019; Mallinson et al. 2018). The hydrolase tannase (EC 3.1.1.20) releases the laccase substrate gallic acid from gallotannins (also referred to as hydrolyzable tannins) (Mäkelä et al. 2015). Finally, also enzymes of the CAZy AA3 and AA6 families produce laccase substrates by reducing various quinones to their corresponding diphenol (e.g., hydroquinone) counterparts (for details please refer to Sect. 3.2).

5 Conclusions and Outlook

Functional cooperations between laccases and other enzymes could open up new routes for the future development of simultaneously operating multi-enzyme cocktails or sequentially acting enzyme cascades, which potentially may be employed for a wide range of sustainable bioeconomy applications (Martinez et al. 2017; Pezzella et al. 2015; Pollegioni et al. 2015; Roth and Spiess 2015). In some cases, the basic functioning of related enzyme cooperations has already convincingly been demonstrated at lab scale. However, the successful implementation of such novel biocatalytic systems into practical applications mostly still necessitates to overcome issues related to, e.g., operational stability under real process conditions, product yield and purity, and economic as well as legal constraints (Ba et al. 2018; Ba and Vinoth Kumar 2017; Husarcikova and Schallmey 2019; Picart et al. 2017; Roth and Spiess 2015; Schlosser and Hofer 2002). Other potential cooperations between laccases and further enzymes remain hypothetical yet and still have to be proven and further explored with regard to their usability for practical applications. Beyond the use of wild-type enzymes, state-of-the-art techniques such as protein engineering,

CRISPR/Cas9 genome editing, and adaptive evolution are expected to substantially enrich the toolbox of enzymes available for future biotechnological applications with novel and tailored biocatalysts (Kun et al. 2019; Song et al. 2019; Stanzione et al. 2020).

Acknowledgments K. Haghbeen appreciates support for international collaboration provided by the National Institute of Genetic Engineering and Biotechnology of I.R. Iran (project 504). This work as further supported by the Helmholtz Association of German Research Centers within the frame of the integrated project 'Controlling Chemicals' Fate' of the Chemicals In The Environment (CITE) research program (conducted at the Helmholtz Centre for Environmental Research—UFZ).

References

Abdelaziz OY, Brink DP, Prothmann J, Ravi K, Sun M, Garcia-Hidalgo J, Sandahl M, Hulteberg CP, Turner C, Liden G, Gorwa-Grauslund MF (2016) Biological valorization of low molecular weight lignin. Biotechnol Adv 34(8):1318–1346. https://doi.org/10.1016/j.biotechadv.2016.10.001

Agrawal K, Chaturvedi V, Verma P (2018) Fungal laccase discovered but yet undiscovered. Bioresour Bioprocess 5(1):4. https://doi.org/10.1186/s40643-018-0190-z

An H, Wei D, Xiao T (2015) Transcriptional profiles of laccase genes in the brown rot fungus *Postia placenta* MAD-R-698. J Microbiol 53(9):606–615. https://doi.org/10.1007/s12275-015-4705-4

Aranda E, Marco-Urrea E, Caminal G, Arias ME, Garcia-Romera I, Guillen F (2010) Advanced oxidation of benzene, toluene, ethylbenzene and xylene isomers (BTEX) by *Trametes versicolor*. J Hazard Mater 181(1–3):181–186. https://doi.org/10.1016/j.jhazmat.2010.04.114

Arantes V, Milagres AM, Filley TR, Goodell B (2011) Lignocellulosic polysaccharides and lignin degradation by wood decay fungi: the relevance of nonenzymatic Fenton-based reactions. J Ind Microbiol Biotechnol 38(4):541–555. https://doi.org/10.1007/s10295-010-0798-2

Ba S, Vinoth Kumar V (2017) Recent developments in the use of tyrosinase and laccase in environmental applications. Crit Rev Biotechnol 37(7):819–832. https://doi.org/10.1080/07388551.2016.1261081

Ba S, Haroune L, Soumano L, Bellenger J-P, Jones JP, Cabana H (2018) A hybrid bioreactor based on insolubilized tyrosinase and laccase catalysis and microfiltration membrane remove pharmaceuticals from wastewater. Chemosphere 201:749–755. https://doi.org/10.1016/j.chemosphere.2018.03.022

Baldrian P (2006) Fungal laccases–occurrence and properties. FEMS Microbiol Rev 30(2):215–242. https://doi.org/10.1111/j.1574-4976.2005.00010.x

Battaini G, Monzani E, Casella L, Lonardi E, Tepper AW, Canters GW, Bubacco L (2002) Tyrosinase-catalyzed oxidation of fluorophenols. J Biol Chem 277(47):44606–44612. https://doi.org/10.1074/jbc.M207829200

Becker J, Wittmann C (2019) A field of dreams: lignin valorization into chemicals, materials, fuels, and health-care products. Biotechnol Adv 37(6):107360. https://doi.org/10.1016/j.biotechadv.2019.02.016

Bertrand T, Jolivalt C, Briozzo P, Caminade E, Joly N, Madzak C, Mougin C (2002) Crystal structure of a four-copper laccase complexed with an arylamine: insights into substrate recognition and correlation with kinetics. Biochemistry 41(23):7325–7333. https://doi.org/10.1021/bi0201318

Bhuvaneshwari D, Elango K (2006) Correlation analysis of reactivity in the oxidation of anilines by nicotinium dichromate in nonaqueous media. Int J Chem Kin 38(11):657–665. https://doi.org/10.1002/kin.20199

Bissaro B, Varnai A, Rohr AK, Eijsink VGH (2018) Oxidoreductases and reactive oxygen species in conversion of lignocellulosic biomass. Microbiol Mol Biol Rev 82(4):e00029–e00018. https://doi.org/10.1128/MMBR.00029-18
Bormann S, Baraibar AG, Ni Y, Holtmann D, Hollmann F (2015) Specific oxyfunctionalisations catalysed by peroxygenases: opportunities, challenges and solutions. Cat Sci Technol 5 (4):2038–2052. https://doi.org/10.1039/c4cy01477d
Borojerdi SS, Haghbeen K, Karkhane AA, Fazli M, Saboury AA (2004) Successful resonance Raman study of cresolase activity of mushroom tyrosinase. Biochem Biophy Res Commun 314 (4):925–930. https://doi.org/10.1016/j.bbrc.2003.12.197
Brenelli L, Squina FM, Felby C, Cannella D (2018) Laccase-derived lignin compounds boost cellulose oxidative enzymes AA9. Biotechnol Biofuels 11(1):10. https://doi.org/10.1186/s13068-017-0985-8
Bugg TD, Ahmad M, Hardiman EM, Singh R (2011a) The emerging role for bacteria in lignin degradation and bio-product formation. Curr Opin Biotechnol 22(3):394–400. https://doi.org/10.1016/j.copbio.2010.10.009
Bugg TDH, Ahmad M, Hardiman EM, Rahmanpour R (2011b) Pathways for degradation of lignin in bacteria and fungi. Nat Prod Rep 28(12):1883–1896. https://doi.org/10.1039/C1NP00042J
CAZypedia (2019) Auxiliary activity family 3. https://www.cazypedia.org/index.php/Auxiliary_Activity_Family_3. Accessed 10 July 2019
Chang H-K, Zylstra GJ (2008) Examination and expansion of the substrate range of m-hydroxybenzoate hydroxylase. Biochem Biophys Res Commun 371(1):149–153. https://doi.org/10.1016/j.bbrc.2008.04.032
Chioccara F, Chiodini G, Farina F, Orlandi M, Rindone B, Sebastiano R (1995) Synthetic mimics of tyrosinase: catechols from ortho-, meta-and para-substituted phenols and copper (I) complexes. J Mol Catal A 97(3):187–194. https://doi.org/10.1016/1381-1169(94)00064-6
Ciullini I, Tilli S, Scozzafava A, Briganti F (2008) Fungal laccase, cellobiose dehydrogenase, and chemical mediators: combined actions for the decolorization of different classes of textile dyes. Bioresour Technol 99(15):7003–7010. https://doi.org/10.1016/j.biortech.2008.01.019
d'Acunzo F, Galli C, Masci B (2002) Oxidation of phenols by laccase and laccase-mediator systems: solubility and steric issues. Eur J Biochem 269(21):5330–5335. https://doi.org/10.1046/j.1432-1033.2002.03256.x
de Gonzalo G, Colpa DI, Habib MH, Fraaije MW (2016) Bacterial enzymes involved in lignin degradation. J Biotechnol 236:110–119. https://doi.org/10.1016/j.jbiotec.2016.08.011
Eisenhofer G, Tian H, Holmes C, Matsunaga J, Roffler-Tarlov S, Hearing VJ (2003) Tyrosinase: a developmentally specific major determinant of peripheral dopamine. FASEB J 17(10):1248–1255. https://doi.org/10.1096/fj.02-0736com
Enayatzamir K, Tabandeh F, Yakhchali B, Alikhani HA, Rodriguez Couto S (2009) Assessment of the joint effect of laccase and cellobiose dehydrogenase on the decolouration of different synthetic dyes. J Hazard Mater 169(1–3):176–181. https://doi.org/10.1016/j.jhazmat.2009.03.088
Fawal N, Li Q, Savelli B, Brette M, Passaia G, Fabre M, Mathe C, Dunand C (2013) PeroxiBase: a database for large-scale evolutionary analysis of peroxidases. Nucleic Acids Res 41(Database issue):D441–D444. https://doi.org/10.1093/nar/gks1083
Ferraroni M, Westphal AH, Borsari M, Tamayo-Ramos JA, Briganti F, de Graaff LH, van Berkel WJ (2017) Structure and function of *Aspergillus niger* laccase McoG. Biocatalysis 3(1):1–21. https://doi.org/10.1515/boca-2017-0001
Floudas D, Binder M, Riley R, Barry K, Blanchette RA, Henrissat B, Martinez AT, Otillar R, Spatafora JW, Yadav JS, Aerts A, Benoit I, Boyd A, Carlson A, Copeland A, Coutinho PM, de Vries RP, Ferreira P, Findley K, Foster B, Gaskell J, Glotzer D, Gorecki P, Heitman J, Hesse C, Hori C, Igarashi K, Jurgens JA, Kallen N, Kersten P, Kohler A, Kues U, Kumar TK, Kuo A, LaButti K, Larrondo LF, Lindquist E, Ling A, Lombard V, Lucas S, Lundell T, Martin R, McLaughlin DJ, Morgenstern I, Morin E, Murat C, Nagy LG, Nolan M, Ohm RA, Patyshakuliyeva A, Rokas A, Ruiz-Duenas FJ, Sabat G, Salamov A, Samejima M, Schmutz J, Slot JC, St John F, Stenlid J, Sun H, Sun S, Syed K, Tsang A, Wiebenga A, Young D, Pisabarro A, Eastwood DC, Martin F, Cullen D, Grigoriev IV, Hibbett DS (2012) The Paleozoic

origin of enzymatic lignin decomposition reconstructed from 31 fungal genomes. Science 336 (6089):1715–1719. https://doi.org/10.1126/science.1221748

Flurkey WH, Ratcliff B, Lopez L, Kuglin J, Dawley RM (1995) Differentiation of fungal tyrosinases and laccases using selective inhibitors and substrates. In: Lee CY, Whitaker JR (eds) Enzymatic browning and its prevention. ACS, Washington, DC, pp 81–89. https://doi.org/10.1021/bk-1995-0600.ch006

Frasconi M, Favero G, Boer H, Koivula A, Mazzei F (2010) Kinetic and biochemical properties of high and low redox potential laccases from fungal and plant origin. Biochim Biophys Acta 1804 (4):899–908. https://doi.org/10.1016/j.bbapap.2009.12.018

Freire RS, Thongngamdee S, Durán N, Wang J, Kubota LT (2002) Mixed enzyme (laccase/tyrosinase)-based remote electrochemical biosensor for monitoring phenolic compounds. Analyst 127(2):258–261. https://doi.org/10.1039/B110011D

Frommhagen M, Mutte SK, Westphal AH, Koetsier MJ, Hinz SWA, Visser J, Vincken JP, Weijers D, van Berkel WJH, Gruppen H, Kabel MA (2017) Boosting LPMO-driven lignocellulose degradation by polyphenol oxidase-activated lignin building blocks. Biotechnol Biofuels 10(1):121. https://doi.org/10.1186/s13068-017-0810-4

Fuentes S, Mendez V, Aguila P, Seeger M (2014) Bioremediation of petroleum hydrocarbons: catabolic genes, microbial communities, and applications. Appl Microbiol Biotechnol 98 (11):4781–4794. https://doi.org/10.1007/s00253-014-5684-9

Gandía-Herrero F, García-Carmona F (2013) Biosynthesis of betalains: yellow and violet plant pigments. Trends Plant Sci 18(6):334–343. https://doi.org/10.1016/j.tplants.2013.01.003

Gandía-Herrero F, Escribano J, García-Carmona F (2005) Betaxanthins as substrates for tyrosinase. An approach to the role of tyrosinase in the biosynthetic pathway of betalains. Plant Physiol 138 (1):421–432. https://doi.org/10.1104/pp.104.057992

García-Molina F, Hiner AN, Fenoll LG, Rodríguez-Lopez JN, García-Ruiz PA, García-Cánovas F, Tudela J (2005) Mushroom tyrosinase: catalase activity, inhibition, and suicide inactivation. J Agri Food Chem 53(9):3702–3709. https://doi.org/10.1021/jf048340h

Giardina P, Faraco V, Pezzella C, Piscitelli A, Vanhulle S, Sannia G (2010) Laccases: a never-ending story. Cell Mol Life Sci 67(3):369–385. https://doi.org/10.1007/s00018-009-0169-1

Glazunova OA, Trushkin NA, Moiseenko KV, Filimonov IS, Fedorova TV (2018) Catalytic efficiency of basidiomycete laccases: redox potential versus substrate-binding pocket structure. Catalysts 8(4):152. https://doi.org/10.3390/catal8040152

Gomez-Toribio V, Garcia-Martin AB, Martinez MJ, Martinez AT, Guillen F (2009a) Enhancing the production of hydroxyl radicals by *Pleurotus eryngii* via quinone redox cycling for pollutant removal. Appl Environ Microbiol 75(12):3954–3962. https://doi.org/10.1128/AEM.02138-08

Gomez-Toribio V, Garcia-Martin AB, Martinez MJ, Martinez AT, Guillen F (2009b) Induction of extracellular hydroxyl radical production by white-rot fungi through quinone redox cycling. Appl Environ Microbiol 75(12):3944–3953. https://doi.org/10.1128/AEM.02137-08

Gorbacheva M, Morozova O, Shumakovich G, Streltsov A, Shleev S, Yaropolov A (2009) Enzymatic oxidation of manganese ions catalysed by laccase. Bioorg Chem 37(1):1–5. https://doi.org/10.1016/j.bioorg.2008.09.002

Granja-Travez RS, Wilkinson RC, Persinoti GF, Squina FM, Fulop V, Bugg TDH (2018) Structural and functional characterisation of multi-copper oxidase CueO from lignin-degrading bacterium *Ochrobactrum* sp. reveal its activity towards lignin model compounds and lignosulfonate. FEBS J 285(9):1684–1700. https://doi.org/10.1111/febs.14437

Gray HB, Ellis WR Jr (1994) Electron transfer. In: Bertini I, Gray HB (eds) Bioinorganic chemistry. University Science Books, Sausalito, pp 315–363

Guillen F, Martinez MJ, Munoz C, Martinez AT (1997) Quinone redox cycling in the ligninolytic fungus *Pleurotus eryngii* leading to extracellular production of superoxide anion radical. Arch Biochem Biophys 339(1):190–199. https://doi.org/10.1006/abbi.1996.9834

Guillen F, Munoz C, Gomez-Toribio V, Martinez AT, Jesus Martinez M (2000) Oxygen activation during oxidation of methoxyhydroquinones by laccase from *Pleurotus eryngii*. Appl Environ Microbiol 66(1):170–175. https://doi.org/10.1128/aem.66.1.170-175.2000

Haghbeen K, Tan EW (1998) Facile synthesis of catechol azo dyes. J Org Chem 63(13):4503–4505. https://doi.org/10.1021/jo972151z

Haghbeen K, Tan EW (2003) Direct spectrophotometric assay of monooxygenase and oxidase activities of mushroom tyrosinase in the presence of synthetic and natural substrates. Anal Biochem 312(1):23–32. https://doi.org/10.1016/s0003-2697(02)00408-6

Hakulinen N, Kiiskinen LL, Kruus K, Saloheimo M, Paananen A, Koivula A, Rouvinen J (2002) Crystal structure of a laccase from Melanocarpus albomyces with an intact trinuclear copper site. Nat Struct Mol Biol 9(8):601–605. https://doi.org/10.1038/nsb823

Harms H, Schlosser D, Wick LY (2011) Untapped potential: exploiting fungi in bioremediation of hazardous chemicals. Nat Rev Microbiol 9(3):177–192. https://doi.org/10.1038/nrmicro2519

Harris PV, Welner D, McFarland KC, Re E, Navarro Poulsen JC, Brown K, Salbo R, Ding H, Vlasenko E, Merino S, Xu F, Cherry J, Larsen S, Lo Leggio L (2010) Stimulation of lignocellulosic biomass hydrolysis by proteins of glycoside hydrolase family 61: structure and function of a large, enigmatic family. Biochemistry 49(15):3305–3316. https://doi.org/10.1021/bi100009p

Harrison W, Whisler WW, Ko S (1967) Detection and study by fluorescence spectrometry of stereospecificity in mushroom tyrosinase-catalyzed oxidations proposal of a copper-containing reaction rate control site. J Biol Chem 242(7):1660–1667

Hassani S, Gharechaei B, Nikfard S, Fazli M, Gheibi N, Hardré R, Legge RL, Haghbeen K (2018) New insight into the allosteric effect of L-tyrosine on mushroom tyrosinase during L-dopa production. Int J Biol Macromol 114:821–829. https://doi.org/10.1016/j.ijbiomac.2018.03.185

Hofer C, Schlosser D (1999) Novel enzymatic oxidation of Mn^{2+} to Mn^{3+} catalyzed by a fungal laccase. FEBS Lett 451(2):186–190. https://doi.org/10.1016/s0014-5793(99)00566-9

Hofrichter M, Ullrich R (2014) Oxidations catalyzed by fungal peroxygenases. Curr Opin Chem Biol 19:116–125. https://doi.org/10.1016/j.cbpa.2014.01.015

Hofrichter M, Ziegenhagen D, Vares T, Friedrich M, Jager MG, Fritsche W, Hatakka A (1998) Oxidative decomposition of malonic acid as basis for the action of manganese peroxidase in the absence of hydrogen peroxide. FEBS Lett 434(3):362–366. https://doi.org/10.1016/S0014-5793(98)01023-0

Hofrichter M, Ullrich R, Pecyna MJ, Liers C, Lundell T (2010) New and classic families of secreted fungal heme peroxidases. Appl Microbiol Biotechnol 87(3):871–897. https://doi.org/10.1007/s00253-010-2633-0

Hofrichter M, Kellner H, Pecyna MJ, Ullrich R (2015) Fungal unspecific peroxygenases: heme-thiolate proteins that combine peroxidase and cytochrome P450 properties. In: Hrycay EG, Bandiera SM (eds) Monooxygenase, peroxidase and peroxygenase properties and mechanisms of cytochrome P450. Springer International, Cham, pp 341–368. https://doi.org/10.1007/978-3-319-16009-2_13

Husarcikova J, Schallmey A (2019) Whole-cell cascade for the preparation of enantiopure beta-*O*-4 aryl ether compounds with glutathione recycling. J Biotechnol 293:1–7. https://doi.org/10.1016/j.jbiotec.2019.01.002

Husarcikova J, Voss H, Dominguez de Maria P, Schallmey A (2018) Microbial beta-etherases and glutathione lyases for lignin valorisation in biorefineries: current state and future perspectives. Appl Microbiol Biotechnol 102(13):5391–5401. https://doi.org/10.1007/s00253-018-9040-3

Hussain R, Ahmed M, Khan TA, Akhter Y (2020) Fungal P450 monooxygenases – the diversity in catalysis and their promising roles in biocontrol activity. Appl Microbiol Biotechnol 104(3):989–999. https://doi.org/10.1007/s00253-019-10305-3

Isaschar-Ovdat S, Fishman A (2018) Crosslinking of food proteins mediated by oxidative enzymes– a review. Trends Food Sci Technol 72:134–143. https://doi.org/10.1016/j.tifs.2017.12.011

Jensen KA Jr, Ryan ZC, Vanden Wymelenberg A, Cullen D, Hammel KE (2002) An NADH:quinone oxidoreductase active during biodegradation by the brown-rot basidiomycete *Gloeophyllum trabeum*. Appl Environ Microbiol 68(6):2699–2703. https://doi.org/10.1128/aem.68.6.2699-2703.2002

Kallio J, Auer S, Jänis J, Andberg M, Kruus K, Rouvinen J, Koivula A, Hakulinen N (2009) Structure–function studies of a *Melanocarpus albomyces* laccase suggest a pathway for oxidation of phenolic compounds. J Mol Biol 392(4):895–909. https://doi.org/10.1016/j.jmb.2009.06.053

Kallio JP, Gasparetti C, Andberg M, Boer H, Koivula A, Kruus K, Rouvinen J, Hakulinen N (2011) Crystal structure of an ascomycete fungal laccase from *Thielavia arenaria*–common structural features of asco-laccases. FEBS J 278(13):2283–2295. https://doi.org/10.1111/j.1742-4658.2011.08146.x

Kanteev M, Goldfeder M, Fishman A (2015) Structure–function correlations in tyrosinases. Protein Sci 24(9):1360–1369. https://doi.org/10.1002/pro.2734

Karich A, Scheibner K, Ullrich R, Hofrichter M (2016) Exploring the catalase activity of unspecific peroxygenases and the mechanism of peroxide-dependent heme destruction. J Mol Catal B Enzym 134:238–246. https://doi.org/10.1016/j.molcatb.2016.10.014

Karich A, Ullrich R, Scheibner K, Hofrichter M (2017) Fungal unspecific peroxygenases oxidize the majority of organic EPA priority pollutants. Front Microbiol 8:1463. https://doi.org/10.3389/fmicb.2017.01463

Kerem Z, Hammel, Hammel KE (1999) Biodegradative mechanism of the brown rot basidiomycete *Gloeophyllum trabeum*: evidence for an extracellular hydroquinone-driven fenton reaction. FEBS Lett 446(1):49–54. https://doi.org/10.1016/s0014-5793(99)00180-5

Kinne M, Poraj-Kobielska M, Ullrich R, Nousiainen P, Sipilä J, Scheibner K, Hammel KE, Hofrichter M (2011) Oxidative cleavage of non-phenolic *ß*-O-4 lignin model dimers by an extracellular aromatic peroxygenase. Holzforschung 65:673–679

Kracher D, Scheiblbrandner S, Felice AK, Breslmayr E, Preims M, Ludwicka K, Haltrich D, Eijsink VG, Ludwig R (2016) Extracellular electron transfer systems fuel cellulose oxidative degradation. Science 352(6289):1098–1101. https://doi.org/10.1126/science.aaf3165

Kramer C, Kreisel G, Fahr K, Kassbohrer J, Schlosser D (2004) Degradation of 2-fluorophenol by the brown-rot fungus *Gloeophyllum striatum*: evidence for the involvement of extracellular Fenton chemistry. Appl Microbiol Biotechnol 64(3):387–395. https://doi.org/10.1007/s00253-003-1445-x

Krueger MC, Hofmann U, Moeder M, Schlosser D (2015) Potential of wood-rotting fungi to attack polystyrene sulfonate and its depolymerisation by *Gloeophyllum trabeum* via hydroquinone-driven Fenton chemistry. PLoS One 10(7):e0131773. https://doi.org/10.1371/journal.pone.0131773

Krueger MC, Bergmann M, Schlosser D (2016) Widespread ability of fungi to drive quinone redox cycling for biodegradation. FEMS Microbiol Lett 363(11). https://doi.org/10.1093/femsle/fnw105

Kues U (2015) Fungal enzymes for environmental management. Curr Opin Biotechnol 33:268–278. https://doi.org/10.1016/j.copbio.2015.03.006

Kun RS, Gomes ACS, Hilden KS, Salazar Cerezo S, Makela MR, de Vries RP (2019) Developments and opportunities in fungal strain engineering for the production of novel enzymes and enzyme cocktails for plant biomass degradation. Biotechnol Adv 37(6):107361. https://doi.org/10.1016/j.biotechadv.2019.02.017

Larrondo LF, Vicuña R, Cullen D (2005) *Phanerochaete chrysosporium* genomics. In: Arora DK, Berka RM (eds) Applied mycology and biotechnology, vol 5. Elsevier, Amsterdam, pp 315–352. https://doi.org/10.1016/S1874-5334(05)80016-4

Lee SS, Moon DS, Choi HT, Song HG (2007) Purification and characterization of an intracellular NADH: quinone reductase from *Trametes versicolor*. J Microbiol 45(4):333–338

Levasseur A, Drula E, Lombard V, Coutinho PM, Henrissat B (2013) Expansion of the enzymatic repertoire of the CAZy database to integrate auxiliary redox enzymes. Biotechnol Biofuels 6 (1):41. https://doi.org/10.1186/1754-6834-6-41

Lombard V, Golaconda Ramulu H, Drula E, Coutinho PM, Henrissat B (2014) The carbohydrate-active enzymes database (CAZy) in 2013. Nucleic Acids Res 42(Database issue):D490–D495. https://doi.org/10.1093/nar/gkt1178

Lubbers RJM, Dilokpimol A, Visser J, Makela MR, Hilden KS, de Vries RP (2019) A comparison between the homocyclic aromatic metabolic pathways from plant-derived compounds by bacteria and fungi. Biotechnol Adv 37(7):107396. https://doi.org/10.1016/j.biotechadv.2019.05.002

Majeau JA, Brar SK, Tyagi RD (2010) Laccases for removal of recalcitrant and emerging pollutants. Bioresour Technol 101(7):2331–2350. https://doi.org/10.1016/j.biortech.2009.10.087

Mäkelä MR, Marinović M, Nousiainen P, Liwanag AJM, Benoit I, Sipilä J, Hatakka A, de Vries RP, Hildén KS (2015) Aromatic metabolism of filamentous fungi in relation to the presence of aromatic compounds in plant biomass. In: Sima S, Geoffrey Michael G (eds) Advances in applied microbiology, vol 91. Academic, Cambridge, pp 63–137. https://doi.org/10.1016/bs.aambs.2014.12.001

Mallinson SJB, Machovina MM, Silveira RL, Garcia-Borras M, Gallup N, Johnson CW, Allen MD, Skaf MS, Crowley MF, Neidle EL, Houk KN, Beckham GT, DuBois JL, McGeehan JE (2018) A promiscuous cytochrome P450 aromatic O-demethylase for lignin bioconversion. Nat Commun 9(1):2487. https://doi.org/10.1038/s41467-018-04878-2

Marco-Urrea E, Radjenovic J, Caminal G, Petrovic M, Vicent T, Barcelo D (2010) Oxidation of atenolol, propranolol, carbamazepine and clofibric acid by a biological Fenton-like system mediated by the white-rot fungus *Trametes versicolor*. Water Res 44(2):521–532. https://doi.org/10.1016/j.watres.2009.09.049

Marinovic M, Nousiainen P, Dilokpimol A, Kontro J, Moore R, Sipila J, de Vries RP, Makela MR, Hilden K (2018) Selective cleavage of lignin *β-O*-4 aryl ether bond by *β*-etherase of the white-rot fungus *Dichomitus squalens*. ACS Sustain Chem Eng 6(3):2878–2882. https://doi.org/10.1021/acssuschemeng.7b03619

Martinez AT (2016) CHEMISTRY. How to break down crystalline cellulose. Science 352 (6289):1050–1051. https://doi.org/10.1126/science.aaf8920

Martinez D, Challacombe J, Morgenstern I, Hibbett D, Schmoll M, Kubicek CP, Ferreira P, Ruiz-Duenas FJ, Martinez AT, Kersten P, Hammel KE, Vanden Wymelenberg A, Gaskell J, Lindquist E, Sabat G, Bondurant SS, Larrondo LF, Canessa P, Vicuna R, Yadav J, Doddapaneni H, Subramanian V, Pisabarro AG, Lavin JL, Oguiza JA, Master E, Henrissat B, Coutinho PM, Harris P, Magnuson JK, Baker SE, Bruno K, Kenealy W, Hoegger PJ, Kues U, Ramaiya P, Lucas S, Salamov A, Shapiro H, Tu H, Chee CL, Misra M, Xie G, Teter S, Yaver D, James T, Mokrejs M, Pospisek M, Grigoriev IV, Brettin T, Rokhsar D, Berka R, Cullen D (2009) Genome, transcriptome, and secretome analysis of wood decay fungus *Postia placenta* supports unique mechanisms of lignocellulose conversion. Proc Natl Acad Sci USA 106 (6):1954–1959. https://doi.org/10.1073/pnas.0809575106

Martinez AT, Ruiz-Duenas FJ, Camarero S, Serrano A, Linde D, Lund H, Vind J, Tovborg M, Herold-Majumdar OM, Hofrichter M, Liers C, Ullrich R, Scheibner K, Sannia G, Piscitelli A, Pezzella C, Sener ME, Kilic S, van Berkel WJH, Guallar V, Lucas MF, Zuhse R, Ludwig R, Hollmann F, Fernandez-Fueyo E, Record E, Faulds CB, Tortajada M, Winckelmann I, Rasmussen JA, Gelo-Pujic M, Gutierrez A, Del Rio JC, Rencoret J, Alcalde M (2017) Oxidoreductases on their way to industrial biotransformations. Biotechnol Adv 35(6):815–831. https://doi.org/10.1016/j.biotechadv.2017.06.003

Mathieu Y, Piumi F, Valli R, Aramburu JC, Ferreira P, Faulds CB, Record E (2016) Activities of secreted aryl alcohol Quinone oxidoreductases from *Pycnoporus cinnabarinus* provide insights into fungal degradation of plant biomass. Appl Environ Microbiol 82(8):2411–2423. https://doi.org/10.1128/AEM.03761-15

Matoba Y, Kihara S, Bando N, Yoshitsu H, Sakaguchi M, Ke K, Yanagisawa S, Ogura T, Sugiyama M (2018) Catalytic mechanism of the tyrosinase reaction toward the Tyr98 residue in the caddie protein. PLoS Biol 16(12):e3000077. https://doi.org/10.1371/journal.pbio.3000077

Mirazizi F, Bahrami A, Asl SS, Zaribafan A, Haghbeen K, Aminzadeh S (2018) Evaluation of oxidative enzymes for efficient oxidation of aniline and phenolic pollutants. Int J Environ Sci Technol 15(8):1679–1686. https://doi.org/10.1007/s13762-017-1493-x

Morel M, Meux E, Mathieu Y, Thuillier A, Chibani K, Harvengt L, Jacquot JP, Gelhaye E (2013) Xenomic networks variability and adaptation traits in wood decaying fungi. Microb Biotechnol 6(3):248–263. https://doi.org/10.1111/1751-7915.12015

Morin E, Kohler A, Baker AR, Foulongne-Oriol M, Lombard V, Nagy LG, Ohm RA, Patyshakuliyeva A, Brun A, Aerts AL, Bailey AM, Billette C, Coutinho PM, Deakin G, Doddapaneni H, Floudas D, Grimwood J, Hilden K, Kues U, Labutti KM, Lapidus A, Lindquist EA, Lucas SM, Murat C, Riley RW, Salamov AA, Schmutz J, Subramanian V, Wosten HA, Xu J, Eastwood DC, Foster GD, Sonnenberg AS, Cullen D, de Vries RP, Lundell T, Hibbett DS, Henrissat B, Burton KS, Kerrigan RW, Challen MP, Grigoriev IV, Martin F (2012) Genome sequence of the button mushroom *Agaricus bisporus* reveals mechanisms governing adaptation to a humic-rich ecological niche. Proc Natl Acad Sci USA 109(43):17501–17506. https://doi.org/10.1073/pnas.1206847109

Moshtaghioun SM, Haghbeen K, Lotfi Sahebghadam A, Legge RL, Khoshneviszadeh R, Farhadi S (2011) Direct spectrophotometric assay of laccase using diazo derivatives of guaiacol. Anal Chem 83(11):4200–4205. https://doi.org/10.1021/ac200501w

Munk L, Sitarz AK, Kalyani DC, Mikkelsen JD, Meyer AS (2015) Can laccases catalyze bond cleavage in lignin? Biotechnol Adv 33(1):13–24. https://doi.org/10.1016/j.biotechadv.2014.12.008

Otsuka Y, Sonoki T, Ikeda S, Kajita S, Nakamura M, Katayama Y (2003) Detection and characterization of a novel extracellular fungal enzyme that catalyzes the specific and hydrolytic cleavage of lignin guaiacylglycerol beta-aryl ether linkages. Eur J Biochem 270 (11):2353–2362. https://doi.org/10.1046/j.1432-1033.2003.03545.x

Pardo I, Santiago G, Gentili P, Lucas F, Monza E, Medrano FJ, Galli C, Martínez AT, Guallar V, Camarero S (2016) Re-designing the substrate binding pocket of laccase for enhanced oxidation of sinapic acid. Cat Sci Technol 6(11):3900–3910. https://doi.org/10.1039/C5CY01725D

Peng RH, Xiong AS, Xue Y, Fu XY, Gao F, Zhao W, Tian YS, Yao QH (2010) A profile of ring-hydroxylating oxygenases that degrade aromatic pollutants. Rev Environ Contam Toxicol 206:65–94. https://doi.org/10.1007/978-1-4419-6260-7_4

Perez-Gilabert M, García Carmona F (2000) Characterization of catecholase and cresolase activities of eggplant polyphenol oxidase. J Agric Food Chem 48(3):695–700. https://doi.org/10.1021/jf990292r

Pezzella C, Guarino L, Piscitelli A (2015) How to enjoy laccases. Cell Mol Life Sci 72(5):923–940. https://doi.org/10.1007/s00018-014-1823-9

Picart P, Liu H, Grande PM, Anders N, Zhu L, Klankermayer J, Leitner W, Dominguez de Maria P, Schwaneberg U, Schallmey A (2017) Multi-step biocatalytic depolymerization of lignin. Appl Microbiol Biotechnol 101(15):6277–6287. https://doi.org/10.1007/s00253-017-8360-z

Piontek K, Antorini M, Choinowski T (2002) Crystal structure of a laccase from the fungus Trametes versicolor at 1.90-Å resolution containing a full complement of coppers. J Biol Chem 277(40):37663–37669. https://doi.org/10.1074/jbc.M204571200

Plonka PM, Grabacka M (2006) Melanin synthesis in microorganisms—biotechnological and medical aspects. Acta Biochim Pol 53(3):429–443. https://doi.org/10.18388/abp.2006_3314

Pollegioni L, Tonin F, Rosini E (2015) Lignin-degrading enzymes. FEBS J 282(7):1190–1213. https://doi.org/10.1111/febs.13224

Pretzler M, Rompel A (2018) What causes the different functionality in type-III-copper enzymes? A state of the art perspective. Inorg Chim Acta 481:25–31. https://doi.org/10.1016/j.ica.2017.04.041

Rescigno A, Sollai F, Pisu B, Rinaldi A, Sanjust E (2002) Tyrosinase inhibition: general and applied aspects. J Enzyme Inhib Med Chem 17(4):207–218. https://doi.org/10.1080/14756360210000010923

Rodgers CJ, Blanford CF, Giddens SR, Skamnioti P, Armstrong FA, Gurr SJ (2010) Designer laccases: a vogue for high-potential fungal enzymes? Trends Biotechnol 28(2):63–72. https://doi.org/10.1016/j.tibtech.2009.11.001

Roth S, Spiess AC (2015) Laccases for biorefinery applications: a critical review on challenges and perspectives. Bioprocess Biosyst Eng 38(12):2285–2313. https://doi.org/10.1007/s00449-015-1475-7

Roy LE, Jakubikova E, Guthrie MG, Batista ER (2009) Calculation of one-electron redox potentials revisited. Is it possible to calculate accurate potentials with density functional methods? J Phys Chem A 113(24):6745–6750. https://doi.org/10.1021/jp811388w

Saboury A, Zolghadri S, Haghbeen K, Moosavi-Movahedi A (2006) The inhibitory effect of benzenethiol on the cresolase and catecholase activities of mushroom tyrosinase. J Enzyme Inhib Med Chem 21(6):711–717. https://doi.org/10.1080/14756360600810787

Schlosser D, Hofer C (2002) Laccase-catalyzed oxidation of Mn^{2+} in the presence of natural Mn^{3+} chelators as a novel source of extracellular H_2O_2 production and its impact on manganese peroxidase. Appl Environ Microbiol 68(7):3514–3521. https://doi.org/10.1128/aem.68.7.3514-3521.2002

Selinheimo E, NiEidhin D, Steffensen C, Nielsen J, Lomascolo A, Halaouli S, Record E, O'Beirne D, Buchert J, Kruus K (2007) Comparison of the characteristics of fungal and plant tyrosinases. J Biotechnol 130(4):471–480. https://doi.org/10.1016/j.jbiotec.2007.05.018

Shamsipur M, Maddah B, Hemmateenejad B, Rouhani S, Haghbeen K, Alizadeh K (2008) Multiwavelength spectrophotometric determination of acidity constants of some azo dyes. Spectrochim Acta A Mol Biomol Spectrosc 70(1):1–6. https://doi.org/10.1016/j.saa.2007.07.006

Shleev S, Morozova O, Nikitina O, Gorshina E, Rusinova T, Serezhenkov V, Burbaev D, Gazaryan I, Yaropolov A (2004) Comparison of physico-chemical characteristics of four laccases from different basidiomycetes. Biochimie 86(9–10):693–703. https://doi.org/10.1016/j.biochi.2004.08.005

Solem E, Tuczek F, Decker H (2016) Tyrosinase versus catechol oxidase: one asparagine makes the difference. Angew Chem Int Ed 55(8):2884–2888. https://doi.org/10.1002/anie.201508534

Solomon EI, Heppner DE, Johnston EM, Ginsbach JW, Cirera J, Qayyum M, Kieber-Emmons MT, Kjaergaard CH, Hadt RG, Tian L (2014) Copper active sites in biology. Chem Rev 114 (7):3659–3853. https://doi.org/10.1021/cr400327t

Song R, Zhai Q, Sun L, Huang E, Zhang Y, Zhu Y, Guo Q, Tian Y, Zhao B, Lu H (2019) CRISPR/Cas9 genome editing technology in filamentous fungi: progress and perspective. Appl Microbiol Biotechnol 103(17):6919–6932. https://doi.org/10.1007/s00253-019-10007-w

Stanzione I, Pezzella C, Giardina P, Sannia G, Piscitelli A (2020) Beyond natural laccases: extension of their potential applications by protein engineering. Appl Microbiol Biotechnol 104(3):915–924. https://doi.org/10.1007/s00253-019-10147-z

Sutzl L, Laurent C, Abrera AT, Schutz G, Ludwig R, Haltrich D (2018) Multiplicity of enzymatic functions in the CAZy AA3 family. Appl Microbiol Biotechnol 102(6):2477–2492. https://doi.org/10.1007/s00253-018-8784-0

Suzuki MR, Hunt CG, Houtman CJ, Dalebroux ZD, Hammel KE (2006) Fungal hydroquinones contribute to brown rot of wood. Environ Microbiol 8(12):2214–2223. https://doi.org/10.1111/j.1462-2920.2006.01160.x

Syed K, Shale K, Pagadala NS, Tuszynski J (2014) Systematic identification and evolutionary analysis of catalytically versatile cytochrome P450 monooxygenase families enriched in model basidiomycete fungi. PLoS One 9(1):e86683. https://doi.org/10.1371/journal.pone.0086683

Theerachat M, Guieysse D, Morel S, Remaud-Simeon M, Chulalaksananukul W (2019) Laccases from marine organisms and their applications in the biodegradation of toxic and environmental pollutants: a review. Appl Biochem Biotechnol 187(2):583–611. https://doi.org/10.1007/s12010-018-2829-9

Thurston CF (1994) The structure and function of fungal laccases. Microbiology 140(1):19–26

Urzua U, Kersten PJ, Vicuna R (1998a) Kinetics of Mn^{3+}-oxalate formation and decay in reactions catalyzed by manganese peroxidase of *Ceriporiopsis subvermispora*. Arch Biochem Biophys 360(2):215–222. https://doi.org/10.1006/abbi.1998.0952

Urzua U, Kersten PJ, Vicuna R (1998b) Manganese peroxidase-dependent oxidation of glyoxylic and oxalic acids synthesized by *Ceriporiopsis subvermispora* produces extracellular hydrogen peroxide. Appl Environ Microbiol 64(1):68–73

Vaaje-Kolstad G, Westereng B, Horn SJ, Liu Z, Zhai H, Sorlie M, Eijsink VG (2010) An oxidative enzyme boosting the enzymatic conversion of recalcitrant polysaccharides. Science 330 (6001):219–222. https://doi.org/10.1126/science.1192231

Vasiliadou IA, Molina R, Pariente MI, Christoforidis KC, Martinez F, Melero JA (2019) Understanding the role of mediators in the efficiency of advanced oxidation processes using white-rot fungi. Chem Eng J 359:1427–1435. https://doi.org/10.1016/j.cej.2018.11.035

Voss H, Heck CA, Schallmey M, Schallmey A (2020) Database mining for novel bacterial beta-etherases, glutathione-dependent lignin-degrading enzymes. Appl Environ Microbiol 86(2). https://doi.org/10.1128/AEM.02026-19

Wang C, Fu Y, Liu L (2010) Theoretical estimation of Hammett σ p constants of organic radical groups. Chin Sci Bull 55(25):2904–2908. https://doi.org/10.1007/s11434-010-4015-5

Wang Y, Li J, Liu A (2017) Oxygen activation by mononuclear nonheme iron dioxygenases involved in the degradation of aromatics. J Biol Inorg Chem 22(2–3):395–405. https://doi.org/10.1007/s00775-017-1436-5

Wei H, Xu Q, Taylor LE 2nd, Baker JO, Tucker MP, Ding SY (2009) Natural paradigms of plant cell wall degradation. Curr Opin Biotechnol 20(3):330–338. https://doi.org/10.1016/j.copbio.2009.05.008

Wei D, Houtman CJ, Kapich AN, Hunt CG, Cullen D, Hammel KE (2010) Laccase and its role in production of extracellular reactive oxygen species during wood decay by the brown rot basidiomycete *Postia placenta*. Appl Environ Microbiol 76(7):2091–2097. https://doi.org/10.1128/AEM.02929-09

Wood JM, Schallreuter KU (1991) Studies on the reactions between human tyrosinase, superoxide anion, hydrogen peroxide and thiols. Biochim Biophys Acta 1074(3):378–385. https://doi.org/10.1016/0304-4165(91)90088-x

Xu F (1996) Oxidation of phenols, anilines, and benzenethiols by fungal laccases: correlation between activity and redox potentials as well as halide inhibition. Biochemistry 35 (23):7608–7614. https://doi.org/10.1021/bi952971a

Xu F (1997) Effects of redox potential and hydroxide inhibition on the pH activity profile of fungal laccases. J Biol Chem 272(2):924–928. https://doi.org/10.1074/jbc.272.2.924

Xu F, Shin W, Brown SH, Wahleithner JA, Sundaram UM, Solomon EI (1996) A study of a series of recombinant fungal laccases and bilirubin oxidase that exhibit significant differences in redox potential, substrate specificity, and stability. Biochim Biophys Acta 1292(2):303–311. https://doi.org/10.1016/0167-4838(95)00210-3

Xu F, Berka RM, Wahleithner JA, Nelson BA, Shuster JR, Brown SH, Palmer AE, Solomon EI (1998) Site-directed mutations in fungal laccase: effect on redox potential, activity and pH profile. Biochem J 334(1):63–70. https://doi.org/10.1042/bj3340063

Xu F, Kulys JJ, Duke K, Li K, Krikstopaitis K, Deussen HJ, Abbate E, Galinyte V, Schneider P (2000) Redox chemistry in laccase-catalyzed oxidation of N-hydroxy compounds. Appl Environ Microbiol 66(5):2052–2056. https://doi.org/10.1128/aem.66.5.2052-2056.2000

Yang J, Li W, Ng TB, Deng X, Lin J, Ye X (2017) Laccases: production, expression regulation, and applications in pharmaceutical biodegradation. Front Microbiol 8:832. https://doi.org/10.3389/fmicb.2017.00832

Yasunobu K, Peterson E, Mason H (1959) The oxidation of tyrosine-containing peptides by tyrosinase. J Biol Chem 234(12):3291–3295

Yelle DJ, Ralph J, Lu F, Hammel KE (2008) Evidence for cleavage of lignin by a brown rot basidiomycete. Environ Microbiol 10(7):1844–1849. https://doi.org/10.1111/j.1462-2920.2008.01605.x

Yoon J, Fujii S, Solomon EI (2009) Geometric and electronic structure differences between the type 3 copper sites of the multicopper oxidases and hemocyanin/tyrosinase. Proc Nat Acad Sci 106 (16):6585–6590. https://doi.org/10.1073/pnas.0902127106

Zamocky M, Ludwig R, Peterbauer C, Hallberg BM, Divne C, Nicholls P, Haltrich D (2006) Cellobiose dehydrogenase—a flavocytochrome from wood-degrading, phytopathogenic and saprotropic fungi. Curr Protein Pept Sci 7(3):255–280. https://doi.org/10.2174/138920306777452367

Immobilized Laccase: A Promising Bioremediation Tool for the Removal of Organic Contaminants in Wastewater

Vasanth Kumar Vaithyanathan, Abiram Karanam Rathankumar, Vinoth Kumar Vaidyanathan, and Hubert Cabana

Contents

V. K. Vaithyanathan · H. Cabana (✉)
University of Sherbrooke Water Research Group (USWRG), Department of Civil and Building Engineering, University of Sherbrooke, Sherbrooke, QC, Canada
e-mail: Hubert.Cabana@USherbrooke.ca

A. K. Rathankumar
Department of Biotechnology, Integrated Bioprocessing Laboratory, SRM Institute of Technology, Chennai, Tamil Nadu, India

V. K. Vaidyanathan
University of Sherbrooke Water Research Group (USWRG), Department of Civil and Building Engineering, University of Sherbrooke, Sherbrooke, QC, Canada

Department of Biotechnology, Integrated Bioprocessing Laboratory, SRM Institute of Technology, Chennai, Tamil Nadu, India

D. Schlosser (ed.), *Laccases in Bioremediation and Waste Valorisation*,
Microbiology Monographs 33, https://doi.org/10.1007/978-3-030-47906-0_5

Abstract Laccase, an incredible enzyme, has a wide prospective in bioremediation processes, mainly due to its relative broad oxidation capacity, the lack of requirement of cofactors, and the use of readily available oxygen as the final electron acceptor. However, the large-scale application of laccases in bioremediation necessitates immobilization/insolubilization of the biocatalysts to enhance their operational stability. With the burgeoning use of laccases in wastewater treatment, several state-of-the-art methods have been developed over the past few years to immobilize laccase, derived from various microbial sources, in order to enhance the selectivity, activity, stability, and reusability. Recent advances in these immobilization methods offer promising solutions to the limitations of soluble enzymes, such as poor reusability due to poor recoverability, low stability, and high costs, to name a few. This article is intended to review the various recent methods employed for immobilization or insolubilization of laccase and its use in treating various types of organic contaminants in wastewaters including those from olive mill, pulp and paper, biorefinery, municipal, hospital, and textile industries. Furthermore, to improve the potential of the laccase-based biocatalytic system against wastewater/pollution treatment, co-immobilization of enzymes such as tyrosinase, peroxidase, and glucose oxidase, with laccase, would serve as a promising bioremediation tool for treating the organic contaminants in industrial and municipal wastewater. The concept and approach of this review also renders knowledge on a yet unexplored focus on the pioneering advances on the development of immobilized laccase-based reusable biocatalysts, which could be employed for treatment of industrial and hospital wastewater.

1 Laccase: A Bioremediation Tool

The global rise in environmental contaminants has led to a reduction in the quality of life, which has become a major concern in recent years. The root of these contaminants can be mainly traced to the waste treatment plants, and the source and type of treatment is determined by the nature of these contaminants (Pal et al. 2010). Typically, wastewater treatment focuses on remediation of the effluent in order to meet the regulations of the effluent rather than using the treated water in the industrial process. Despite the development of several wastewater remediation processes, effective and rapid large-scale water treatment still remains a challenging problem (Dignac et al. 2000). On the other hand, with the advances in enzyme technology over the past decade, an increased interest in the use of enzymes in wastewater treatment has been noted. Among these enzymes, laccase (EC 1.10.3.2) has been widely exploited as a biocatalytic tool for various industrial applications including wastewater treatment due to its ability to utilize a broader substrate range and molecular oxygen for catalysis (Gasser et al. 2014). Moreover, the ability of laccases to transform phenol and non-phenolic compounds by oxidation process makes them an excellent choice for the remediation of a wide range of contaminants

(Couto and Herrera 2006). Laccases are multicopper oxidases, with a potential to oxidize a broad range of phenolic compounds including amino phenols, *ortho-* and *para*-diphenols, polyphenols, and aromatic and aliphatic compounds, which are coupled with electron reduction of O_2 to H_2O (Strong and Claus 2011). Laccases can degrade potential contaminants, making them less toxic and imparting high bioavailability, which can be removed by conventional physical or mechanical methods. During the laccase-mediated bioremediation process, the redox potential ($E°$) difference between the enzyme and the substrate is one of the crucial factors that determine the oxidation rate of the substrate, thus limiting the use of laccase. On the other hand, with the use of redox mediators, which act as a diffusible shuttle from laccase enzyme to substrate, the scope of laccase on the remediation of high redox potential pollutants has broadened to a greater extent (Morozova et al. 2007; Frasconi et al. 2010). The best examples for laccase-mediated removal of contaminants include the removal of PAHs such as anthracene or benzopyrene (Zeng et al. 2016; Wu et al. 2008), recalcitrant dyes (Kumar et al. 2014), and organophosphorus compounds, such as nerve agents VX or Russian VX (Amitai et al. 1998). However, large-scale treatment processes require a large amount of the enzyme which is not considered economical (Osma et al. 2011). Moreover, several factors such as pH and chemical compounds present in wastewater hinder the enzyme activity by inhibiting the activity site of the enzyme or inducing modification in the amino acid residues or chelation of the copper atoms of the enzyme, leading to the reduction or inhibition of the enzyme activities (Johannes and Majcherczyk 2000; Bollag and Leonowicz 1984). In addition, the high ionic strength imparted due to the presence of halide groups (e.g., F^-, Cl^-, Br^-), heavy metals (e.g., Hg^{2+}, sn^{2+}, Zn^{2+}, Fe^{2+}, Fe^{3+}), surfactants and some organic solvents (acetone, acetonitrile, dimethyl sulfoxide), or proteases may inhibit the activity of laccase (Madhavi and Lele 2009; Yaropolov et al. 1994; Couto and Herrera 2006; Cabana et al. 2007). The difficulty in the recovery and reusability of the enzymes has also posed as a major limitation in the large-scale application of laccase. Thus, development of an immobilized laccase system not only overcomes the drawbacks but also enhances the stability of the enzyme to a greater extent.

2 Laccase Immobilization

Irrespective of the general definition of immobilization, the prime functionality of the immobilization process addresses two main functionalities: first, to aid in effectively stabilizing the functionality of the enzyme within a desired time and space (catalytic function) and, second, to ease the separation of the enzyme, in turn enabling reusability of the biocatalytic system (non-catalytic function). The enzyme immobilization process depends on the interaction of two functional compounds, the enzyme and the carrier (carrier-based immobilization). The amino acid and the functional groups on the side chain of laccases play a pivotal role in enhancing the stability. Moreover, the presence of the side chain influences most of the surface properties of the enzyme, including the net charge, resultant of the individual amino acid dissociation constants (pKa) that dictate the interaction of the enzyme with the carrier particle (Secundo 2013; Halling et al. 2005). In fungal laccases, the three main amino acid residues that are reported to be present are histidine (His), cysteine (Cys), and methionine (Met). As Met is a non-polar amino acid with a sulfur moiety, it is not involved in the cross-linking reaction in contrast to His and Cys (Ba et al. 2013; Habeeb and Hiramoto 1968). On the other hand, cysteine along with histidine, lysine, and tyrosine residues is found to interact with the functional group present on the carrier molecule (Ba et al. 2013). In addition, the stability and flexibility of the immobilized particle can be varied by modifying the chemical bonds between the enzyme and carrier particle, for example, the introduction of a covalent bond by cross-linking agents. In most cases, the efficiency of the immobilized enzyme mostly depends on the properties of the carrier particle (Sheldon and van Pelt 2013). Considering this, it is generally accepted that the carrier should have a large accessible surface area (i.e., $>100\ m^2\ g^{-1}$) combined with a pore size of approximately three times larger than the average diameter of the enzyme (i.e., >30 nm). The latter results in better enzyme loading, retention, and reduced substrate and cofactor diffusion constraints (Cao 2005).

Ideally, the carrier material should be inert and prepared to correlate with the surface property of the enzymes, with a large porous surface area to accommodate a large number of enzyme molecules. Moreover, the carrier particles should be chemically and mechanically stable. For wastewater application, the carrier particles should also be economically feasible, environmentally friendly, and more widely available, which makes their selection a pivotal step in the immobilization process. However, the presence of carrier particles may hinder the mass transfer rate leading to a decrease in catalytic activity, which has led to the development of a carrier-free immobilization technique by cross-linking the biocatalyst with cross-linking agents such as glutaraldehyde (Sheldon and van Pelt 2013).

3 Immobilization Methods

Considering the surface properties of enzymes and their application, several immobilization methods have been introduced (Fig. 1). These methods have been broadly divided into five main types for ease of understanding:

1. Covalent immobilization
2. Cross-linked enzyme immobilization
3. Adsorption
4. Entrapment
5. Encapsulation

This chapter presents the various laccase immobilization technologies that exist for bioremediation of wastewater (Table 1), for which most of the studies were conducted either at a lab-scale or at a pilot-scale level.

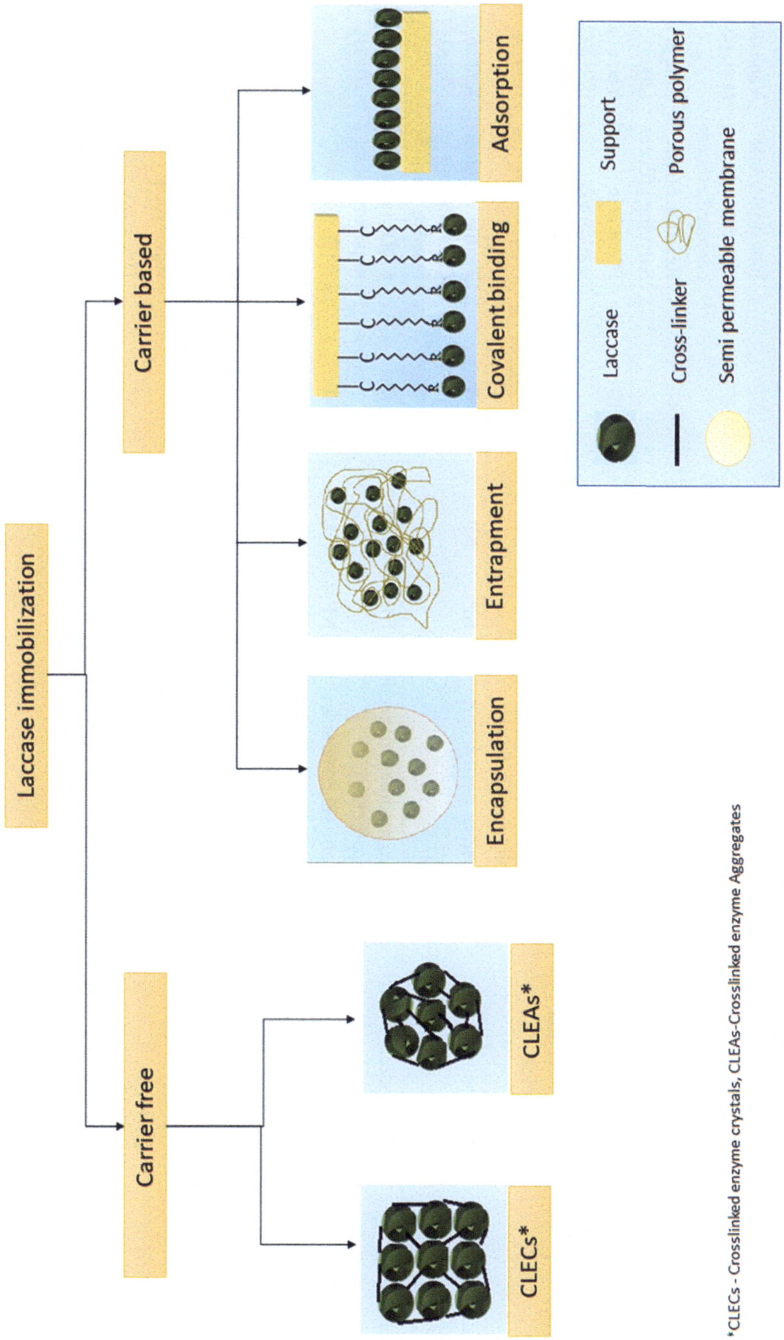

Fig. 1 Schematic representation of types of immobilization

Table 1 Laccase-based immobilized systems used for wastewater remediation

Microorganism	Immobilization process	Support	Yield (%)	Biocatalytic load (U/mg)	Reusability (cycle)	Application	References
White-rot fungi family	Adsorption	Woodchips	NP[a]	NP	NP	Industrial crude effluent and ozonated effluent	Vanhulle et al. (2005)
T. villosa	Adsorption with carrier modifier	Aminopropyltriethoxysilane alumina spherical particle	~50	NP	4	Reactive black 5 dyeing effluent	Zille et al. (2003)
T. versicolor	Physical adsorption	Halloysite nanotubes and chitosan	NP	123.1 mg/g	10	Phenols in wastewater	Yao et al. (2015)
Coriolus versicolor	Adsorption and entrapment	Laccase adsorbed on activated carbon entrapped in calcium alginate bead	NP	NP	8	Dichlorophenols in wastewater	Zhang et al. (2006)
Lactarius volemus	Adsorption	Microsilica	NP	NP	NP	Textile wastewater containing reactive black 5	Kalkan et al. (2014)
T. versicolor	Adsorption	Surface modified magnetic silica particles	NP	58.3	NP	COD[b] reduction in papermaking wastewater	Liu and Wang (2014)
		Magnetic silica particles	NP	34.7	NP		
Paraconiothyrium variabile	Entrapment	Gelatin–alginate	NP	NP	5	Decolorization of industrial synthetic dyes	Mogharabi et al. (2012)
T. versicolor	Entrapment	Copper alginate and iron oxide	85.5	NP	NP	Synthetic pollutants in wastewater	Le et al. (2016)
Trametes sp.	Entrapment	Manganese ferrite ($MnFe_2O_4$)	NP	0.0165	NP	Dye wastewater	Shojaat et al. (2016)
Pleurotus ostreatus IBL-02	Encapsulation	Hydrophobic gel entrapment	NP	NP	NP	Decolorization of different dyes and local textile wastewaters	Asgher et al. (2012)

T. versicolor	Encapsulation	Multi-walled carbon nanotube	88.9	NP	NP	Phenolic organics from water	Dai et al. (2016)
Ganoderma lucidum	Sol gel	Trimethoxysilane and propyltrimethoxysilane	~90	NP	NP	Wastewater effluent	Irshad et al. (2012)
Lentinula edodes	Covalent	Eupergit	45	NP	8	Phenols in olive mill wastewater	D'Annibale et al. (2000)
L. edodes	*Covalent*	Chitosan	NP	0.520	NP	Polyphenols, *ortho*-diphenols, and dye in olive mill wastewater	D'Annibale et al. (1999)
T. hirsuta	Covalent	Aminopropyltriethoxysilane alumina	68	0.14 mg/g	NP	Textile dyes and dying effluent	Abadulla et al. (2000)
Pycnoporus coccineus	Covalent	Acrylic epoxy-activated resins, Eupergit C 250 L	NP	0.110	NP	Olive oil mill wastewater phenolic compounds	Berrio et al. (2007)
Coriolopsis polyzona	Covalent	Amino-modified silica nanoparticles	61.4	2.67	NP	Real-time wastewater	Zimmermann et al. (2011)
Phoma sp. UHH 5-1-03	Cross-linking	PVDF membranes by electron beam irradiation	NP	NP	NP	PhAC in wastewater	Jahangiri et al. (2018)
T. versicolor	CLEAs[c]	Magnetic mesoporous silica microbeads (MMSMB)	39	1.53	NP	PhAC in wastewater	Arca-Ramos et al. (2016b)
T. versicolor (laccase) and *Myceliophthora thermophila* (laccase)	Co-immobilization	Fumed silica nanoparticles	89.9	1.10	10	Micropollutants from wastewater	Arca-Ramos et al. (2016a)
T. versicolor (Laccase) and *M. thermophila* (Laccase)	Combi-CLEAs	Functionalized chitosan	23.3 (Lac) 115.6 (Try)	NP	NP	COD of paper mill wastewater	Ba et al. (2012)

(continued)

Table 1 (continued)

Microorganism	Immobilization process	Support	Yield (%)	Biocatalytic load (U/mg)	Reusability (cycle)	Application	References
T. versicolor laccase, *Bjerkandera adusta* versatile peroxidase (VP), *Aspergillus niger* glucose oxidase (GOD)	Combi-CLEAs	Chitosan	NP	NP	NP	Pharmaceuticals from urban wastewater	Touahar et al. (2014)
T. versicolor laccase (TvL) *Mushroom tyrosinase (Tyr)*	Combi-CLEAs	Chitosan	10.6 (lac) 61.8 (Tyr)	NP	NP	Acetaminophen in hospital and municipal wastewater	Ba et al. (2014a)

[a]*NP* not provided
[b]*COD* chemical oxygen demand
[c]*CLEAs* cross-linked enzyme aggregates

3.1 Covalent Binding

Covalent immobilization provides strong bonding and helps in overcoming the drawbacks of leaching, specifically when working in aqueous media consisting of denaturing factors (wastewater). This robustness of the enzyme could be due to the restriction in the conformation changes by resisting the enzyme folding and thermal vibrations by multiple bonding between the carrier and enzyme molecule (Brady and Jordaan 2009). The covalent association of laccase with the support occurs due to the presence of the amino acids in the side chain (eg. methionine, cysteine, lysine, histidine), whose reactivity depends on the functional groups, such as hydroxyl, imide, phenolic, indolyl, etc. (Sheldon and van Pelt 2013; Singh et al. 2015). The specific activity of the immobilized laccase depends on the surface area of the support material, which acts as a decisive factor. Similar to adsorption, several synthetic and naturally occurring carriers can be used for the covalent immobilization of laccase. The most common synthetic carriers, which include magnetic nanoparticles (Wang et al. 2012), mesoporous silica (Salis et al. 2009), silica (Zawisza et al. 2006), Eupergit® (Lloret et al. 2012), etc. along with the naturally occurring carrier molecules such as chitosan (Kalkan et al. 2012), chitin, activated coal (Davis and Burns 1992), and coconut fiber (de Souza Bezerra et al. 2015), were commonly used for immobilization and subsequent bioremediation purposes. The occurrence of steric hindrance, a major limitation, due to direct coupling of support with enzymes can be overcome by the introduction of spacer arms such as glutaraldehyde, which separate the enzyme from the support molecule. On the contrary, the carrier molecules containing carboxylic acid, such as polymers of acrylic acid, require activators, such as carbodiimide, which under slightly acidic conditions react with the carboxylic acid group to form highly reactive *O*-acylisourea derivatives. Among many commercially available water soluble carbodiimides, cyclohexyl-3-(2-morpholino-ethyl)-carbodiimide (CMC), 1-ethyl-3-(3-dimethylamino propyl)-carbodiimide (EDC), and 1-cyclohexyl-3-(2-morpholino-ethyl)-carbodiimide (CMC) are commonly reported. Among the other functional groups, amine-bearing supports were widely used as carrier particles for covalent immobilization. The amino functionalization can be introduced by either aminosilane attachment or polyethyleneimine coating.

Coupling of laccases with the modified support can be achieved by various methods, which include the use of bifunctional groups such as dialdehydes, di-imidate esters, and diisocyanates. Among the various bifunctional groups, glutaraldehyde was the most commonly used cross-linking agent during covalent immobilization. Glutaraldehyde forms a complex Schiff base with an amine functional group on the carrier to form α,β-unsaturated carbonyl groups on which enzyme may attach. Arıca et al. (2009) used a bifunctional glutaraldehyde for the immobilization of laccase on non-porous poly glycidyl methacrylate and ethylene glycol dimethacrylate beads with a maximum biocatalytic load of 4.9 mg/g along with bioremediation application of the textile industrial dye Reactive Red 120 (Arıca et al. 2009). Over the past decade, an increase in the number of covalently immobilized laccase particle-based reports for bioremediation application reflects the popularity of the covalent immobilization technique. One such report includes the use of covalently immobilized *Lentinula edodes* laccase on Eupergit (D'Annibale et al. 1999), which proves its effective removal of phenolics from olive mill wastewater in lab scale (D'Annibale et al. 2000). Despite its merits, diffusion limitation of the substrate, intense aggregation, and settling along with the possibility of structural alternation due to covalent bond formation on the active site of laccase limits the use of covalent-based laccase immobilization.

3.2 Cross-Linking of Laccase

Enzyme immobilization by cross-linking the reactive NH_2 group of the enzyme using bi- or multifunctional cross-linking agents resulted in the strategical development of carrier-free immobilization. Carrier-free immobilized biocatalysts can be produced directly by cross-linked enzyme crystals (CLECs) or cross-linked enzyme aggregates (CLEAs). However, as CLECs require highly purified crystalline enzyme, CLEAs established itself as a potential alternative in developing a stable biocatalyst that gears up activity to the maximum level possible. CLEAs is a rapid, gentle, and cost-effective method for the production of carrier-free biocatalysts (Sheldon 2011). In recent years, CLEAs have become a novel and potential biocatalytic system on both the lab and industrial scale and are generally considered to be the next-generation biocatalyst for their applicability in the elimination of emerging contaminants (Cabana et al. 2009; Ba et al. 2012, 2014a; Kumar et al. 2012). The main advantage associated with CLEAs includes simultaneous purification and immobilization of an enzyme particle in a simpler method in a short duration of time, with the aid of a salting-out agent, ammonium sulfate, in the presence of organic solvents, such as polyethylene glycol (Cabana et al. 2009), tert-butanol (Kumar et al. 2012), and isopropanol (Matijošytė et al. 2010) under optimum physiological conditions of the enzyme, along with cross-linking agents, such as glutaraldehyde (Cabana et al. 2009). The resultant aggregates are separated by simple centrifugation followed by washing with a buffer solution. Formation of linking bonds among lysine amino acid residues by the action of dialdehyde group (glutaraldehyde) was reported to be a cause for the formation of smart biocatalysts, CLEA particles, with tunable physical properties (Cabana et al. 2009). Since the immobilization technique does not require a carrier particle as support, high biocatalytic activity can be achieved at a lower production cost (Kumar et al. 2012). Another additional benefit of the CLEAs technology is that it stabilizes the quaternary structure of multimeric enzymes and enhances the operational stability of the biocatalytic system (Kumar et al. 2012). Generally, the resultant CLEAs exhibit stability in both aqueous and solvent phases, making them suitable immobilization processes (Sheldon 2011; Ba et al. 2012).

However, the gelatinous nature of CLEA particles leads to low reproducibility and low mechanical stability. As a result, the CLEAs are often bound to a carrier to improve the operational stability of CLEAs or to ease the separation process. Thus, the process can be upgraded by introducing carrier particles such as mesoporous silica particle or magnetic nanoparticle (Kumar et al. 2014). CLEAs have been extensively used in various applications, including bioremediation. With the use of modern nano-engineered technologies, a creative fusion of biocatalysis, chemical engineering fundamentals, and nanotechnology has opened up attractive horizons towards immobilization techniques. This amalgamation resulted in the development of magnetic nanoparticle-based CLEAs of laccase to bolster long-term mechanical stability and could be recycled by the application of a magnetic field (Kumar et al. 2014). The operational stability of the CLEAs was enhanced by coating with a chitosan and 3-aminopropyltriethoxysilane polymer network (Hassani et al. 2013). Due to the limitation of restricted mass transfer of substrate in CLEAs, porous cross-linked aggregates (*p*-CLEAs) with enhanced biocatalytic activity were developed and proved to be effective with recyclability up to 15 cycles (Kumar et al. 2012).

Consequently, CLEAs are considered as the most attractive carrier-free method of immobilization for the effective treatment of wastewater. As an application in real matrix, magnetic CLEAs (mCLEAs) from *T. versicolor*, which is known for its potential in the effective biotransformation of pharmaceutically active compounds—acetaminophen, mefenamic acid, fenofibrate, and indomethacin—claimed to exhibit an enhanced stability against chemical denaturants in real wastewater matrix (Arca-Ramos et al. 2016b). The potential application of CLEAs laccase in wastewater was widely discussed by Ba et al. (2013). Despite scientific advances in immobilization by cross-linking, most of the application of CLEA particles was restricted to simulated wastewater treatment rather than real matrix, which includes mCLEAs of laccase for decolorization of dye (Kumar et al. 2014) and antibiotics (Yang et al. 2017) and porous CLEAs for reactive dyes (Kumar et al. 2012). However, low yield of the enzyme and absence of required mechanical strength due to the lack of carrier particle act as a major drawback for a CLEA-based laccase system.

3.3 Adsorption

Adsorption of enzymes involves physical interaction of the enzyme with the supports through dipole–dipole, hydrogen bonding, electrostatic, or hydrophobic and hydrophilic interactions (Jesionowski et al. 2014). For adsorption, the desired adsorbate (support) is made to contact with the laccase for a fixed time period at suitable conditions, and the unabsorbed enzymes are removed by washing. In spite of the weak interaction of the enzyme and adsorbent, high catalytic activity, less or no requirement of chemicals, and provision for reusability of support make this technique popular among others. Depending on nature and occurrence, the carrier molecules can be categorized into organic carriers, which include naturally occurring compounds such as chitosan, chitin, cellulose, alginates, etc. and inorganic carriers such as silica, titania, hydroxyapatite, etc. (Jesionowski et al. 2014). Laccase generally exhibits higher affinity towards organic carriers in comparison with the inorganic carrier molecule (Jesionowski et al. 2014) and is found to be more selective towards chitosan due to the enhanced ionic interactions. The work with chitosan as a carrier molecule along with itaconic acid and Cu(II) as a carrier modifier displayed an enhanced catalytic loading, which in turn improved the efficiency bioremediation of contaminants (Bayramoglu et al. 2012). Similarly, laccase immobilized on various organic carriers—chitosan (Yang et al. 2006), wood fiber (Saarinen et al. 2008), inorganic carrier—alumina (Abadulla et al. 2000), quartz (Saarinen et al. 2008), metals (zirconium) (Li et al. 2018), and porous support—mesoporous silica (Shao et al. 2009), mesoporous molecular sieve MCM-41 (Fernández-Fernández et al. 2013) was reported for its subsequent application towards bioremediation of various contaminants. On the other hand, use of magnetic nanoparticle as a carrier opened a new window in the metal affinity adsorption-based immobilization method. Metal ions, such as poly crystal gold, metal (gold and silver)-coated electrodes, indium tin oxide films, magnetic nanoparticle, and metal chelates have been tested as ion exchange supports for laccase immobilization based on adsorption.

Among the metal chelates, application of laccase-based copper-chelated magnetic nanoparticles for the treatment of coking wastewater was considered as the best example for metal affinity adsorption (Wang et al. 2012). The other metal-based adsorptions include the use of either magnetic copper nanoparticles (Wang et al. 2008) or magnetic chitosan microspheres (Fang et al. 2009). The affinity of the enzyme and carrier is considered as the most important criterion for adsorption and its use in wastewater treatment. In most cases, the weaker interaction of carrier particle can be enhanced using intermediate agents (carrier modifiers) or by modification of the carrier surface. The histidine functional group containing support molecules was reported to exhibit a favorable adsorption condition for laccase immobilization (Fernández-Fernández et al. 2013). The bifunctional (glutaraldehyde and ethylene glycol-*N*-hydroxysuccinimide) and monofunctional (citraconic anhydride) groups were reported as widely used surface modification agents to introduce functional group on the surface of carrier for the enhanced adsorption of laccase (Fernández-Fernández et al. 2013). Liu and co-workers modified the magnetic nanoparticle surface using a bifunctional agent and showed a two-fold improvement in the biocatalytical loading with application in reduction of COD levels in real-time papermaking (pulp and paper) effluent (Liu and Wang 2014). On the other hand, with global awareness on eco-friendly processes, researchers are working on renewable carrier supports such as coconut fibers with good water holding along with cation exchange properties, modified Kaolin with chemical acetylation, etc. As an important aspect, it should be noted that enhanced operational stability was observed in most of the support's molecules, with the exception of some supports such as aluminum hydroxide (Ahn et al. 2007). Despite its relatively simple and inexpensive method, with industrial potential, the presence of a weak interaction leading to desorption of laccase and reduction in the biocatalytic loading under the harsh conditions of wastewater treatment acts as a major limitation. The exposure of laccase to the microbial attack that is present in the wastewater environment acts as another major drawback. On the other hand, the competitive adsorption of the pollutants to the adsorbent reduces the catalytic loading of laccase in real matrix (Alshabib and Onaizi 2019). Moreover, the adsorbed enzymes are shielded by the hydrophobic interface which hinders the interactions with substrates (Spahn and Minteer 2008).

3.4 Encapsulation

Encapsulation is considered as the simplest and the most straightforward method involving entrapment of the biomolecules in a polymer matrix without any association of bonding between the network and the biomolecules (Mohidem and Mat 2012). Similar to entrapment, this method has the advantage of reusability and permeability of the substrates. The widely applied sol–gel technology and bioencapsulation are considered as two different encapsulation methods with an application in bioremediation (Irshad et al. 2012). The sol–gel is chemically inert and exhibits a remarkable half-life period along operational stability making this method popular (Mohidem and Mat 2012). Among many immobilization methods, laccase encapsulation was quite popular in bioremediation along with its other applications, which include as selective coating for optical and electrochemical biosensors and as a stationary phase of an affinity chromatography. With advances in polymer science, several studies have been reported on laccase immobilization with a wide range of natural and synthetic polymers. For enhanced shelf life period, doping of additives was introduced to protect laccase from denaturation by minimizing the nucleophilic attraction of water and also maintained the hydration levels by modifying the microenvironment of laccase (Fernández-Fernández et al. 2013). Mohidem and co-workers demonstrated the effect of additive agents polyvinyl alcohol, polyethylene glycol, and (3-aminopropyl)triethoxysilane in the enhancement of enzyme stability (Mohidem and Mat 2012). Addition of these agents as additives during the pre-gelation process is aimed at preserving the laccase catalytic activity during the gelation process. The literature regarding laccase encapsulation for the elimination of real-time pollutants in wastewater is limited. Encapsulated laccase in hydrogels by the sol–gel method exhibited an enhanced degradation of the dye in industrial effluent in comparison with its free counterpart from *P. ostreatus* (Asgher et al. 2012). Furthermore, with the invention of the electrospinning method, several eco-friendly and recoverable fibrous membranes were developed, which in turn were used for the entrapment of laccase using in situ emulsion electrospinning immobilization technology (Niu et al. 2013). Even though encapsulation of laccase exhibits enhanced operational stability along with resistance against inactivating agents, the leakage of the enzyme and diffusion limitation hindered their industrial usage (Ba et al. 2013).

3.5 Entrapment

Enzyme entrapment is considered a simple immobilization method achieved using a polymer network such as a natural or organic polymer. The classic example of enzyme entrapment is the polymerization of sodium alginate with calcium chloride, which leads to the formation of interfacial polymerization with the precipitation of calcium alginates (Fraser and Bickerstaff 1997). The gel obtained is considered to be chemically inert and mechanically stable, which protects the enzyme from the external environment (Fraser and Bickerstaff 1997; Bickerstaff 1997). In addition, the open lattice structure of the immobilized beads allows a greater mass transfer due to high porosity ranging from 200 to 150 nm in diameter (Fraser and Bickerstaff 1997). In addition, laccase entrapment can be performed through thermo-reversal polymerization by using either natural or synthetic polymers such as alginate, gelatine, polyvinyl acetate, beta-carrageenan, and acrylic acid (Fraser and Bickerstaff 1997). As a wastewater treatment application, this immobilization method is limited due to the presence of chelating agents such as EDTA, citrate, Mg^{2+}, and phosphate in the wastewater, which may destabilize the calcium alginate beads (Fraser and Bickerstaff 1997). Some techniques, such as the use of stabilizing agents or stabilizing by cross-linking with other polymers such as chitosan (Lu et al. 2007), were found to increase the stability of laccase. For instance, alginate was reported as a stabilized agent with gelatine as a natural polymer and demonstrated an enhanced stability of up to 85% of residual activity after five successive cycles of decolorization of industrial synthetic dye (Mogharabi et al. 2012). In parallel with the knowledge on the laccase and their dependence on the copper molecule, Teerapatsakul et al. investigated the use of copper sulfate as an alternative to calcium chloride for the effective formation of copper alginates, with enhanced laccase activity. In comparison with calcium alginate, copper alginate proved to be a better support with enhanced residual activity and reusability (Teerapatsakul et al. 2008). In spite of several advancements in the entrapment of laccases, their usage is limited due to low catalytic loading and excessive leakage of laccases during the process (Spahn and Minteer 2008; Cao 2005). Moreover, due to the high thermal conductivity of the polymers used in the entrapment process, a rapid reduction in the thermal stability of the laccase during the bioremediation process was assumed to be another common drawback of the entrapment process (Cao 2005).

3.6 Co-immobilization

The knowledge about enzyme chemistry along with studies on their reaction mechanisms led to the strategic development of a novel immobilization technique where more than one enzyme is used during immobilization. This process is called co-immobilization. Among the co-immobilization strategies, multipurpose CLEAs involving more than one enzyme, multi-CLEAs or combi-CLEAs, are gaining prominence for their sole process, the multi-enzyme cascade process. By definition, combi-CLEAs are defined as CLEAs of different proteins/enzymes. In addition, requirement of less space with high economic and eco-friendly benefits of the combi-CLEAs made this technique preferable. Due to substrate limitations, changing operating conditions, and the synergistic action that results from the combination of different enzymes, it is preferable to use, for example, multiple oxidoreductase enzymes like laccase (Lac), tyrosinase (Tyr), peroxidase (VP), and glucose oxidase (GO) to eliminate a wide range of trace level organic contaminants like endocrine-disrupting compounds and pharmaceuticals (Anderson et al. 2018).

These versatile biocatalysts can be comprised of oxidative and supporting enzymes that interact in a cascade of reactions, which makes them a perfect bioremediation tool for dye decolorization, lignin bioprocessing, elimination of pharmaceuticals, and endocrine-disrupting compounds (Taboada-Puig et al. 2011; Abadulla et al. 2000; Ba et al. 2012, 2013, 2014a; Touahar et al. 2014). Even though several enzymes have been immobilized by this technique, immobilization of laccase along with other oxidoreductase enzymes for bioremediation application is limited. Ba et al. used a bi-enzymatic system, with laccase and tyrosinase immobilized as combi-CLEAs that showed improved efficiency in the transformation of acetaminophen from municipal and hospital wastewaters, respectively, at pH 7.0 (Ba et al. 2014a). Immobilization of the versatile tri-enzymatic system consisting of laccase, versatile peroxidase, and glucose oxidase by CLEAs technique (Touahar et al. 2014) exhibited an enhanced stability of the enzymes. The resultant biocatalyst showed an increase in the oxidation spectrum of 14 pharmaceutical compounds such as acetaminophen, diclofenac, mefenamic acid, atenolol, epoxy carbamazepine, fenofibrate, diazepam, trimethoprim, ketoprofen, indomethacin, carbamazepine, caffeine, and naproxen in synthetic wastewater, which was more efficient than the free enzymes (Touahar et al. 2014). By exploring the advantage of laccase with varying redox potential, a multi-laccase system from *Thielavia* genus, *Coriolopsis polyzona*, *Cerrena unicolor*, *Pleurotus ostreatus*, and *Trametes versicolor* was immobilized onto fumed silica nanoparticles for the oxidation of trace organic contaminants, namely, carbamazepine, diclofenac, sulfamethoxazole, ibuprofen, gemfibrozil, benzophenone-2, benzophenone-4, and bisphenol A (Ammann et al. 2014). On the other hand, the pollutants, gemfibrozil and benzophenone-2, which resist oxidation by the sole action of *T. versicolor* or *C. polyzona* laccase, were reported to be successfully remediated by co-immobilized *C. polyzona* and *T. versicolor* laccase nano-biocatalyst. This reflects the importance of the co-immobilization technique in bioremediation. Due to the variation in the pH of the wastewater, a multipurpose immobilized biocatalyst was developed with subunits of laccase from *C. polyzona*, with an acidic pH optimum, and from *C. cinerea*, with a neutral pH optimum, which showed biocatalytic activity through a broad pH range, implying their suitability for the treatment of micropollutant-contaminated real wastewaters of varying pH (Agathos 2012). In the same context, Ba et al. insolubilized two commercial laccases, from a fungus and a bacterium, using combi-CLEAs, and claimed to achieve a 70% reduction in total COD of a pulp and paper mill wastewater (Ba et al. 2012).

4 Immobilized Laccase-Based Reactors for Wastewater Treatment

The selection of the reactors for the immobilized laccase and establishment of its operation strategy for the efficient treatment of wastewater is considered a major challenge. Depending on the mode of immobilization and treatment strategies of wastewater, various bioreactors, along with multifunctional hybrid reactors, have been designed, which makes the categorization of reactors a difficult task. However, based on the design, the immobilized enzyme-based bioreactors are grouped into:

1. Stirred tank reactors (STR)
2. Membrane-based reactors (MBR)
3. Packed bed reactors (PBR)
4. Fluidized bed reactors (FBR)

These enzyme reactors generally operate either in batch, fed batch, or continuous mode. However, the batch reactor is considered a more versatile tool in determining the degradation kinetics of wastewater, which makes it a prime choice of reactor design at a laboratory scale or pilot scale.

Despite the extensive work and potential applications of immobilized laccase in degradation of contaminants, limited studies were reported on the remediation of real-time wastewater using immobilized laccase in a bioreactor (Table 2). Generally, packed bed reactor, fluidized bed reactor, and membrane reactors are considered to be ideal reactors for immobilized enzyme systems (Wanga and Zhonga 2011; Stanbury et al. 2013; Messing 2012). However, with technological advancements, some bioreactors are reported to have a suspension system, which includes stirred tank reactors, airlift, and bubble column bioreactors. For instance, Osma et al. used a continuous stirred tank reactor (CSTR) for the degradation of textile dye in simulated wastewater using covalently immobilized laccase (Osma et al. 2010).

The selection of a bioreactor generally depends on some key fundamental principles, such as transport phenomena, which include mass transfer along with the efficiency to maintain the optimum conditions required by the immobilized laccase (Stanbury et al. 2013). However, in most cases, it is difficult to design an ideal reactor that meets all the requirements, and hence some of these requirements can be packed accordingly. For example, it is crucial to balance the agitation speed and the mass transfer in STR in order to avoid the reduction in biocatalytic load due to shear force (Wanga and Zhonga 2011). Accordingly, several bioreactors and hybrid reactors have been designed for wastewater treatment which is briefly discussed in the next sections.

Table 2 Various strategies of laccase immobilized based reactors for the remediation of contaminants in wastewater

Microorganism	Immobilization strategy	Reactor	Design	Flow rate (ml/min)	Hydraulic retention time HRT (h)	Application	References
Genetically modified *Aspergillus oryzae*	Physical adsorption on granular activated carbon	Packed bed reactor	Width: 1 cm Height: 22 cm Bed volume: 17 mL	2.4	NP[a]	Micropollutants in simulated water and wastewater	Nguyen et al. (2016)
T. versicolor	Metal affinity adsorption on copper-chelated mesoporous silica nanoparticle	Magnetically stabilized fluidized bed bioreactor	Glass column Height: 20 cm Width: 1.5 cm Power supply: 0–10 A	5.5–10.8	NP	Phenols from coking wastewater	Wang et al. (2012)
T. versicolor	Calcium alginate bead-based entrapment	Packed bed reactor	Glass column Height: 50 cm Width: 2 cm Temperature: 30 ± 2 °C	1	NP	Simulated paper industrial effluent	Niladevi and Prema (2008)
T. versicolor	Encapsulation on poly(D,L-lactide-co-glycolide) (PDLGA) by emulsification of laccase	Laccase-loading spider-type reactor (LSTR)	Spider-type reactor	NP	NP	PAHs in wastewater	Niu et al. (2013)
Coriolopsis gallica	Covalently immobilized on mesoporous silica particle	Continuous fed batch reactor	Working volume: 50 mL	0.6	1.25	Organic contaminant in wastewater	Nair et al. (2013)
Myceliophthora	Covalently immobilized on silanized alumina pellets	Fixed bed reactor	Working volume: 31.4 ml Width: 2 cm Length: 55 cm	10	NP	Maillard products from distillery wastewater	Singh et al. (2015)

(continued)

Table 2 (continued)

Microorganism	Immobilization strategy	Reactor	Design	Flow rate (ml/min)	Hydraulic retention time HRT (h)	Application	References
P. ostreatus	Covalently immobilized on TiO_2 particle	Hybrid membrane reactor	rpm: 150 Temperature: 25 °C	~40	24	Bisphenol-A and carbamazepine in sewage	Ji et al. (2017)
T. pubescens	Covalently immobilized on surface-modified alumina spheres	Fluidized bed bioreactor	Glass column Height: 20 cm Width: 4.5 cm Working volume: 200 mL Air flow: 0.5 vvm	NP	NP	Simulated textile effluent	Osma et al. (2010)
		CSTR	Glass column Height: 20 cm Width: 7 cm Working volume: 200 mL Air flow: 0.5 vvm	NP	33		
T. versicolor	Covalently immobilized on hydrophilic PVDF microfiltration membrane	Microfilter membrane reactor	Feed solution: 200 mL Air flow: 1.26 mMol/L Temperature: 25 °C	6	NP	Phenylurea pesticide in wastewater	Jolivalt et al. (2000)

C. versicolor	Covalently immobilized on functionalized activated carbon	Packed bed reactor	–	30	12	Pulp mill and bleach plant effluent	Davis and Burns (1992)
T. versicolor	Covalently immobilized on TiO_2 modified particles with APTES	Membrane hybrid reactor	–	NP	96	Carbamazepine in effluent	Ji et al. (2016)
T. versicolor	Particle CLEAs on chitosan	Hybrid bioreactors	Reactor volume: 1 L rpm: 300	0.9	NP	Aromatic pharmaceutical from wastewater	Ba et al. (2014b)

[a] *NP* not provided

4.1 Stirred Tank Reactor

An STR is considered as the most commonly used bioreactor design in bioprocess industries for its ease in fabrication and simple operation with minimal process controls (Fig. 2a). An STR consists of a simple tank containing a sparger for aeration along with an agitator or impeller with a wide range of functionalities like heat and mass transfer (Messing 2012). Depending on the transport phenomena, power consumption, and fluid dynamics, numerous impellers were designed, which were grouped into axial and radial flow impellers. For a shear-sensitive immobilized laccase system, the conventional impeller may adversely affect the system due to either high hydrodynamic shear force or air bubble produced by the gas sparger (Messing 2012). On this aspect, a number of alternative forms of impeller, like Intermig, Prochem Maxflo T, and Scaba 6SRGT, were developed with improved mixing efficiency at low tip speed (Wanga and Zhonga 2011), which are assumed to be suitable for immobilized laccase. Besides the impeller, the design of STR for wastewater treatment follows the standard geometrical specifications that are required in conventional STR, which were widely reported by some authors (Stanbury et al. 2013; Kargi and Shuler 1992).

Depending on the mode of operation, a STR can be operated either in batch or in continuous mode. In batch mode, the recovery of the immobilized laccase is often accompanied by centrifugation or filtration. However, the formation of clumps, especially in CLEAs, leads to a reduction in the efficiency of immobilized particles, which acts as a major limitation in the treatment of wastewater (Arca-Ramos et al. 2016b). To address this setback, Arca-Ramos et al. developed laccase cross-linked magnetic silica microbeads for successful treatment of pharmaceutically active compound (PhAC) in wastewater (Arca-Ramos et al. 2016b), which was based on the design of magnetic-assisted laccase immobilization reported (Kumar et al. 2014). In the continuous stirred tank reactor (CSTR), hydraulic retention time (HRT) is considered a key design variable, which is defined as the amount of time spent by the wastewater in the reactor (Osma et al. 2010). Similar to a batch reactor, CSTR can also be operated for effective wastewater treatment. However, reduction in the biocatalytic load due to shear force and requirement of further downstream processing for the separation of the immobilized laccase in CSTR increases the overall process cost of wastewater treatment using immobilized laccase, thus limiting its usage on a large scale.

4.2 Membrane Bioreactor

A membrane bioreactor is a simple device that combines the concept of biocatalytic conversion with membrane separation (Fig. 2b) (Stanbury et al. 2013). Restriction of the enzymes within the reactor, which further reduces the washing loss of the enzymes, is considered a major advantage of a membrane bioreactor (Wanga and

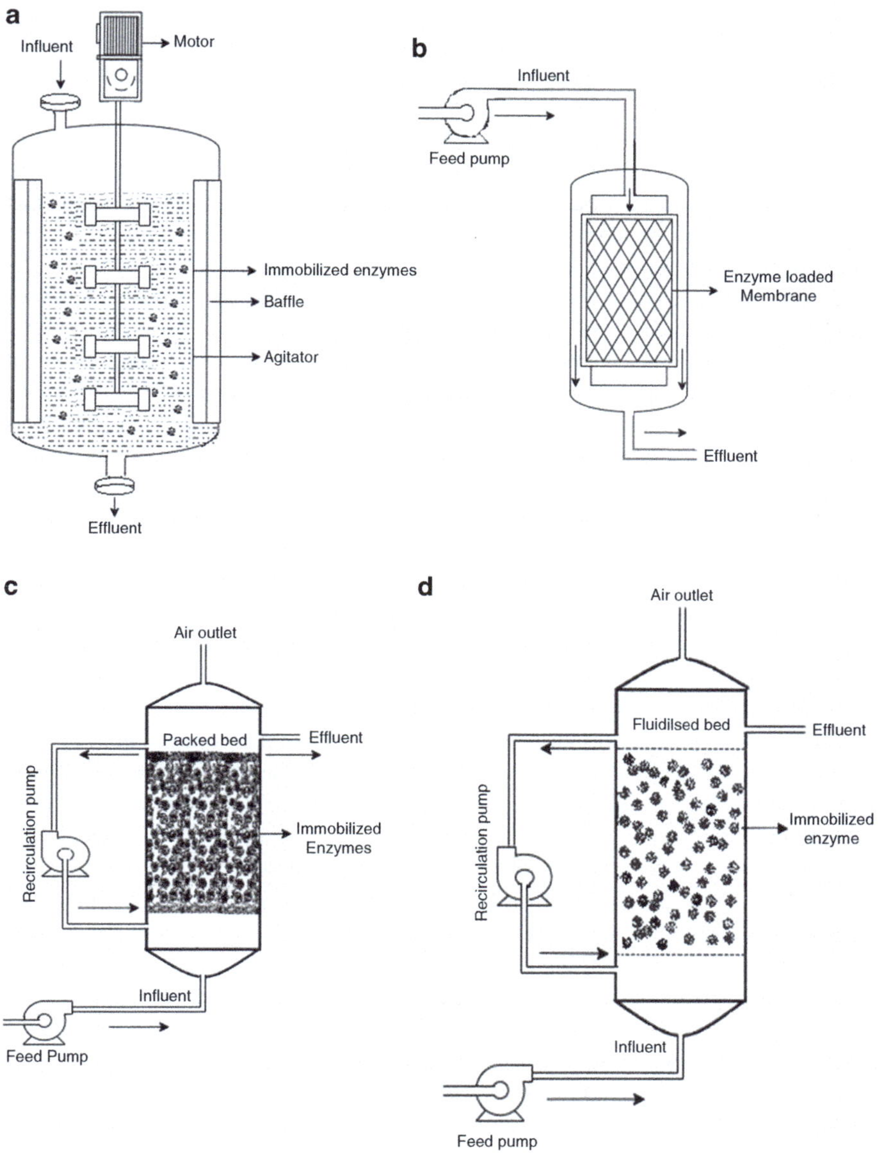

Fig. 2 Schematic representation of (**a**) stirred tank reactors (STRs), (**b**) membrane-based reactors (MBR), (**c**) packed bed reactors (PBR), and (**d**) fluidized bed reactors (FBR)

Zhonga 2011). With the advancement in immobilization techniques, biocatalytic-dense membranes have successfully replaced conventional membranes and are considered a promising technology in the field of wastewater treatment using a membrane bioreactor. Additionally, obligatory interaction of the pollutants with the immobilized laccase during filtration makes this reactor an obvious choice in the treatment of pollutants (De Cazes et al. 2014). The membrane is selected depending on the application and the immobilization strategy, and some of the common membranes include polysulfone, cellulose, polytetrafluoroethylene, ceramic, polyvinylidene difluoride (PVDF), and polypropylene. On the other hand, pore size of the membrane acts as a decisive property in arbitrating the mass transfer rate of the substrate into the membrane (Kargi and Shuler 1992; Stanbury et al. 2013). Depending on the mode of membrane packing, these membrane reactors can be grouped into any of the following modules: plate and sheet, spiral wound, and tubular or hollow-fiber modules.

The biocatalyst, laccase, can be either be entrapped inside the porous structure or adsorbed onto the surface of the membrane. The immobilization techniques used in a membrane bioreactor are mainly classified as non-covalent immobilization, covalent immobilization on the surface of the membrane, and entrapment of laccase in the membrane (Arca-Ramos et al. 2018). Entrapment of laccase in the membrane was reported to be a benign approach when compared to the other immobilization techniques, as it shields the enzyme from harsh conditions that avail in the wastewater treatment process (Wanga and Zhonga 2011). The work on laccase-mediated remediation of PAHs in real-time wastewater matrix, where laccase was encapsulated on poly(D,L-lactide-co-glycolide) in spider type bioreactors, showed the effectiveness of the entrapped laccase system. This immobilized laccase exhibited enhanced stability when compared to the free enzymes and retained more than 80% of its initial activity for about 60 days of incubation (Niu et al. 2013).

On the contrary, during entrapment, the limited volume porosity leads to a reduction in the biocatalytic load, which was considered a common limitation. As an alternative, immobilization on the surface of the membrane, specifically by covalent immobilization, was considered as a strategic solution. For instance, Jolivalt et al. and Ji et al. covalently immobilized laccase and showed an enhanced remediation of micropollutant in sewage water in a membrane reactor (Ji et al. 2016, 2017; Jolivalt et al. 2000). In spite of several advantages, such as low maintenance and low sludge generation with a higher degradation rate of organic pollutants, MBR has limited use due to the fouling of membrane and limited contact time between laccase and contaminants. This phenomenon may decrease the permeate flux or surge in the transmembrane pressure, which increases the overall operational cost and decreases the membrane's life span. Additionally, most of the previous studies reported were mostly on simulated wastewater bioremediation using laccase-assisted membrane reactors, among which only limited reports have been published on real wastewater matrices (Table 2).

4.3 Packed Bed Reactor

A packed bed reactor (PBR) also known as a fixed bed reactor is the most common type of bioreactor used for an immobilized enzyme system owing to its simplicity and high reaction rates (Fig. 2c). The conventional PBR consists of a vertical column packed with appropriate carrier particles loaded with biocatalysts, forming a submerged bed. The feed can be introduced either from the top or from the bottom, and in most laboratory scale reactors, the feed is given from the bottom due to the ease in maintaining the liquid levels above the bed (Sondhi et al. 2018). However, on an industrial scale, to avoid the use of a pump and to reduce power consumption, the feed is often supplied from the top under the influence of gravitational force. Reversal of the feed flow has the advantage of maintaining the compactness of the bed and also aids in clearing the loaded biocatalyst (Illanes 2008). Ideally, the flow of the liquid is parallel to the reactor axis, thereby avoiding the back mixing (Arca-Ramos et al. 2018). Due to the ease in operation and simple design, these reactors are widely used as model reactors for the treatment of wastewater, irrespective of the laccase immobilization strategy (Niladevi and Prema 2008; Nguyen et al. 2016; Davis and Burns 1992). Furthermore, with the development of co-immobilization technology, Krastanov developed a simple biocatalytic system by co-immobilizing *Pyricularia oryzae* laccase and mushroom tyrosinase on Mikroperl support for the removal of various phenolic compounds, such as 4-methoxyphenol, 2,6 dimethoxyphenol, 2,4 dichlorophenol, 4-choloro-3-methylphenol, 4-chlorophenol, naphthol, chlorogenic acid, 3-methoxyphenol, 2-chlorophenol, guaiacol, *m*-cresol, *p*-cresol, *o*-cresol, 3-chlorophenol, phenol, catechol, catechin, and DOPA in a packed/fixed bed tubular bioreactor (Krastanov 2000).

However, relatively poor mass and heat transfer due to low liquid velocity are considered the major drawback of PBR (Wanga and Zhonga 2011). Another drawback of PBR includes accumulation of stagnant gas packets, which leads to poor transfer phenomena (Wanga and Zhonga 2011). As a solution to this setback, Shiotanni and Yamane proposed a shallow horizontal packed bed reactor with free head space, and the designed reactor was successful in removing the produced carbon dioxide accumulated in the head space above the packed bed due its buoyancy. This design can be used as a better alternative in wastewater treatment applications to avoid the formation of stagnant gas pockets during the remediation process.

4.4 Fluidized Bed Reactor

A fluidized bed reactor (FBR) is considered an effective reactor used in the wastewater treatment process due to its low cost, high mass transfer rates, and uniform mixing (Fig. 2d) (Bello et al. 2017). The FBR is specifically used for high viscous effluent or low-soluble pollutant that creates clogging problems in PBR (Arca-Ramos et al. 2018; Lema Rodicio et al. 2014). The basic principle of FBR involves the passage of the fluid through the static bed of immobilized biocatalysts with a sufficient superficial velocity, such that the immobilized particles behave like a fluid. When the fluid velocity is less than the drag and gravitational force of the bed, the reactor behaves as a packed bed reactor. However, at particle velocity, the applied fluid velocity balances the weight of the particle and, thus, completely suspends the particles in the fluid. The fluid velocity required to move the bed is termed as minimum fluidization velocity (Bello et al. 2017). Depending on the minimum fluidization velocity, the FBR is categorized as smooth fluidization, slugging fluidization, bubbling fluidization, turbulent fluidization, or pneumatic fluidization (Yang 2003).

Based on the application and required transport phenomena, the FBR can operate either in a two-phase flow system (solid–gas or solid–liquid) or in a three-phase flow system (solid–liquid–gas) (Grace and Bi 1997). However, in wastewater treatment, a solid–liquid system is commonly used, where the solid phase can be immobilized laccase and the liquid phase wastewater effluent. This configuration was successfully applied by Osma et al., for the remediation of simulated textile wastewater using laccase covalently immobilized on silica particle (Osma et al. 2010). Additionally, the reports on remediation of pollutants from coking wastewater claimed more than 95% remediation of phenols using laccase immobilized on magnetic mesoporous silica particle (Wang et al. 2012). Similarly, Cabana et al. used CLEAs of laccase from *Coriolopsis polyzona* for the effective elimination of BPA, nonylphenol, and triclosan in a fluidized bed reactor (Cabana et al. 2007). However, lack of proper oxygen transfer to the immobilized laccase in a two-phase flow system can be a major limitation (Chao et al. 2011). In this situation, the three-phase flow system, which includes an inlet for aeration or oxygenation, acts as a suitable alternative. In addition, the fluidization can be achieved by providing either current or countercurrent of either effluent or gas, which makes it a suitable reactor for the wastewater treatment using immobilized laccase (Chao et al. 2011; Wanga and Zhonga 2011; Grace and Bi 1997). Despite these advantages, limited research has been reported with the three-phase flow FBR due to high shear force and elutriation of particles at a high flow rate.

4.5 Other Miscellaneous Reactors

A perfusion bioreactor was designed to overcome the disadvantages of mechanical shear in reactors and reduction in the reusability of the particles (Fig. 3). Based on metallic filtrations like membranes, a perfusion basket reactor, filled with immobilized laccase particles, was developed for the effective degradation of endocrine-disrupting chemicals (EDCs) (Cabana et al. 2009). This unbaffled and 3-blade marine propeller showed more than 80% of degradation efficiency of EDCs in a hydraulic retention time (HRT) of 325 min (Cabana et al. 2009). Applying a similar concept, a perfusion basket reactor equipped with mCLEAs was developed for continuous degradation of synthetic dyes (Kumar et al. 2014). This reactor was reported to perform in continuous mode and could be used for the effective treatment of wastewater on a large scale.

Based on conventional bioreactor functionalities, several hybrid reactors were introduced for the effective remediation of wastewater, which combined two or more functionalities of different reactors. One such hybrid reactor was designed by Ba

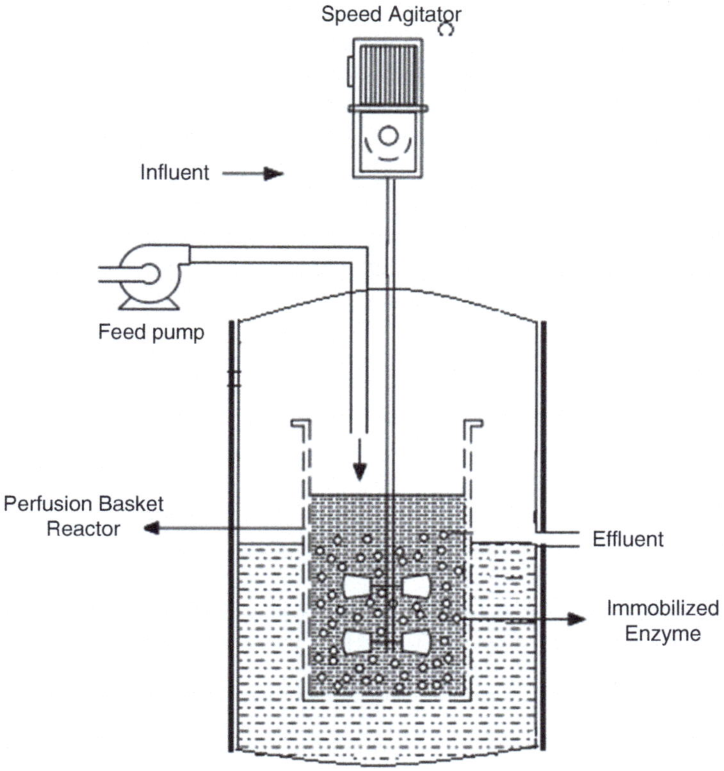

Fig. 3 Schematic representation of perfusion basket reactor with immobilized enzyme

et al. with a combination of simple STR (loaded with CLEAs of laccase) and a membrane filter unit (Ba et al. 2014b). This hybrid reactor reported more than 93% elimination of carbamazepine in 72 h along with complete removal of acetaminophen and mefenamic acid from municipal wastewater (Ba et al. 2014b). Although the proof of concept of these reactors was successful in laboratory scale, lack of extensive research on these reactors (with immobilized laccase) for the treatment of wastewater acts as a major drawback in commercialization of these reactors.

5 Conclusion

Laccase has emerged as a potential biocatalytic tool in the degradation of a broader spectrum of contaminants due to its high redox potential. However, the lack of sufficient availability of the enzyme impedes its commercial application for the treatment of wastewater, which can be overcome by the insolubilization process. An understanding of different types of immobilization strategies, along with their interactions with the amino acid moiety of laccase, is considered highly indispensable for enhancing the effectiveness of the immobilization strategies for their use in large-scale applications. This biocatalytic system generally shows a greater potential for the biotransformation of contaminants into innocuous products when used in a suitable bioreactor. Despite the advancements in wastewater treatment, numerous challenges still exist that have to be addressed in order to make laccase-mediated systems a commercially competitive technology. However, with a clear understanding of the challenges, it will be possible in the future to design a desired laccase system without compromising its stability and activity, for the continuous remediation of wastewater in an environmentally friendly and economically feasible way.

Acknowledgment This work was supported by grants from the Fonds de recherche du Québec—Nature et Technologies, the Natural Sciences and Engineering Research Council of Canada, and Sri Ramaswamy Memorial Institute of Science and Technology.

References

Abadulla E, Tzanov T, Costa S, Robra K-H, Cavaco-Paulo A, Gübitz GM (2000) Decolorization and detoxification of textile dyes with a laccase from *Trametes hirsuta*. Appl Environ Microbiol 66(8):3357–3362

Agathos SN (2012) Biocatalytic weapons against micropollutants. Environ Eng Manag J 11:SS99–SS104

Ahn M-Y, Zimmerman AR, Martínez CE, Archibald DD, Bollag J-M, Dec J (2007) Characteristics of *Trametes villosa* laccase adsorbed on aluminum hydroxide. Enzyme Microb Technol 41 (1–2):141–148

Alshabib M, Onaizi SA (2019) A review on phenolic wastewater remediation using homogeneous and heterogeneous enzymatic processes: current status and potential challenges. Sep Purif Technol 219:186

Amitai G, Adani R, Sod-Moriah G, Rabinovitz I, Vincze A, Leader H, Chefetz B, Leibovitz-Persky L, Friesem D, Hadar Y (1998) Oxidative biodegradation of phosphorothiolates by fungal laccase. FEBS Lett 438(3):195–200

Ammann EM, Gasser CA, Hommes G, Corvini PF-X (2014) Immobilization of defined laccase combinations for enhanced oxidation of phenolic contaminants. Appl Microbiol Biotechnol 98 (3):1397–1406

Anderson G, Bleakley B, McDaniel R, Mardani S, Jahandideh A (2018) Fungal treatment of pharmaceuticals in effluents: current state, perspectives, limitations, and opportunities. In: Life cycle assessment of wastewater treatment. CRC, Boca Raton, pp 139–170

Arca-Ramos A, Ammann E, Gasser C, Nastold P, Eibes G, Feijoo G, Lema J, Moreira M, Corvini P-X (2016a) Assessing the use of nanoimmobilized laccases to remove micropollutants from wastewater. Environ Sci Pollut Res 23(4):3217–3228

Arca-Ramos A, Kumar V, Eibes G, Moreira M, Cabana H (2016b) Recyclable cross-linked laccase aggregates coupled to magnetic silica microbeads for elimination of pharmaceuticals from municipal wastewater. Environ Sci Pollut Res 23(9):8929–8939

Arca-Ramos A, Eibes G, Feijoo G, Lema JM, Moreira MT (2018) Enzymatic reactors for the removal of recalcitrant compounds in wastewater. Biocatal Biotransform 36(3):195–215

Arıca MY, Altıntas B, Bayramoğlu G (2009) Immobilization of laccase onto spacer-arm attached non-porous poly (GMA/EGDMA) beads: application for textile dye degradation. Bioresour Technol 100(2):665–669

Asgher M, Kamal S, Iqbal HMN (2012) Improvement of catalytic efficiency, thermo-stability and dye decolorization capability of *Pleurotus ostreatus* IBL-02 laccase by hydrophobic sol gel entrapment. Chem Cent J 6(1):110

Ba S, Mialon A, Cabana H, Jones J (2012) Laccase insolubilized as combination of crosslinked enzyme aggregates: characterization and potential application in wastewater treatment. J Enzym Res 3:23–28

Ba S, Arsenault A, Hassani T, Jones JP, Cabana H (2013) Laccase immobilization and insolubilization: from fundamentals to applications for the elimination of emerging contaminants in wastewater treatment. Crit Rev Biotechnol 33(4):404–418

Ba S, Haroune L, Cruz-Morató C, Jacquet C, Touahar IE, Bellenger J-P, Legault CY, Jones JP, Cabana H (2014a) Synthesis and characterization of combined cross-linked laccase and tyrosinase aggregates transforming acetaminophen as a model phenolic compound in wastewaters. Sci Total Environ 487:748–755

Ba S, Jones JP, Cabana H (2014b) Hybrid bioreactor (HBR) of hollow fiber microfilter membrane and cross-linked laccase aggregates eliminate aromatic pharmaceuticals in wastewaters. J Hazard Mater 280:662–670

Bayramoglu G, Gursel I, Yilmaz M, Arica MY (2012) Immobilization of laccase on itaconic acid grafted and cu (II) ion chelated chitosan membrane for bioremediation of hazardous materials. J Chem Technol Biotechnol 87(4):530–539

Bello MM, Raman AAA, Purushothaman M (2017) Applications of fluidized bed reactors in wastewater treatment–a review of the major design and operational parameters. J Clean Prod 141:1492–1514

Berrio J, Plou FJ, Ballesteros A, Martínez ÁT, Martínez MJ (2007) Immobilization of *Pycnoporus coccineus* laccase on Eupergit C: stabilization and treatment of olive oil mill wastewaters. Biocatal Biotransform 25(2–4):130–134

Bickerstaff GF (1997) Immobilization of enzymes and cells. In: Immobilization of enzymes and cells. Springer, New York, pp 1–11

Bollag J-M, Leonowicz A (1984) Comparative studies of extracellular fungal laccases. Appl Environ Microbiol 48(4):849–854

Brady D, Jordaan J (2009) Advances in enzyme immobilisation. Biotechnol Lett 31(11):1639

Cabana H, Jones JP, Agathos SN (2007) Preparation and characterization of cross-linked laccase aggregates and their application to the elimination of endocrine disrupting chemicals. J Biotechnol 132(1):23–31

Cabana H, Jones JP, Agathos SN (2009) Utilization of cross-linked laccase aggregates in a perfusion basket reactor for the continuous elimination of endocrine-disrupting chemicals. Biotechnol Bioeng 102(6):1582–1592

Cao L (2005) Immobilised enzymes: science or art? Curr Opin Chem Biol 9(2):217–226

Chao Z, Wang Y, Jakobsen JP, Fernandino M, Jakobsen HA (2011) Derivation and validation of a binary multi-fluid Eulerian model for fluidized beds. Chem Eng Sci 66(16):3605–3616

Couto SR, Herrera JLT (2006) Industrial and biotechnological applications of laccases: a review. Biotechnol Adv 24(5):500–513

D'Annibale A, Stazi SR, Vinciguerra V, Di Mattia E, Sermanni GG (1999) Characterization of immobilized laccase from *Lentinula edodes* and its use in olive-mill wastewater treatment. Process Biochem 34(6–7):697–706

D'Annibale A, Stazi SR, Vinciguerra V, Sermanni GG (2000) Oxirane-immobilized *Lentinula edodes* laccase: stability and phenolics removal efficiency in olive mill wastewater. J Biotechnol 77(2–3):265–273

Dai Y, Yao J, Song Y, Wang S, Yuan Y (2016) Enhanced adsorption and degradation of phenolic pollutants in water by carbon nanotube modified laccase-carrying electrospun fibrous membranes. Environ Sci 3(4):857–868

Davis S, Burns RG (1992) Covalent immobilization of laccase on activated carbon for phenolic effluent treatment. Appl Microbiol Biotechnol 37(4):474–479

De Cazes M, Abejón R, Belleville M-P, Sanchez-Marcano J (2014) Membrane bioprocesses for pharmaceutical micropollutant removal from waters. Membranes 4(4):692–729

de Souza Bezerra TM, Bassan JC, de Oliveira Santos VT, Ferraz A, Monti R (2015) Covalent immobilization of laccase in green coconut fiber and use in clarification of apple juice. Process Biochem 50(3):417–423

Dignac M-F, Ginestet P, Rybacki D, Bruchet A, Urbain V, Scribe P (2000) Fate of wastewater organic pollution during activated sludge treatment: nature of residual organic matter. Water Res 34(17):4185–4194

Fang H, Huang J, Ding L, Li M, Chen Z (2009) Preparation of magnetic chitosan nanoparticles and immobilization of laccase. J Wuhan Univ Technol Mater Sci Ed 24(1):42–47

Fernández-Fernández M, Sanromán MÁ, Moldes D (2013) Recent developments and applications of immobilized laccase. Biotechnol Adv 31(8):1808–1825

Frasconi M, Favero G, Boer H, Koivula A, Mazzei F (2010) Kinetic and biochemical properties of high and low redox potential laccases from fungal and plant origin. Biochim Biophys Acta 1804 (4):899–908

Fraser JE, Bickerstaff GF (1997) Entrapment in calcium alginate. In: Immobilization of enzymes and cells. Springer, New York, pp 61–66

Gasser CA, Ammann EM, Shahgaldian P, Corvini PF-X (2014) Laccases to take on the challenge of emerging organic contaminants in wastewater. Appl Microbiol Biotechnol 98(24):9931–9952

Grace JR, Bi H (1997) Introduction to circulating fluidized beds. In: Circulating fluidized beds. Springer, New York, pp 1–20

Habeeb A, Hiramoto R (1968) Reaction of proteins with glutaraldehyde. Arch Biochem Biophys 126(1):16–26

Halling PJ, Ulijn RV, Flitsch SL (2005) Understanding enzyme action on immobilised substrates. Curr Opin Biotechnol 16(4):385–392

Hassani T, Ba S, Cabana H (2013) Formation of enzyme polymer engineered structure for laccase and cross-linked laccase aggregates stabilization. Bioresour Technol 128:640–645

Illanes A (2008) Enzyme biocatalysis. Principles and applications. Springer, New York

Irshad M, Bahadur BA, Anwar Z, Yaqoob M, Ijaz A, Iqbal HMN (2012) Decolorization applicability of sol-gel matrix-immobilized laccase produced from *Ganoderma lucidum* using agro-industrial waste. BioResour 7(3):4249–4261

Jahangiri E, Thomas I, Schulze A, Seiwert B, Cabana H, Schlosser D (2018) Characterisation of electron beam irradiation-immobilised laccase for application in wastewater treatment. Sci Total Environ 624:309–322

Jesionowski T, Zdarta J, Krajewska B (2014) Enzyme immobilization by adsorption: a review. Adsorption 20(5–6):801–821

Ji C, Hou J, Wang K, Zhang Y, Chen V (2016) Biocatalytic degradation of carbamazepine with immobilized laccase-mediator membrane hybrid reactor. J Membrane Sci 502:11–20

Ji C, Nguyen LN, Hou J, Hai FI, Chen V (2017) Direct immobilization of laccase on titania nanoparticles from crude enzyme extracts of *P. ostreatus* culture for micro-pollutant degradation. Sep Purif Technol 178:215–223

Johannes C, Majcherczyk A (2000) Natural mediators in the oxidation of polycyclic aromatic hydrocarbons by laccase mediator systems. Appl Environ Microbiol 66(2):524–528

Jolivalt C, Brenon S, Caminade E, Mougin C, Pontié M (2000) Immobilization of laccase from *Trametes versicolor* on a modified PVDF microfiltration membrane: characterization of the grafted support and application in removing a phenylurea pesticide in wastewater. J Membrane Sci 180(1):103–113

Kalkan NA, Aksoy S, Aksoy EA, Hasirci N (2012) Preparation of chitosan-coated magnetite nanoparticles and application for immobilization of laccase. J Appl Polym Sci 123(2):707–716

Kalkan E, Nadaroğlu H, Celebi N, Tozsin G (2014) Removal of textile dye reactive black 5 from aqueous solution by adsorption on laccase-modified silica fume. Desalin Water Treat 52 (31–33):6122–6134

Kargi F, Shuler ML (1992) Bioprocess engineering: basic concepts. Prentice-Hall PTR, Englewood Cliffs

Krastanov A (2000) Removal of phenols from mixtures by co-immobilized laccase/tyrosinase and polyclar adsorption. J Ind Microbiol Biotechnol 24(6):383–388

Kumar VV, Kumar MP, Thiruvenkadaravi K, Baskaralingam P, Kumar PS, Sivanesan S (2012) Preparation and characterization of porous cross linked laccase aggregates for the decolorization of triphenyl methane and reactive dyes. Bioresour Technol 119:28–34

Kumar VV, Sivanesan S, Cabana H (2014) Magnetic cross-linked laccase aggregates—bioremediation tool for decolorization of distinct classes of recalcitrant dyes. Sci Total Environ 487:830–839

Le TT, Murugesan K, Lee C-S, Vu CH, Chang Y-S, Jeon J-R (2016) Degradation of synthetic pollutants in real wastewater using laccase encapsulated in core–shell magnetic copper alginate beads. Bioresour Technol 216:203–210

Lema Rodicio JM, Teresa Moreira M, Eibes G, Lu-Chau TA, Lloret L, Taboada R, Arca-Ramos A, Feijoo G (2014) Enzymatic reactors applied for the biotransformation of endocrine disrupting chemicals. Transf Prod Emerg Contam Environ 229–260

Li G, Pang S, Wu Y, Ouyang J (2018) Enhanced removal of hydroquinone by graphene aerogel-Zr-MOF with immobilized laccase. Chem Eng Comm 205(5):698–705

Liu Y, Wang Z (2014) Immobilization of laccase on surface modified magnetic silica particles and its use for the papermaking wastewater. In: Applied mechanics matter. Trans Tech, pp 267–270

Lloret L, Hollmann F, Eibes G, Feijoo G, Moreira M, Lema J (2012) Immobilisation of laccase on Eupergit supports and its application for the removal of endocrine disrupting chemicals in a packed-bed reactor. Biodegradation 23(3):373–386

Lu L, Zhao M, Wang Y (2007) Immobilization of laccase by alginate–chitosan microcapsules and its use in dye decolorization. World J Microbiol Biotechnol 23(2):159–166

Madhavi V, Lele S (2009) Laccase: properties and applications. BioResour 4(4):1694–1717

Matijošytė I, Arends IW, de Vries S, Sheldon RA (2010) Preparation and use of cross-linked enzyme aggregates (CLEAs) of laccases. J Mol Catal B 62(2):142–148

Messing R (2012) Immobilized enzymes for industrial reactors. Elsevier, Oxford

Mogharabi M, Nassiri-Koopaei N, Bozorgi-Koushalshahi M, Nafissi-Varcheh N, Bagherzadeh G, Faramarzi MA (2012) Immobilization of laccase in alginate-gelatin mixed gel and decolorization of synthetic dyes. Bioinorg Chem Appl 2012:1

Mohidem NA, Mat HB (2012) Catalytic activity and stability of laccase entrapped in sol–gel silica with additives. J Sol-Gel Sci Technol 61(1):96–103

Morozova O, Shumakovich G, Shleev S, Yaropolov YI (2007) Laccase-mediator systems and their applications: a review. Appl Biochem Microbiol 43(5):523–535

Nair RR, Demarche P, Agathos SN (2013) Formulation and characterization of an immobilized laccase biocatalyst and its application to eliminate organic micropollutants in wastewater. New Biotechnol 30(6):814–823

Nguyen LN, Hai FI, Dosseto A, Richardson C, Price WE, Nghiem LD (2016) Continuous adsorption and biotransformation of micropollutants by granular activated carbon-bound laccase in a packed-bed enzyme reactor. Bioresour Technol 210:108–116

Niladevi K, Prema P (2008) Immobilization of laccase from *Streptomyces psammoticus* and its application in phenol removal using packed bed reactor. World J Microbiol Biotechnol 24 (7):1215–1222

Niu J, Dai Y, Guo H, Xu J, Shen Z (2013) Adsorption and transformation of PAHs from water by a laccase-loading spider-type reactor. J Hazar Mater 248:254–260

Osma JF, Toca-Herrera JL, Rodríguez-Couto S (2010) Biodegradation of a simulated textile effluent by immobilised-coated laccase in laboratory-scale reactors. Appl Catal A Gen 373 (1–2):147–153

Osma JF, Toca-Herrera JL, Rodríguez-Couto S (2011) Cost analysis in laccase production. J Environ Manag 92(11):2907–2912

Pal A, Gin KY-H, Lin AY-C, Reinhard M (2010) Impacts of emerging organic contaminants on freshwater resources: review of recent occurrences, sources, fate and effects. Sci Total Environ 408(24):6062–6069

Saarinen T, Orelma H, Grönqvist S, Andberg M, Holappa S, Laine J (2008) Adsorption of different laccases on cellulose and lignin surfaces. BioResour 4(1):94–110

Salis A, Pisano M, Monduzzi M, Solinas V, Sanjust E (2009) Laccase from *Pleurotus sajor-caju* on functionalised SBA-15 mesoporous silica: immobilisation and use for the oxidation of phenolic compounds. J Mol Catal B Enzym 58(1–4):175–180

Secundo F (2013) Conformational changes of enzymes upon immobilisation. Chem Soc Rev 42 (15):6250–6261

Shao J-G, Deng W-J, Yang Y-X, Liu X-N, Chen Y-R (2009) Adsorption of laccase onto mesoporous silica prepared with inorganic counterions. Adsorpt Sci Technol 27(2):147–165

Sheldon RA (2011) Characteristic features and biotechnological applications of cross-linked enzyme aggregates (CLEAs). Appl Microbiol Biotechnol 92(3):467–477

Sheldon RA, van Pelt S (2013) Enzyme immobilisation in biocatalysis: why, what and how. Chem Soc Rev 42(15):6223–6235

Shojaat R, Saadatjoo N, Karimi A, Aber S (2016) Simultaneous adsorption–degradation of organic dyes using MnFe2O4/calcium alginate nano-composites coupled with GOx and laccase. J Environ Chem Eng 4(2):1722–1730

Singh N, Basu S, Vankelecom IF, Balakrishnan M (2015) Covalently immobilized laccase for decolourization of glucose-glycine maillard products as colourant of distillery wastewater. Appl Biochem Biotechnol 177(1):76–89

Sondhi S, Kaur R, Kaur S, Kaur PS (2018) Immobilization of laccase-ABTS system for the development of a continuous flow packed bed bioreactor for decolorization of textile effluent. Int J Biol Macromol 117:1093–1100

Spahn C, Minteer SD (2008) Enzyme immobilization in biotechnology. Recent Pat Eng 2 (3):195–200

Stanbury PF, Whitaker A, Hall SJ (2013) Principles of fermentation technology. Elsevier, Oxford

Strong P, Claus H (2011) Laccase: a review of its past and its future in bioremediation. Crit Rev Env Sci Technol 41(4):373–434

Taboada-Puig R, Junghanns C, Demarche P, Moreira M, Feijoo G, Lema J, Agathos SN (2011) Combined cross-linked enzyme aggregates from versatile peroxidase and glucose oxidase: production, partial characterization and application for the elimination of endocrine disruptors. Bioresour Technol 102(11):6593–6599

Teerapatsakul C, Bucke C, Parra R, Keshavarz T, Chitradon L (2008) Dye decolorisation by laccase entrapped in copper alginate. World J Microbiol Biotechnol 24(8):1367–1374

Touahar IE, Haroune L, Ba S, Bellenger J-P, Cabana H (2014) Characterization of combined cross-linked enzyme aggregates from laccase, versatile peroxidase and glucose oxidase, and their utilization for the elimination of pharmaceuticals. Sci Total Environ 481:90–99

Vanhulle S, Lucas M, Mertens V, Gobeaux B, Corbisier A, Bols C-M, Buchon F, Wesenberg D, Agathos S (2005) Sustainable process for the treatment and detoxification of liquid waste. Google Patents

Wang F, Guo C, Liu HZ, Liu CZ (2008) Immobilization of *Pycnoporus sanguineus* laccase by metal affinity adsorption on magnetic chelator particles. J Chem Technol Biotechnol 83 (1):97–104

Wang F, Hu Y, Guo C, Huang W, Liu C-Z (2012) Enhanced phenol degradation in coking wastewater by immobilized laccase on magnetic mesoporous silica nanoparticles in a magnetically stabilized fluidized bed. Bioresour Technol 110:120–124

Wanga S-J, Zhonga J-J (2011) Bioreactor engineering. Bioprocessing for value-added products from renewable resources: new technologies and applications, 131

Wu Y, Teng Y, Li Z, Liao X, Luo Y (2008) Potential role of polycyclic aromatic hydrocarbons (PAHs) oxidation by fungal laccase in the remediation of an aged contaminated soil. Soil Biol Biochem 40(3):789–796

Yang W-C (2003) Bubbling fluidized beds. In: Handbook of fluidization and fluid-particle systems. CRC, Boca Raton, pp 64–123

Yang WY, Wen SX, Jin L, Rong L, Tetsuo M, Bo C (2006) Immobilization and characterization of laccase from Chinese *Rhus vernicifera* on modified chitosan. Process Biochem 41 (6):1378–1382

Yang J, Lin Y, Yang X, Ng TB, Ye X, Lin J (2017) Degradation of tetracycline by immobilized laccase and the proposed transformation pathway. J Hazard Mater 322:525–531

Yao J, Wang Q, Wang Y, Zhang Y, Zhang B, Zhang H (2015) Immobilization of laccase on chitosan–halloysite hybrid porous microspheres for phenols removal. Desalin Water Treat 55 (5):1293–1301

Yaropolov A, Skorobogat'Ko O, Vartanov S, Varfolomeyev S (1994) Laccase. Appl Biochem Biotechnol 49(3):257–280

Zawisza I, Rogalski J, Opallo M (2006) Electrocatalytic reduction of dioxygen by redox mediator and laccase immobilized in silicate thin film. J Electroanal Chem 588(2):244–252

Zeng J, Zhu Q, Wu Y, Lin X (2016) Oxidation of polycyclic aromatic hydrocarbons using *Bacillus subtilis* CotA with high laccase activity and copper independence. Chemosphere 148:1–7

Zhang S, Gao E, Xia L (2006) Dechlorination of dichlorophenol in waste water by immobilized laccase. J Chem Ind Eng 57(2):359

Zille A, Tzanov T, Gübitz GM, Cavaco-Paulo A (2003) Immobilized laccase for decolourization of reactive black 5 dyeing effluent. Biotechnol Lett 25(17):1473–1477

Zimmermann Y-S, Shahgaldian P, Corvini PF, Hommes G (2011) Sorption-assisted surface conjugation: a way to stabilize laccase enzyme. Appl Microbiol Biotechnol 92(1):169

Challenges in Applying Cross-Linked Laccase Aggregates in Bioremediation of Emerging Contaminants from Municipal Wastewater

Arielle Farida Ariste and Hubert Cabana

Contents

Abstract Municipal wastewaters have been recognized as the primary source of emerging contaminants such as pharmaceuticals and personal care products (PPCPs) in the aquatic receiving environment. The impacts of this contamination on the environment and human health are still not fully understood. Laccases are enzymes capable of degrading various PPCPs leading to widespread promising applications of laccases in bioremediation processes. Laccases are immobilized to broaden their applicability. Immobilization improves the enzyme stability regarding denaturing and operational conditions and creates a recyclable catalyst. Because of its high activity retention, stability and simplicity, the cross-linked enzyme aggregate (CLEA) technology has been considered as an effective carrier-free immobilization method. CLEA technology has been successfully applied for laccase insolubilization (lac-CLEA). However, considerable attempts are still needed to improve the stability and reusability of lac-CLEAs, especially in continuous treatment processes. Novel stabilization strategies have been developed to overcome the undesirable limitations of lac-CLEA applications in bioremediation. This chapter focuses on the latest improvements in the preparation and stabilization of lac-CLEAs for the removal of pollutants from waste streams.

A. F. Ariste · H. Cabana (✉)
University of Sherbrooke Water Research Group (USWRG), Department of Civil and Building Engineering, University of Sherbrooke, Sherbrooke, QC, Canada
e-mail: Hubert.Cabana@USherbrooke.ca

D. Schlosser (ed.), *Laccases in Bioremediation and Waste Valorisation*,
Microbiology Monographs 33, https://doi.org/10.1007/978-3-030-47906-0_6

1 Introduction

Pharmaceuticals and personal care products (PPCPs) are among emerging contaminants in water that have been commonly identified in the environment (Wilkinson et al. 2017). Municipal wastewater treatment plant effluents have been recognized as the primary source of PPCPs in the environment (Verlicchi et al. 2012). Hence, existing WWTPs designed in the nineteenth and twentieth centuries to treat household wastewater and industrial effluents are not effective in removing these emerging contaminants (Lofrano and Brown 2010).

The presence of PPCPs in wastewater is of interest in environmental pollution (aquatic and terrestrial ecosystems) because some PPCPs are biologically active even at low concentrations and have bioaccumulation potential (Brausch and Rand 2011; Tanoue et al. 2015; Hopkins and Blaney 2016). PPCP contamination may lead to environmental and health problems (Rasheed et al. 2018). For example, male fish feminization as a result of exposure to steroidal hormones, development of antibiotic-resistant genes due to the release of antibiotic active compounds into the environment and various harmful effects on development or reproduction that may be caused by endocrine-disrupting chemicals (EDCs) (Ortiz de García et al. 2013; Riva et al. 2018; Sharma et al. 2019). The occurrence of these PPCPs in drinking water and irrigation water may also pose health concerns and wastewater management challenges in many regions, particularly in developing countries where wastewater treatment plants (WWTPs) are rare (Drechsel et al. 2015; Gogoi et al. 2017).

Furthermore, to face the problems of water scarcity in some regions of the world, the mobilization of alternative water sources, including wastewater recycling, is increasing (Wilcox et al. 2016). Recycled wastewater is mainly used for non-potable direct and indirect uses, such as irrigation, watering green spaces or industrial processes (Drechsel et al. 2015). The impacts of these practices on the environment and human health are still not very well-known (Kabir et al. 2015; Kumar and Pal 2018; Nilsen et al. 2019). The health risks (chronic exposure to contaminants) and environmental risks (soil degradation, contamination of groundwater) present a gap to be filled to understand the impact of reusing wastewater containing emerging contaminants (Semerjian et al. 2018).

The progress in advanced wastewater treatment technology development has led to the creation of some innovative treatment processes. Bioremediation processes are considered one of the most effective and eco-friendly mechanisms for municipal wastewater treatment. Enzymatic technologies have gained interest because of their specificity and selectivity in removing contaminants from waste streams (Bilal et al. 2019a).

Enzymes are catalysts capable of carrying out specific reactions. Laccases (EC 1.10.3.2) and peroxidases (EC 1.11.1.7) are enzymes capable of degrading various PPCPs (Alneyadi et al. 2018). The effectiveness of enzymes has led to a focus on ligninolytic enzymes as a promising tool for biotransformation processes. Laccases have drawn attention in recent decades because of their high bioremediation potential (Madhavi and Lele 2009). Laccases are copper-containing oxidoreductase glycoprotein widely distributed in fungi, bacteria and plants. They have been broadly applied in numerous biotechnological and industrial fields (Demarche et al. 2012; Gasser et al. 2014). Laccases are versatile catalysts capable of oxidizing a variety of phenolic and non-phenolic compounds and have been used in different operational conditions. They only need oxygen to oxidize their substrates, which is a significant economic advantage. The reaction end products resulting in laccase degradation of phenolic compounds are also one of the advantages of laccase application in bioremediation processes. The resulting metabolites (reactive phenoxy radicals) form a variety of oligomers that have been found to significantly lose their biological activity and toxicity (Ba et al. 2013; Bilal et al. 2019b).

The economic viability of a large-scale application of enzymes in bioremediation is directly affected by the cost contribution of the enzyme to the treatment. So, to be cost-effective, enzymes must be simple to produce and handle, as well as stable and reusable, with high catalytic activity (Naghdi et al. 2018). When suspended in an aqueous phase, enzymes are soluble, making them challenging to reuse. The immobilization techniques consist of physical or chemical interactions to generate insoluble enzymes by confining them in a certain region of space. The immobilized enzymes are then reusable and contribute to reducing the treatment cost (Zdarta et al. 2019).

Physical immobilization methods consist of entrapment or binding enzymes onto a carrier material. The physical support reduces the volumetric activity of the enzymes compared with carrier-free immobilization approaches (Wang et al. 2015). The most valuable carrier-free immobilization technique is the cross-linking enzyme aggregate (CLEA) method. The CLEAs are produced through covalent bonds that keep the enzyme aggregates together. CLEAs appear to have considerable bioremediation potential because of their high activity retention and stability, combined with their simple preparation protocol. CLEAs display a high catalytic property because the insolubilization method does not need physical support (Cao et al. 2003; Sheldon 2011). CLEA technology has been used for laccase insolubilization (lac-CLEA). Lac-CLEAs have been successfully applied to treat pollutants from different waste streams. However, to increase the probability of satisfying the operational requirements for large-scale applications, it is still necessary to improve the existing lac-CLEA preparation methods, to enhance the insolubilized laccase stability, as well as its catalytic properties. Better reactors adapted for the application of lac-CLEAS in real environmental and operational conditions need to be developed (Ba et al. 2018; Bilal et al. 2019a).

This chapter presents an overview of the latest strategies to enhance lac-CLEA formulation and stability enhancement to apply lac-CLEAs as a promising bioremediation tool for the removal of emerging contaminants from municipal wastewater.

2 Cross-Linked Enzyme Aggregate Technology

Enzyme cross-linking technologies were developed in the middle of the twentieth century. Three cross-linking methods using dissolved enzymes (CLEs), crystallized enzymes (CLECs) and aggregated enzymes (CLEAs) have been reported and extensively reviewed (Cao et al. 2003; Sheldon et al. 2005; Velasco-Lozano et al. 2016; Yusof and Khanahmadi 2019). Among these methods, CLEA technology has shown many advantages for large-scale industrial applications. The method is simple and allows rapid optimization of the insolubilization conditions. CLEA technology provides a stable and recyclable catalyst with high enzymatic activity retention. Also, crude enzyme extract can be used to prepare CLEAs, which results in the formation of a low-cost biocatalyst (Sheldon 2011). CLEAs can be applied to a wide range of enzymes (including laccases) that can be combined to form combined cross-linked enzyme aggregates (combi-CLEAs). A detailed description of CLEA technology has been exhaustively reviewed (Sheldon 2011; Velasco-Lozano et al. 2016; Yusof and Khanahmadi 2019).

Briefly, CLEAs are obtained from precipitated enzymes forming aggregates that remain permanently insoluble after covalent binding with a cross-linking agent in a two-step process (Fig. 1).

The first step in CLEA preparation consists of physical precipitation of the enzymes. The precipitation occurs by changing the hydration state of the molecules or by altering the electrostatic constant of the solution (Matijošyte et al. 2010). The common precipitants involved in laccase aggregation are salts, organic solvents or non-ionic polymers, such as ammonium sulphate, polyethylene glycol and tert-butyl alcohol (Matijošyte et al. 2010).

The second step is based on the formation of covalent bindings between the enzymes' primary amino groups of lysine and the reactive groups of the cross-linker (Sheldon 2011). Depending on the enzyme used, different cross-linkers can be used in CLEA formulation. Due to its availability and low cost, glutaraldehyde, a bifunctional aldehyde agent, is the most widely used cross-linker in CLEA preparation (Sheldon 2011). Glutaraldehyde reacts with the amino groups of the enzymes to

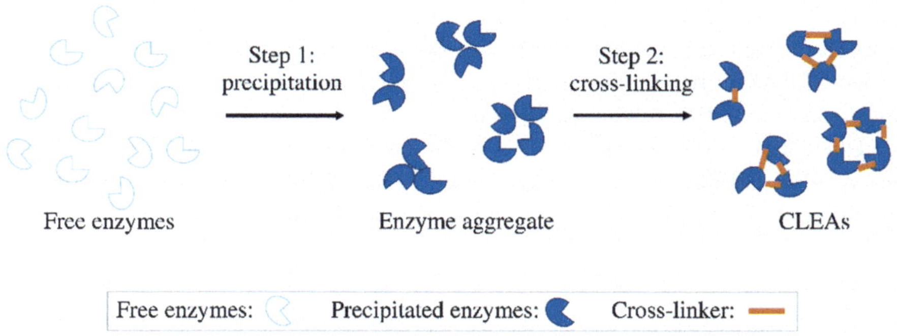

Fig. 1 Schematic representation of cross-linked enzyme aggregate preparation

form an amide bond that links the enzymes together. Other cross-linkers such as glyoxal, dextran polyaldehyde and the biopolymer chitosan can also be used for lac-CLEA preparation (Ba et al. 2013).

The two-step procedure in CLEA preparation is simple. However, the combination of different enzyme properties, as well as the choice of precipitants and cross-linking agents, can significantly turn this simple procedure into a more complex one that can result in several permutations. It is essential therefore to determine the cross-linking condition from the start of the insolubilization process.

3 Determining the Conditions for Laccase Aggregation and Cross-Linking

The two-step CLEA preparation procedure can affect the catalytic properties of the enzymes. These properties depend on the precipitating agent and the cross-linker used, as well as their concentrations. Since the precipitant and cross-linker can affect the conformation and rigidity of the enzyme which then changes the CLEA reactivity (Matijošyte et al. 2010), they must be carefully chosen. This choice depends on the biochemical properties of the enzymes. So, the best conditions must be defined experimentally and optimized to determine the concentrations needed for the enzymes' insolubilization. The optimum ratio of cross-linkers and enzymes is crucial to determine, especially when using an undefined crude enzyme preparation (Cabana et al. 2007b; Matijošyte et al. 2010). Generally, a low concentration of cross-linking agent is known to create insufficient bonds with the enzyme, while at a high concentration, it can increase the enzyme rigidity and prevent the substrate from reaching the active site resulting in lower apparent enzyme activity (Cabana et al. 2007b).

To reduce these adverse effects on the catalytic properties of lac-CLEAs, it is crucial to optimize the formulation conditions. Mathematical models using experimental designs are helpful in process optimization. However, these empirical models do not allow for an in-depth characterization of the insolubilized enzymes. Other alternatives exist, such as the addition of polymers containing primary amino groups during the precipitation step to avoid excessive cross-linking of the enzyme. Bovine serum albumin (BSA) is mostly used as a source of protein and amino groups during lac-CLEA preparation. This technology can help increase the stability of the immobilized enzyme (Cabana et al. 2007b).

Optimization helps to reduce the use of the reagents and their undesirable effects on lac-CLEA properties and can also help to reduce the cost of the resulting insolubilized biocatalyst by reducing the reagent quantities to the minimum necessary, avoiding waste. During the optimization process, it is important to determine an equilibrium between the stability and apparent activity of the resulting lac-CLEAs, not to favour one parameter more than the others (Arsenault et al. 2011).

In some cases, laccase insolubilization might lead to a partial or complete loss of laccase activity because the reagents used during lac-CLEA preparation might interfere with the active site of the enzyme. Although glutaraldehyde remains an inexpensive and widely used cross-linker in laccase insolubilization, it is well-known that it can contribute to enzyme inactivation during CLEA preparation. Glutaraldehyde can modify the active site residues of the enzyme. So, it is crucial to find solutions to prevent the cross-linker from entering the active site or inducing extreme conformational changes. Mild insolubilization conditions must be employed, or the reagents must be replaced, to prevent such problems.

Only a few studies report perspectives about the investigation of new cross-linking agents in lac-CLEA formulation for a sustainable biocatalyst preparation. A dextran-based cross-linker has been developed and successfully applied in CLEA preparation from several enzymes (penicillin G acylase, hydroxynitrile lyase, alcohol dehydrogenase and two different nitrilases) (Mateo et al. 2004). The dextran molecules are too large to penetrate the active site of the enzyme, and CLEAs exhibited a higher enzyme activity. This technique has not been applied in lac-CLEA preparation yet.

Recently, chitosan was successfully tested as a cross-linking agent for laccase from *C. polyzona* insolubilization. The results showed hyperactivation of the lac-CLEA-based chitosan conjugation formed. This overexpression can come from the conformational changes in laccase during the conjugation reaction by locking the enzyme into a more favourable conformation (Arsenault et al. 2011; Cabana et al. 2011).

Most of the studies about lac-CLEAs have focused on characterization and batch experiment applications (Table 1). The shape and size of the CLEA particles are essential characteristics of a direct effect on the biocatalyst catalytic properties and recyclability. The structural organization of CLEAs is influenced by the cross-linking time. Generally, the particle size of CLEAs varies between 0.1 and 200 μm (Velasco-Lozano et al. 2016). Although CLEA structure is known to significantly influence the effectiveness of the insolubilized biocatalyst, this parameter is not often evaluated, and information on the lac-CLEA structure is still missing (Tables 1 and 2).

4 Lac-CLEAs in Bioremediation: Promises, Efficiency and Drawbacks

In addition to being a low-cost process, one of the benefits of bioremediation is that it results in the generation of less toxic reaction end products compared to chemical treatment processes (Varga et al. 2019). A significant number of PPCPs are phenolic compounds, oxidable by laccases. The oxidation reaction involving laccases leads to the production of water and free reactive phenoxy radicals. These radicals are likely to polymerize and precipitate from the reaction solution (Ba et al. 2014a). The

Table 1 Lac-CLEA preparation and efficiency from contaminant bioremediation from waste streams

Laccase sources	Precipitants/cross-linkers	Lac-CLEA activity (substrate)	Structures	Experiment conditions and pollutant removal rate.	References
Crude laccases from *C. polyzona*	Polyethylene glycol/ glutaraldehyde	148 U/g (ABTS)	BSA-free: 1–5 μm With BSA: 100–200 μm	Continuous elimination: fluidized bed reactor Elimination after 50 min: nonylphenol = 90%; bisphenol A = 30%; triclosan = 90% After 150 min: >95% elimination for all compounds	Cabana et al. (2007b)
Crude laccases from *C. polyzona*	Polyethylene glycol/ glutaraldehyde	148 U/g (ABTS)	NP[a]	Continuous elimination: perfusion basket reactor (BR) EDC elimination after 325 min: 85% Stable performances of the BR over 7 days	Cabana et al. (2008)
Laccases from *T. versicolor*, *T. villosa* and *A. bisporus*	Different precipitants/glutaraldehyde Best precipitants: polyethylene glycol 3400 (PEG 3400); ammonium sulphate AS); 1,2 dimethoxyethane (DME)	(Metol) *T. versicolor:* 17.6 U/mg (PEG 34000 at 20 °C) *T. villosa*: 9.6 U/mg (AS at 4 °C) *A. bisporus:* 3.6 U/mg (DME and AS at 4 °C)	NP	A batch experiment of *T. versicolor* lac-CLEAs with TEMPO as a mediator. After 5 h: benzyl alcohol, 80%; 1-pentanol, 7%; 1-hexanol, 15% After 20 h: 1-heptanol, 55%; 1-octanol, 34%; 1-nonanol, 13%; 1-decanol, 21%	Matijošyte et al. (2010)
Laccases from *T. versicolor*	Tris buffer/chitosan/EDC	626 U/g (ABTS)	NP	6 h batch experiment Triclosan removal: free laccase = 60%; Lac-CLEAs = 100%	Cabana et al. (2011)
Crude laccases from *C. polyzona*	Ammonium sulphate/ chitosan/EDC	737 U/g (ABTS)	1694.3–2313.7 nm	–	Arsenault et al. (2011)

(continued)

Table 1 (continued)

Laccase sources	Precipitants/cross-linkers	Lac-CLEA activity (substrate)	Structures	Experiment conditions and pollutant removal rate.	References
Laccases from *S. putrefaciens*	Ammonium sulphate/ glutaraldehyde	(ABTS)	NP	24 h batch experiment Malachite green dye decolorization. Free laccase and Lac-CLEAs ≈90%	Sinirlioglu et al. (2013)
Three oxidative enzymes: laccase (Lac) from *T. versicolor*, versatile peroxidase (VP) from *B. adusta* and glucose oxidase (GOD) from *A. niger*	Ammonium sulphate/ chitosan_EDC	30% of initial laccase activity was recovered along with 40% for each of VP and GOD Lac (ABTS); VP (veratryl alcohol); GOD (qualitatively estimated through the increase of Mn-oxidizing activity of VP)	NP	Batch experiments with free enzyme ability to degrade a pharmaceutical compound (acetaminophen, naproxen, mefenamic acid, indomethacin, diclofenac, ketoprofen, caffeine, diazepam, ciprofloxacin, trimethoprim, fenofibrate and bezafibrate, carbamazepine and its by-products 10–11 epoxy-carbamazepine) in synthetic wastewater More than 80% elimination for the first five compounds Combi-CLEA-mediated removal reached up to 25% contained in real municipal wastewater effluent	Touahar et al. (2014)
Laccases from *T. versicolor* (TvL) and mushroom tyrosinase (Tyr)	Ammonium sulphate/ chitosan_EDC	Combi-CLEAs-TvL: 12.3 U/ g (ABTS) Combi-CLEAs-Tyr: 167.4 U/ g (L-DOPA)	NP	Batch experiment: transformation of 80% to ≈100% of acetaminophen from the municipal wastewater and more than 90% from a hospital wastewater Acetaminophen metabolites: oligomers as dimers, trimers and tetramers due to the laccase and 3-hydroxyacetaminophen due to the tyrosinase	Ba et al. (2014a)

Laccases from *T. versicolor* (TvL)	Ammonium sulphate/ chitosan_EDC	(ABTS)	NP	Batch experiment + hybrid bio-reactor Acetaminophen and mefenamic acid ≈100% in 24 h; carbamazepine 93% in 72 h, from wastewater	Ba et al. (2014b)
Laccases from *Cerrena* sp.	Ammonium sulphate/ glutaraldehyde	Activity recovery: 68.1% (ABTS)	NP	With 80 mM NaCl, over 90% decolourization was obtained with free and immobilized laccases in 4 and 2 h, respectively	Yang et al. (2016)
Laccases from *F. fomentarius* and *T. versicolor*	Ammonium sulphate/ glutaraldehyde	–	NP	Batch experiment: decoulorization of malachite green, bromothymol and methyl red dyes within 10 h *T. versicolor* lac-CLEAs: 95% and *F. fomentarius* lac-CLEAs: 90%	Vršanská et al. (2018)
Laccases from *T. versicolor* (TvL) and mushroom tyrosinase (Tyr)	Ammonium sulphate/ chitosan_EDC	Combi-CLEAs-TvL: 12.3 U/ g (ABTS) Combi-CLEAs-Tyr: 167.4 U/ g (L-DOPA)	NP	Continuous elimination of 14 pharmaceutical compounds (PPs) from municipal wastewater using a hybrid bioreactor (HBR) Complete removal of all PPs after a 5-day continuous operation. The combi-CLEAs retained ≈70% of its initial enzymatic activity after the treatment	Ba et al. (2018)

[a]*NP* not provided

Table 2 Developed strategies for the improvement of catalytic properties as well as the stability of lac-CLEAs

Enzymes	Immobilization method Precipitant/cross-linker	Laccase activity (substrate)	Particle size	Pollutants/removal rate	Stability	References
Laccase from *P. ostreatus*	p-CLEAs Three-phase partitioning co-precipitation technique of laccase and starch, followed by cross-linking with glutaraldehyde	16 U/mg (ABTS)	Spherical shape with pores p-CLEAs with 0.2% starch ≈0–81 nm	Batch experiment elimination of triphenylmethane and reactive dyes by 60–70%, respectively	The half-life of free laccase at 55 °C was calculated to be 1.3 h, while p-CLEAs did not lose any activity even after 14 h p-CLEAs exhibited improved storage stability and catalytic efficiency and could be recycled 15 times with 60% loss of activity	Kumar et al. (2012)
Laccase from *T. versicolor*	EPES-lac and EPES-CLEAs Free laccase and lac-CLEA stabilization through a polymeric network of chitosan and 3-aminopropyltriethoxysilane (APTES) Ammonium sulphate/ chitosan_EDC	Apparent specific activity of laccase and CLEAs dropped from 1110.6 U/g to 34.3 U/g and 55.7 U/g to 11.5 U/g, respectively (ABTS)	EPES-lac ≈100 nm (spherical shape) EPES-CLEAs was amorphous and relaxed rather than compact and uniform	NP[a]	Optimal pH and temperature Free laccase: pH 3/60 °C EPES-lac: pH 4/20–60 °C CLEAs and EPES-CLEAs: pH 4/60 °C Residual activities After 22 days: EPES-lac and EPES-CLEAs ~300%. Free laccase and CLEAs lost 10% and 25%, respectively, after 24 h of incubation and were	Hassani et al. (2013)

					completely denatured after 7 and 9 days In the presence of organic solvents, EPES-lac and EPES-CLEAs showed more than 100% of their initial activity, while free laccase and CLEAs lost 95% and about 30% of their initial activity, respectively In the presence of $ZnCl_2$ and EDTA, free laccase, CLEAs and EPES-lac activities dropped to 5, 35 and 50%, respectively, while EPES-CLEAs maintained nearly 100% of their activity	
Laccase from *T. versicolor*	M-CLEAs Amino-functionalized magnetic nanoparticles bonded to lac-CLEAs Chilled propanol/ glutaraldehyde	Activity recovery ~32% (ABTS)	M-CLEAs = 48–74 nm with few agglomerated particles	A laboratory-scale perfusion basket reactor (BR) application for continuous decolourization of dyes 61–96% of brilliant blue R, malachite green and reactive	Optimal temperature and pH Free laccase: 40 °C/ pH 4 Lac-CLEAs: 4 °C/ pH 5 M-CLEAs: 30 °C/ pH 6	Kumar et al. (2014)

(continued)

Table 2 (continued)

Enzymes	Immobilization method Precipitant/cross-linker	Laccase activity (substrate)	Particle size	Pollutants/removal rate	Stability	References
				black 5 removal in 10 h		
Laccase from *T. versicolor*	h-CLEAs Preparation in a millifluidic reactor with two coaxial laminar flows Acetonitrile/glutaraldehyde	0.26 U/mg (ABTS)	~220 nm	Trypan blue dye was decolourized within 270 min	Free laccase optimum pH = 4. Free laccase retained around 42% and 13.9% of its optimal activity at pH 7.0 and 10.0, respectively h-CLEAs retain about 90.4% and 80.9% of its activity at pH 7.0 and 10.0, respectively h-CLEAs laccase can be trapped in a membrane for continuous degradation of trypan blue up to 96 h without losing any activity	Nguyen et al. (2017)
Crude laccase from *T. pubescens*	E-CLEAs (entrapped CLEAs) E-lac (entrapped laccase) Laccase was cross-linked with glutaraldehyde before entrapment into Ca-alginate beads	Immobilization yield (>72%) than that of the laccase immobilized only by entrapment (61.6%), (ABTS)	NP	E-CLEAs 99% removal of bisphenol A (BPA) from an aqueous solution in 2 h in a batch experiment BPA was removed in ten successive batches with higher than 70% efficiency at the end of the last batch	Thus, the free laccase: pH 3.0 and E-CLEAs: pH 4.0. Both catalysts showed optimal activity at 40 °C At 70 °C, E-lac and E-CLEAs showed a relative activity of 49.1% and 54.6%, respectively, while the free laccase retained	Lassouane et al. (2019)

					only 16.8% of its initial activity	
Laccase from *T. versicolor*	Entrapped lac-CLEAs: E-CLEAs. Entrapment carrier: porous silica Ammonium sulphate/glutaraldehyde	E-CLEAs: 6.29 U/mg (ABTS) ≈50% specific activity loss upon the enzyme insolubilization	Granulated appearance	40 min batch experiment with complete removal of total phenol in solution	Thermal and pH stability and activity retention in hydrophobic and hydrophilic solvents Good operational stability and reusability: >79% of initial activity after 20 cycles of successive operations	Fathali et al. (2019)

[a]*NP* not provided

polymerization implies the inactivation of their biological activity and inhibits their action on living organisms (Cabana et al. 2007a).

Acetaminophen, a model phenolic pharmaceutical compound, has been frequently employed to study this reaction mechanism in order to propose reaction pathways involving laccases in the degradation of phenolic PPCPs (Ba et al. 2014a; Jahangiri et al. 2018). Laccases have low redox potential, which prevents them from degrading non-phenolic compounds. However, some small molecules can act as an electron shuffle (laccase-mediator system) and allow the transformation of non-phenolic compounds. Cross-coupling phenomena between the reactive phenoxy radicals and the non-phenolic compounds can also contribute to the elimination of non-phenolic compounds by laccases (Hachi et al. 2017; Apriceno et al. 2019). These reaction mechanisms expand the range of contaminants that can be oxidized by laccases, making this biocatalyst a promising tool for bioremediation processes.

Lac-CLEAs have been prepared from a wide variety of laccases in the past decades and efficiently applied to remove a variety of contaminants (Table 1). PPCPs have been successfully removed from aqueous solutions in batches and continuous experiments using lac-CLEAs. A crude enzyme extract from *C. polyzona* lac-CLEAs was prepared with glutaraldehyde and applied in a fluidized bed reactor to treat EDCs. The biocatalyst removed more than 95% of EDCs in 150 min (Cabana et al. 2007b). Later, a new cross-linking method was proposed using chitosan to prepare lac-CLEAs (Arsenault et al. 2011; Cabana et al. 2011; Ba et al. 2014a). In a 6-h batch experiment, Cabana et al. (2011) showed that the chitosan-based lac-CLEAs were able to remove 100% of triclosan from an aqueous solution, while the free laccases were less efficient with a 60% removal rate. The same CLEA technology was applied in batches and continuous experiments to remove pharmaceutical compounds from wastewater. Acetaminophen and mefenamic acid were eliminated from the filtered wastewater in 24 h, and up to 93% of carbamazepine, a more recalcitrant compound, was eliminated after 72 h (Ba et al. 2014b). Lac-CLEAs have also been efficiently applied for PPCP elimination from wastewaters and other biotechnological applications such as dye decolourization from solutions (Table 1).

The efficiency of the laccase-based treatment process can be improved by enhancing its selectivity using the combi-CLEA method (Kong et al. 2016). Combi-CLEAs have the same economic and environmental benefits as CLEAs, with the difference that combi-CLEAs can target a greater variety of substrates, specific to each co-immobilized enzyme (Table 1). However, it is crucial to rigorously optimize the insolubilization conditions to respect the specificity of each enzyme. Laccases were successfully co-immobilized with two other oxidative enzymes, namely, versatile peroxidase and glucose oxidase (Touahar et al. 2014). This combi-CLEA contributes to degrading a plethora of pharmaceutical compounds in laboratory conditions in solution and removes 25% of detected acetaminophen in real municipal wastewater in a batch experiment. Ba et al. (2014a) tested the efficiency of a combi-CLEA of laccase and tyrosinase for the transformation of acetaminophen in wastewaters. In batch mode, the combi-CLEA transformed from 80 to 100% of acetaminophen from the municipal wastewater and more than 90%

from a hospital wastewater. Mass spectrometry detection of acetaminophen transformation products showed the formation of its oligomers as dimers, trimers and tetramers due to the laccase and 3-hydroxyacetaminophen due to the tyrosinase.

Lac-CLEAs are efficient in eliminating PPCPs from aqueous solutions and have many advantages for bioremediation applications (Table 1). However, the lack of a biocatalyst production cost estimate as well as the lack of understanding of the heterogeneous catalytic reaction mechanism and the need for the development of suitable reactors are some drawbacks for a large-scale application of lac-CLEAs.

Laccase insolubilization has a significant influence on the catalytic properties of the biocatalyst. Contrary to free laccases, lac-CLEAs are heterogeneous. The reaction occurs where the laccases are located (on the surface or within the insoluble structure of the CLEAs), while the substrates (contaminants) and the reaction products are dissolved in the reaction solution (Alshabib and Onaizi 2019). The catalytic parameters of free and insolubilized laccases are generally measured by determination of Michaelis-Menten kinetic constants with 2,2′-azino-bis (3-éthylbenzothiazoline-6-sulphonique) (ABTS) as substrate (Table 1). Syringaldazine is another model substrate used for laccase activity assay (Leonowicz and Grzywnowicz 1981). The assays using these model substrates are simple, accurate and reproducible. However, for an efficient characterization of the biocatalyst, none of them reflect the laccase activity appropriately compared to a real substrate (e.g. the targeted PPCPs). The choice of a more effective method and substrate for laccase activity assay could help to better understand how the biocatalyst reacts to the target contaminants.

In the case of oxidase catalysts, the activity can be assessed by monitoring the change in pressure in the reaction system (Klis et al. 2007). Recently, this method was developed by Demarche (2014), to measure laccase activity, and provides an alternative to the commonly used spectrophotometric assays. However, the necessity to use high substrate concentrations with high molar extinction coefficient, and the use of a substrate with interfering biocatalytic products in the target UV-vis spectrum, are some of the limitations to the method (Demarche 2014). Another promising alternative consists of assessing the enzyme kinetics using mass spectrometry combined with liquid chromatography (HPLC-MS). The latest method developed using this technique provided an accurate kinetic parameter measurement of laccases from *T. versicolor* and *G. lucidum*. Natural phenolic compounds (p-coumaric, ferulic and sinapic acid, and a lignin model OH-dilignol) were employed as substrates for laccase kinetic determination by HPLC-MS. The new method has been reported to be highly suitable and accurate for assaying laccase activity. This method also has the advantage of concomitantly examining laccase oxidation product profiles (Perna et al. 2018). These examples are some interesting ways that must be explored more in order to better characterize lac-CLEAs.

In addition to the importance of choosing the best method for measuring laccase activity, it is also important to take into consideration that the kinetic properties of the insolubilized lac-CLEAs are influenced by their structure (shape and size), which can induce mass-transfer limitation phenomena. Most of the time, these diffusion phenomena are mingled with the enzyme inhibition caused by the insolubilization process, as discussed in the previous section. The lac-CLEA activity reported in the literature reflects the apparent activity. The intrinsic activity can be approximatively obtained by increasing the substrate diffusion through the heterogeneous structure, by promoting nano-sized aggregates. CLEA structure and size can be tuned during the insolubilization process to overcome this problem. The stirring speed and time during the biocatalyst washing steps can be optimized to avoid CLEA aggregation to form "clusters" (CLEAs size >1 μm). Aggregates can also be formed during the CLEA recuperation by centrifugation or filtration. The aggregation increases the size of the CLEAs and then increases mass-transfer limitations (Velasco-Lozano et al. 2016).

One strategy consisting of the use of a magnetic field to facilitate CLEA recovery has been developed and proposed as an alternative to avoid the drawback of aggregation caused by centrifugation or filtration. Kumar et al. (2014) successfully immobilized laccase to form a stable and rigid structure of magnetic CLEAs (M-CLEAs). The catalytic efficiency of laccase was significantly improved, and the M-CLEAs were effectively applied in batch and continuous experiments of dye degradation in solution. The results showed more than 60% and 90% decolorization, respectively. More recently, Arca-Ramos et al. (2016) also demonstrated the catalytic potential of M-CLEAs to transform various phenolic and non-phenolic pharmaceutical active compounds present in solution. The developed biocatalyst showed good operational stability after ten successive batch reactions, maintaining up to 70% of the initial activity. These results showed that M-CLEA technology provides a remarkable means to efficiently recover and maintain the catalytic properties of lac-CLEAs.

Nguyen et al. (2017) developed a novel strategy consisting of a multifluidic reactor carrying two coaxial laminar flows to create hollow cross-linked aggregates of laccases (h-CLEAs), by using acetonitrile and an aqueous solution. This method creates uniform h-CLEA sizes (220 nm) and highly stabilizes the laccase. The h-CLEAs were able to degrade trypan blue dye molecules under an alkaline condition in a batch and continuous membrane treatment.

In addition to all these newly developed technologies to enhance lac-CLEA catalytic properties, a simple method consisting of co-precipitation of the enzyme with starch, followed by the cross-linking and removal of the starch by α-amylase to form porous CLEAs (p-CLEAs), was proposed. This method helps to decrease mass-transfer limitations (Wang et al. 2011). A p-CLEA of laccase was successively prepared using a three-partitioning method. The resulting p-CLEA exhibited improved storage stability. The porous biocatalyst was successfully recycled 15 times (Kumar et al. 2012).

All these preparation methods succeed in improving lac-CLEA non-catalytic (shape, size) and catalytic (activity, recyclability) properties for bioremediation (Table 2). Furthermore, to obtain a robust biocatalyst for waste stream bioremediation, lac-CLEAs must be highly stable.

5 Enhancement of Lac-CLEA Stability for Bioremediation Applications

CLEA technology improved the stability of laccases regarding denaturing conditions such as pH and temperature variation, as well as the presence of salts and organic solvents (Table 2). The effectiveness of lac-CLEAs for PPCP elimination from aqueous solutions has been widely proved. However, large-scale applications do not exist yet. One of the reasons is that efforts are still needed to improve lac-CLEA stability. A robust biocatalyst is necessary for the long continuous operation process in denaturing conditions like the conditions found in wastewater.

Besides the methods previously presented, new strategies focused on the stabilization of lac-CLEAs have been developed. CLEAs are compatible with some existing technologies such as encapsulation, silica coating or polymer-engineered structure techniques that could provide resistance to operational stress factors (Cui and Jia 2015).

Hence, the well-known and easy method of immobilization of enzymes by entrapment in a solid polymeric matrix is suitable for CLEA stabilization and easy recovery. A crude laccase extract from *T. pubescens* was cross-linked with glutaraldehyde before entrapment into Ca-alginate beads (Lassouane et al. 2019). The resulting entrapped lac-CLEAs contributed to increasing the immobilization yield by 30% and reducing the catalyst leakage from the beads by sevenfold compared to entrapped free laccases. The entrapped lac-CLEAs were able to remove 99% of bisphenol A from an aqueous solution in batch experiments. The efficiency of the biocatalyst remained high (70%) after ten successive batch treatment cycles.

Another strategy, recently developed for lac-CLEA stability enhancement, consists of the formation of a polymeric network to stabilize the insolubilized biocatalyst. The resulting biocatalysts were named EPES-lac and EPES-CLEAs for enzyme polymer-engineered structure using free laccase and lac-CLEAs (Hassani et al. 2013), respectively. This strategy was developed based on the sustainable method of lac-CLEA formulation using the biocompatible cross-linking agent chitosan which was successively developed and optimized by Cabana et al. (2011) and Arsenault et al. (2011). The resulting EPESs were made from a commercial laccase from *T. versicolor*. The enzyme was surrounded by a polymeric network made of chitosan and 3-aminopropyltriethoxysilane (APTES). This stabilization method contributed to improving the thermoresistance of EPES-lac and EPES-CLEAs 30 times more than free laccases and lac-CLEAs. The stabilized biocatalyst also showed higher resistance than the free laccase and lac-CLEAs to denaturing conditions involving salts and organic solvents. However, the effectiveness of EPESs to treat contaminants in solutions has not been studied yet.

It is also interesting to consider the structure of the EPESs. The scanning microscopy of EPES-lac and EPES-CLEAs revealed a spherical shape for EPES-lac with an average size of 100 μm, while the EPES-CLEAs showed an amorphous shape, which is consistent with the previous works about lac-CLEAs prepared with chitosan (Arsenault et al. 2011; Cabana et al. 2011). These results showed that the formulation conditions influence the structure of the EPESs.

The described approach for lac-CLEA stabilization ensured a robust and easily recoverable catalyst, with excellent potential for bioremediation processes. However, as is well-known, the final goal of immobilizing enzyme is to apply the recyclable biocatalyst for operations involving reactors. Further studies are needed to understand the behaviour, particularly the mass-transfer limitations induced by the stabilization methods and resistance of lac-CLEAs, in a continuous reactor operation.

6 Continuous Reactors Designed for Lac-CLEA-Based Bioremediation

For economic and technical reasons, enzyme-based processes require reuse of the biocatalyst in long-time continuous operations. Different reactor configurations ranging from stirred tank to column-type reactors can be applied to enzyme reaction processes either in sequential batch operation or continuously. In batch mode operation, the biocatalyst is recovered after each batch process by centrifugation or filtration. If combined with a membrane filtration system, the biocatalyst can be directly recycled back to the batch reactor for a continuous operation. Despite this wide choice, there are very few studies on the continuous applications of lac-CLEAs. The trend in lac-CLEA-based technology focused on the development and stabilization of the biocatalysts (Table 2). These developments must ensure a high catalytic and operational stability to lac-CLEAs for an efficient continuous application. Nevertheless, few attempts to apply lac-CLEAs in continuous reactors have been proposed.

A perfusion basket reactor (PBR) was developed by Cabana et al. (2008) for the continuous operation of lac-CLEAs. The PBR consisted of an unbaffled basket made of a metallic filtration compartment filled with the biocatalyst and the reaction solution, which were continuously agitated (Fig. 2). The continuous elimination of some EDCs (nonylphenol, bisphenol A and triclosan) was monitored at an initial concentration of 5 mg/L for each EDC. The results showed that most of the EDCs were removed from the solution (>85% removal) with a hydraulic retention time of 325 min. The performance of the PBR was stable over a week of continuous running (Cabana et al. 2008).

Another reactor system was proposed for the continuous application of lac-CLEAs. The developed hybrid bioreactor (HBR) proposed by Ba et al. (2014b) consists of a stirred tank reactor containing lac-CLEAs, combined with a polysulfone hollow fibre microfiltration membrane. The membrane filtration system confined the lac-CLEAs and recycled them back to the reaction solution (Fig. 3). The HBR was operated continuously for pharmaceutical compounds (acetaminophen, mefenamic acid, carbamazepine) removal in aqueous solution at an initial

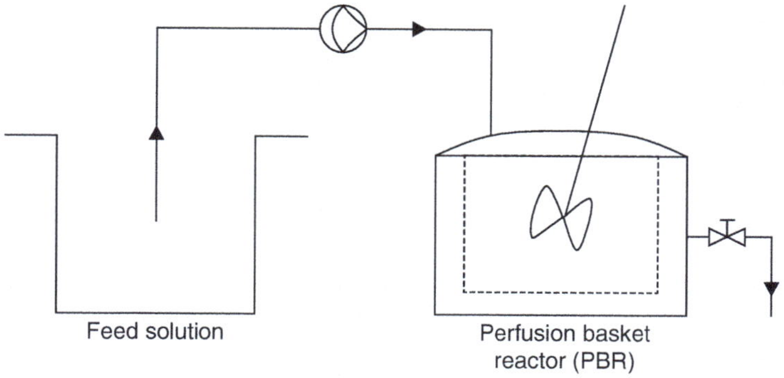

Fig. 2 Schematic representation of a continuous perfusion basket reactor system. Adapted from Cabana H, Jones JP and Agathos SN, Utilization of cross-linked laccase aggregates in a perfusion basket reactor for the continuous elimination of endocrine-disrupting chemicals, Biotechnol Bioeng 102: 1582–1592 (2009), with permission from John Wiley & Sons. (copyright © 2008 Wiley Periodicals, Inc.)

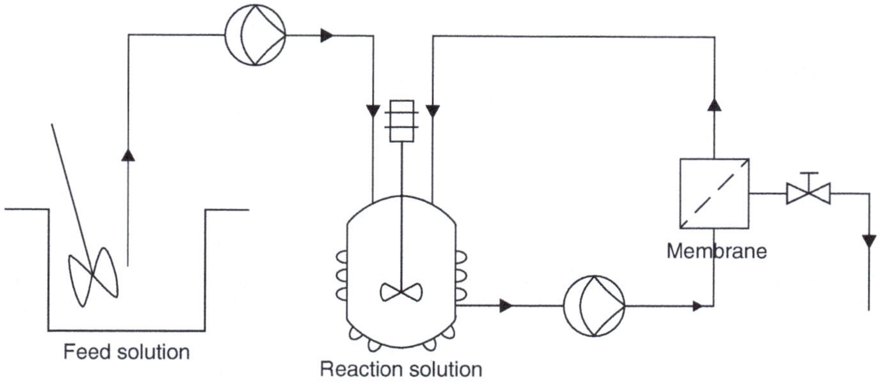

Fig. 3 Experimental setup of the hybrid membrane bioreactor. Reprinted from J Hazard Mater, 280, Ba S, Jones JP and Cabana A, Hybrid bioreactor (HBR) of hollow fibre microfilter membrane and cross-linked laccase aggregates eliminate aromatic pharmaceuticals in wastewaters, 662–670, copyright © 2014, with permission from Elsevier

concentration of 100 μg/L for each compound. The results showed that the synergistic effects of the biocatalyst and the membrane filtration achieved complete elimination of acetaminophen and mefenamic acid and over 90% elimination of carbamazepine from wastewater. In addition, the system efficiently recycled the lac-CLEAs during the experiment with negligible fouling membrane phenomena (Ba et al. 2014b). Later, the same HBR system was operated using a combi-CLEA made of laccase and tyrosinase for a continuous elimination of a mixture of 14 pharmaceutical compounds from municipal wastewater at an environmentally relevant concentration of 10 μg/L (Ba et al. 2018). The results showed complete removal of all active compounds after 5 days of operation. After the treatment, the combi-CLEA retained ~70% of its initial enzymatic activity.

These two examples showed that the continuous operation of lac-CLEAs in reactors is possible. However, the lack of a suitable design restricts the large-scale applications of lac-CLEAs. From these first attempts, it was shown that further investigation is needed to characterize the reactor design in depth and evaluate both parameters (catalytic and operational) for lac-CLEA applications in wastewater bioremediation.

The reactor design must be carefully chosen while taking into consideration the properties of the insolubilized biocatalyst. When applied to large-scale processes, lac-CLEAs will have low volumetric productivity which is influenced by the process volumetric flow. So, we believe that for real application of lac-CLEAs in continuous mode, favoured approaches, which are more compatible with potential streamflow, such as the one proposed by Ba et al. (2018), will be appropriate. Another consideration when designing a suitable reactor for lac-CLEAs application is to understand and develop strategies to deal with the gradual decrease in enzyme activity during the operation processes (Tufvesson et al. 2010).

All in all, the developed continuous reactors should be simple to implement for large-scale application purposes, adapted to current wastewater treatment plants and be economically sustainable (Illanes et al. 2008).

7 Conclusions

Laccases are oxidoreductases, which show excellent potential for the elimination of emerging contaminants from municipal, industrial and hospital wastewaters. This potential has led to a focus on laccase immobilization and insolubilization techniques in recent decades. The insolubilization method using CLEA technology has been shown to apply to a wide range of enzymes, turning it into a useful tool for bioremediation processes.

The simplicity of preparation, lower production costs from crude laccase extract and the possibility of combining laccases with other enzymes are some of the advantages of CLEAs. Also, laccases only need oxygen to oxidize substrates, which is another significant economic advantage for large-scale bioremediation.

Recently, lac-CLEA technology has been improved with some strategies, such as a combination of formulation methods (e.g. physical entrapment of CLEAs) and the formation of polymer-engineered structures. The developed strategies help to enhance the stability of lac-CLEAs and increase their applicability for wastewater bioremediation.

As the impact of PPCPs in the environment and the health risks associated with wastewater reuse are still not well-established, the growing interest in developing lac-CLEAs for wastewater bioremediation purposes gives hope for the refinement of current wastewater treatment. Lac-CLEAs may contribute to reducing the environmental risks associated with some PPCPs, which could help to ease and diversify wastewater reuse.

At this time, efforts are still needed to understand the phenomenon governing the behaviour and performance of lac-CLEA technology. In addition, to overcome the current limitations related to lac-CLEA operations in continuous processes, more scaling-up experiments and applications in real conditions are needed. The choice of lac-CLEA technology for bioremediation of PPCPs from wastewater of other industrial biotechnological applications can be accelerated by the economic viability assessment of the biocatalyst production. To the best of our knowledge, no cost estimation of insolubilized lac-CLEA production has been published. Information on the profitability of lac-CLEA production could help industries and decision-makers analyse the feasibility of these sustainable technologies in bioremediation processes. Also, the lack of a suitable reactor design for the continuous application of insolubilized laccases also limits the lac-CLEA application in large-scale processes.

Acknowledgements This project was supported by a grant from the Fonds de recherche du Québec—Nature et Technologies.

References

Alneyadi AH, Rauf MA, Ashraf SS (2018) Oxidoreductases for the remediation of organic pollutants in water–a critical review. Crit Rev Biotechnol 38:971–988. https://doi.org/10.1080/07388551.2017.1423275

Alshabib M, Onaizi SA (2019) A review on phenolic wastewater remediation using homogeneous and heterogeneous enzymatic processes: current status and potential challenges. Sep Purif Technol 219:186–207. https://doi.org/10.1016/j.seppur.2019.03.028

Apriceno A, Astolfi ML, Girelli AM, Scuto FR (2019) A new laccase-mediator system facing the biodegradation challenge: insight into the NSAIDs removal. Chemosphere 215:535–542. https://doi.org/10.1016/j.chemosphere.2018.10.086

Arca-Ramos A, Kumar VV, Eibes G et al (2016) Recyclable cross-linked laccase aggregates coupled to magnetic silica microbeads for elimination of pharmaceuticals from municipal wastewater. Environ Sci Pollut Res 23:8929–8939. https://doi.org/10.1007/s11356-016-6139-x

Arsenault A, Cabana H, Jones JP (2011) Laccase-based CLEAs: Chitosan as a novel cross-linking agent. Enzyme Res. https://doi.org/10.4061/2011/376015

Ba S, Arsenault A, Hassani T et al (2013) Laccase immobilization and insolubilization: from fundamentals to applications for the elimination of emerging contaminants in wastewater treatment. Crit Rev Biotechnol 33:404–418. https://doi.org/10.3109/07388551.2012.725390

Ba S, Haroune L, Cruz-Morató C et al (2014a) Synthesis and characterization of combined cross-linked laccase and tyrosinase aggregates transforming acetaminophen as a model phenolic compound in wastewaters. Sci Total Environ 487:748–755. https://doi.org/10.1016/j.scitotenv.2013.10.004

Ba S, Jones JP, Cabana H (2014b) Hybrid bioreactor (HBR) of hollow fiber microfilter membrane and cross-linked laccase aggregates eliminate aromatic pharmaceuticals in wastewaters. J Hazard Mater 280:662–670. https://doi.org/10.1016/j.jhazmat.2014.08.062

Ba S, Haroune L, Soumano L et al (2018) A hybrid bioreactor based on insolubilized tyrosinase and laccase catalysis and microfiltration membrane remove pharmaceuticals from wastewater. Chemosphere 201:749–755. https://doi.org/10.1016/j.chemosphere.2018.03.022

Bilal M, Adeel M, Rasheed T et al (2019a) Emerging contaminants of high concern and their enzyme-assisted biodegradation – a review. Environ Int 124:336–353. https://doi.org/10.1016/j.envint.2019.01.011

Bilal M, Rasheed T, Nabeel F et al (2019b) Hazardous contaminants in the environment and their laccase-assisted degradation – a review. J Environ Manag 234:253–264. https://doi.org/10.1016/j.jenvman.2019.01.001

Brausch JM, Rand GM (2011) A review of personal care products in the aquatic environment: environmental concentrations and toxicity. Chemosphere 82:1518–1532. https://doi.org/10.1016/j.chemosphere.2010.11.018

Cabana H, Jones JP, Agathos SN (2007a) Elimination of endocrine disrupting chemicals using white rot fungi and their lignin modifying enzymes: a review. Chemosphere 67:770. https://doi.org/10.1016/j.chemosphere.2006.10.037

Cabana H, Jones JP, Agathos SN (2007b) Preparation and characterization of cross-linked laccase aggregates and their application to the elimination of endocrine disrupting chemicals. J Biotechnol 132:23–31. https://doi.org/10.1016/j.jbiotec.2007.07.948

Cabana H, Jones JP, Agathos SN (2008) Utilization of cross-linked laccase aggregates in a perfusion basket reactor for the continuous elimination of endocrine-disrupting chemicals. Biotechnol Bioeng 102:1582–1592. https://doi.org/10.1002/bit.22198

Cabana H, Ahamed A, Leduc R (2011) Conjugation of laccase from the white rot fungus *Trametes versicolor* to chitosan and its utilization for the elimination of triclosan. Bioresour Technol 102:1656–1662. https://doi.org/10.1016/j.biortech.2010.09.080

Cao L, van Langen L, Sheldon RA (2003) Immobilised enzymes: carrier-bound or carrier-free? Curr Opin Biotechnol 14:387–394. https://doi.org/10.1016/S0958-1669(03)00096-X

Cui JD, Jia SR (2015) Optimization protocols and improved strategies of cross-linked enzyme aggregates technology: current development and future challenges. Crit Rev Biotechnol 35:15–28. https://doi.org/10.3109/07388551.2013.795516

Demarche P (2014) Immobilized phenoloxidases in reactors for the oxidation of aquatic contaminants of emerging concern: toward a scalable industrial system

Demarche P, Junghanns C, Nair RR, Agathos SN (2012) Harnessing the power of enzymes for environmental stewardship. Biotechnol Adv 30:933–953. https://doi.org/10.1016/j.biotechadv.2011.05.013

Drechsel P, Mahjoub O, Keraita B (2015) Social and cultural dimensions in wastewater use. In: Drechsel P, Qadir M, Wichelns D (eds) Economic asset in an Urbanizing world. Springer, Dordrecht, pp 75–92

Fathali Z, Rezaei S, Faramarzi MA, Habibi-Rezaei M (2019) Catalytic phenol removal using entrapped cross-linked laccase aggregates. Int J Biol Macromol 122:359–366. https://doi.org/10.1016/j.ijbiomac.2018.10.147

Gasser CA, Ammann EM, Shahgaldian P, Corvini PFX (2014) Laccases to take on the challenge of emerging organic contaminants in wastewater. Appl Microbiol Biotechnol 98:9931–9952. https://doi.org/10.1007/s00253-014-6177-6

Gogoi A, Mazumder P, Tyagi VK et al (2017) Occurrence and fate of emerging contaminants in water environment: a review. Ground Sustain Dev 6:169–180. https://doi.org/10.1016/j.gsd.2017.12.009

Hachi M, Chergui A, Yeddou AR et al (2017) Removal of acetaminophen and carbamazepine in single and binary systems with immobilized laccase from *Trametes hirsuta*. Biocatal Biotransformation 35:51–62. https://doi.org/10.1080/10242422.2017.1280032

Hassani T, Ba S, Cabana H (2013) Formation of enzyme polymer engineered structure for laccase and cross-linked laccase aggregates stabilization. Bioresour Technol 128:640–645. https://doi.org/10.1016/j.biortech.2012.10.058

Hopkins ZR, Blaney L (2016) An aggregate analysis of personal care products in the environment: identifying the distribution of environmentally-relevant concentrations. Environ Int 92–93:301–316. https://doi.org/10.1016/j.envint.2016.04.026

Illanes A, Altamirano C, Wilson L (2008) Homogeneous enzyme kinetics. In: Enzyme biocatalysis: principles and applications. Springer, Dordrecht, pp 107–153

Jahangiri E, Thomas I, Schulze A et al (2018) Characterisation of electron beam irradiation-immobilised laccase for application in wastewater treatment. Sci Total Environ 624:309–322. https://doi.org/10.1016/j.scitotenv.2017.12.127

Kabir ER, Rahman MS, Rahman I (2015) A review on endocrine disruptors and their possible impacts on human health. Environ Toxicol Pharmacol 40:241–258. https://doi.org/10.1016/j.etap.2015.06.009

Klis M, Rogalski J, Bilewicz R (2007) Voltammetric determination of catalytic reaction parameters of laccase based on electrooxidation of hydroquinone and ABTS. Bioelectrochemistry 71:2–7. https://doi.org/10.1016/J.BIOELECHEM.2006.09.008

Kong W, Chen H, Lyu S et al (2016) Characterization of a novel manganese peroxidase from white-rot fungus *Echinodontium taxodii 2538*, and its use for the degradation of lignin-related compounds. Process Biochem 51:1776–1783. https://doi.org/10.1016/j.procbio.2016.01.007

Kumar A, Pal D (2018) Antibiotic resistance and wastewater: correlation, impact and critical human health challenges. J Environ Chem Eng 6:52–58. https://doi.org/10.1016/j.jece.2017.11.059

Kumar VV, Prem Kumar MP, Thiruvenkadaravi KV et al (2012) Preparation and characterization of porous cross linked laccase aggregates for the decolorization of triphenyl methane and reactive dyes. Bioresour Technol 119:28–34. https://doi.org/10.1016/j.biortech.2012.05.078

Kumar VV, Sivanesan S, Cabana H (2014) Magnetic cross-linked laccase aggregates – bioremediation tool for decolorization of distinct classes of recalcitrant dyes. Sci Total Environ 487:830–839. https://doi.org/10.1016/j.scitotenv.2014.04.009

Lassouane F, Aït-Amar H, Amrani S, Rodriguez-Couto S (2019) A promising laccase immobilization approach for bisphenol a removal from aqueous solutions. Bioresour Technol 271:360–367. https://doi.org/10.1016/j.biortech.2018.09.129

Leonowicz A, Grzywnowicz K (1981) Quantitative estimation of laccase forms in some white-rot fungi using syringaldazine as a substrate. Enzyme Microb Technol 3:55–58

Lofrano G, Brown J (2010) Wastewater management through the ages: a history of mankind. Sci Total Environ 408:5254–5264. https://doi.org/10.1016/j.scitotenv.2010.07.062

Madhavi V, Lele SS (2009) Laccase properties, use. Bioresources 4:1694–1717

Mateo C, Palomo JM, van Langen LM et al (2004) A new, mild cross-linking methodology to prepare cross-linked enzyme aggregates. Biotechnol Bioeng 86:273–276. https://doi.org/10.1002/bit.20033

Matijošyte I, Arends IWCE, de Vries S, Sheldon RA (2010) Preparation and use of cross-linked enzyme aggregates (CLEAs) of laccases. J Mol Catal B Enzym 62:142–148. https://doi.org/10.1016/j.molcatb.2009.09.019

Naghdi M, Taheran M, Brar SK et al (2018) Removal of pharmaceutical compounds in water and wastewater using fungal oxidoreductase enzymes. Environ Pollut 234:190–213. https://doi.org/10.1016/j.envpol.2017.11.060

Nguyen LT, Seow N, Yang KL (2017) Hollow cross-linked enzyme aggregates (h-CLEA) of laccase with high uniformity and activity. Colloids Surf B Biointerfaces 151:88–94. https://doi.org/10.1016/j.colsurfb.2016.12.006

Nilsen E, Smalling KL, Ahrens L et al (2019) Critical review: grand challenges in assessing the adverse effects of contaminants of emerging concern on aquatic food webs. Environ Toxicol Chem 38:46–60

Ortiz de García S, Pinto Pinto G, García Encina P, Irusta Mata R (2013) Consumption and occurrence of pharmaceutical and personal care products in the aquatic environment in Spain. Sci Total Environ 444:451–465. https://doi.org/10.1016/j.scitotenv.2012.11.057

Perna V, Agger JW, Holck J, Meyer AS (2018) Multiple reaction monitoring for quantitative laccase kinetics by LC-MS. Sci Rep 8:1–9. https://doi.org/10.1038/s41598-018-26523-0

Rasheed T, Bilal M, Nabeel F et al (2018) Environmentally-related contaminants of high concern: potential sources and analytical modalities for detection, quantification, and treatment. Environ Int 122:52–66. https://doi.org/10.1016/j.envint.2018.11.038

Riva F, Castiglioni S, Fattore E et al (2018) Monitoring emerging contaminants in the drinking water of Milan and assessment of the human risk. Int J Hyg Environ Health 221:451–457. https://doi.org/10.1016/j.ijheh.2018.01.008

Semerjian L, Shanableh A, Semreen MH, Samarai M (2018) Human health risk assessment of pharmaceuticals in treated wastewater reused for non-potable applications in Sharjah, United Arab Emirates. Environ Int 121:325–331. https://doi.org/10.1016/j.envint.2018.08.048

Sharma BM, Bečanová J, Scheringer M et al (2019) Health and ecological risk assessment of emerging contaminants (pharmaceuticals, personal care products, and artificial sweeteners) in surface and groundwater (drinking water) in the Ganges River Basin, India. Sci Total Environ 646:1459–1467. https://doi.org/10.1016/j.scitotenv.2018.07.235

Sheldon RA (2011) Characteristic features and biotechnological applications of cross-linked enzyme aggregates (CLEAs). Appl Microbiol Biotechnol 92:467–477. https://doi.org/10.1007/s00253-011-3554-2

Sheldon RA, Schoevaart R, Van Langen LM (2005) Cross-linked enzyme aggregates (CLEAs): a novel and versatile method for enzyme immobilization (a review). Biocatal Biotransformation 23:141–147. https://doi.org/10.1080/10242420500183378

Sinirlioglu ZA, Sinirlioglu D, Akbas F (2013) Preparation and characterization of stable cross-linked enzyme aggregates of novel laccase enzyme from *Shewanella putrefaciens* and using malachite green decolorization. Bioresour Technol 146:807–811. https://doi.org/10.1016/j.biortech.2013.08.032

Tanoue R, Nomiyama K, Nakamura H et al (2015) Uptake and tissue distribution of pharmaceuticals and personal care products in wild fish from treated-wastewater-impacted streams. Environ Sci Technol 49:11649–11658. https://doi.org/10.1021/acs.est.5b02478

Touahar IE, Haroune L, Ba S et al (2014) Characterization of combined cross-linked enzyme aggregates from laccase, versatile peroxidase and glucose oxidase, and their utilization for the elimination of pharmaceuticals. Sci Total Environ 481:90–99. https://doi.org/10.1016/j.scitotenv.2014.01.132

Tufvesson P, Fu W, Jensen JS, Woodley JM (2010) Process considerations for the scale-up and implementation of biocatalysis. Food Bioprod Process 88:3–11. https://doi.org/10.1016/j.fbp.2010.01.003

Varga B, Somogyi V, Meiczinger M et al (2019) Enzymatic treatment and subsequent toxicity of organic micropollutants using oxidoreductases – a review. J Clean Prod 221:306–322. https://doi.org/10.1016/j.jclepro.2019.02.135

Velasco-Lozano S, López-Gallego F, Mateos-Díaz JC, Favela-Torres E (2016) Cross-linked enzyme aggregates (CLEA) in enzyme improvement – a review. Biocatalysis 1. https://doi.org/10.1515/boca-2015-0012

Verlicchi P, Al Aukidy M, Zambello E (2012) Occurrence of pharmaceutical compounds in urban wastewater: removal, mass load and environmental risk after a secondary treatment – a review. Sci Total Environ 429:123–155. https://doi.org/10.1016/j.scitotenv.2012.04.028

Vršanská M, Voběrková S, Jiménez Jiménez AM et al (2018) Preparation and optimisation of cross-linked enzyme aggregates using native isolate white rot fungi *Trametes versicolor* and *Fomes fomentarius* for the decolourisation of synthetic dyes. Int J Environ Res Public Health 15. https://doi.org/10.3390/ijerph15010023

Wang M, Jia C, Qi W et al (2011) Porous-CLEAs of papain: application to enzymatic hydrolysis of macromolecules. Bioresour Technol 102:3541–3545. https://doi.org/10.1016/j.biortech.2010.08.120

Wang M, Qi W, Su R, He Z (2015) Advances in carrier-bound and carrier-free immobilized nanobiocatalysts. Chem Eng Sci 135:21–32. https://doi.org/10.1016/j.ces.2015.03.051

Wilcox J, Nasiri F, Bell S, Rahaman MS (2016) Urban water reuse: a triple bottom line assessment framework and review. Sustain Cities Soc 27:448–456. https://doi.org/10.1016/j.scs.2016.06.021

Wilkinson J, Hooda PS, Barker J et al (2017) Occurrence, fate and transformation of emerging contaminants in water: an overarching review of the field. Environ Pollut 231:954–970. https://doi.org/10.1016/j.envpol.2017.08.032

Yang J, Xu X, Yang X et al (2016) Cross-linked enzyme aggregates of *Cerrena* laccase: preparation, enhanced NaCl tolerance and decolorization of Remazol brilliant blue reactive. J Taiwan Inst Chem Eng 65:1–7. https://doi.org/10.1016/j.jtice.2016.04.025

Yusof F, Khanahmadi S (2019) Carrier-free enzyme immobilization by cross-linked enzyme aggregates (CLEA) technology. In: Multifaceted protocol in biotechnology. Springer, Singapore, pp 93–107

Zdarta J, Meyer AS, Jesionowski T, Pinelo M (2019) Research review paper multi-faceted strategy based on enzyme immobilization with reactant adsorption and membrane technology for biocatalytic removal of pollutants: a critical review. Biotechnol Adv 37:107401. https://doi.org/10.1016/j.biotechadv.2019.05.007

Laccase-Assisted Cues: State-of-the-Art Analytical Modalities for Detection, Quantification, and Redefining "Removal" of Environmentally Related Contaminants of High Concern

Lizeth Parra-Arroyo, Roberto Parra-Saldivar, Ricardo A. Ramirez-Mendoza, Tajalli Keshavarz, and Hafiz M. N. Iqbal

Contents

Abstract This chapter focuses on state-of-the-art analytical modalities with reference to laccase-assisted cues to detect, quantify, and redefine "removal" of emerging contaminants (ECs) of high concern. The controlled or uncontrolled discharge and long-term persistence of ECs such as micro-pollutants, endocrine disruptors (EDs), pharmaceuticals, hormones, and toxins all pose environmental threats and considerable risks to human health and aquatic life, alike. Thus, there is a dire need to engineer new tools and adopt strategic measures since the conventional wastewater treatment technologies are proved ineffective to detect or remove ECs. The current boom in biotechnology and environmental engineering offers opportunities to engineer biosensing and biocatalytic prototypes, for various applications, to yield requisite measurements to meet specific requirements. Such prototypes can further help in real-time and in situ monitoring of environmental pollutants. Besides that, on-site monitoring can also reduce the excessive consumption of potentially hazardous chemicals and harsh reagents prior to release into the environment. Thus, the current

L. Parra-Arroyo · R. Parra-Saldivar · R. A. Ramirez-Mendoza · H. M. N. Iqbal (✉)
Tecnologico de Monterrey, School of Engineering and Sciences, Monterrey, Mexico
e-mail: hafiz.iqbal@tec.mx

T. Keshavarz
School of Life sciences, College of Liberal Arts and Sciences, University of Westminster, Westminster, London, UK

D. Schlosser (ed.), *Laccases in Bioremediation and Waste Valorisation*, Microbiology Monographs 33, https://doi.org/10.1007/978-3-030-47906-0_7

advancement in designing and developing biosensing prototypes and biocatalytic cues have gained high research interest and inevitably focused on enzymes. Laccase-assisted sensing prototypes and biocatalytic cues have gained attention and can be an excellent fit for monitoring, quantification, and removal of ECs. This chapter extend the current knowledge on new modalities to detect, quantify, and treat the samples efficiently. Useful information is also given on laccase-assisted sensors, immobilized laccase-based cues, and their role in the pollution detection and redefining "removal" processes. In-depth understanding of critical emergencies and their needful alternatives to preserve the inherent physicochemical, structural, and functional features of the engineered sensing cues is essential for success.

1 Introduction: Problem Statement and Opportunities

Around the globe, environmental pollution concern has been intensified in the past several years. This is mainly because of the controlled or uncontrolled release/disposal of various types of hazardous elements/compounds into all key spheres, including the atmosphere, hydrosphere, biosphere, lithosphere, and anthroposphere, for instance, the release of toxic heavy elements, toxic greenhouse gases, hazardous contaminants from the various industrial sectors, persistence of synthetic dyes and dye-based wastewater effluents, and emerging contaminants (ECs) of high concern such as endocrine disruptors (EDs), pharmaceuticals, hormones, and toxins, all posturing serious threats to the environment (Gavrilescu 2010; Bilal et al. 2017; Barrios-Estrada et al. 2018a; Rasheed et al. 2018a, b; Bilal and Iqbal 2019a; Bilal et al. 2019a; Rasheed et al. 2019c, b). The term ECs has been broadly defined as "contaminants of emerging concerns which have appeared in the environmental matrices recently either due to anthropogenic inputs, i.e., human-made activities, or re-emergence from the contaminants which have been persistent in the environment for a long time." Owing to this rising issue of environmental pollution, people are now more concern about their living ecosystem. Thus, there is a huge demand for legislative authorities to make environmental friendlier policies and pollutant handling regulations to tackle environmental insecurity effectively.

ECs of high concern are a vast and expanding array of anthropogenic compounds that include synthetic chemicals used in various industrial practices, around the globe (Hernandez-Vargas et al. 2018; Rasheed et al. 2019a). Most of such ECs are commonly present in water, but only recently has been identified as significant water pollutants (Gomes et al. 2018). These contaminants contain a variety of extensively used synthetic compounds that are indispensable in modern society around the globe. It is estimated that the worldwide production of these contaminants has been increased from 1 million to 500 million tons per year (Thomaidis et al. 2012). However, these naturally or synthetically occurring substances (i.e., ECs) have been associated with exerting known or suspected adverse consequences on human's health and the environment (Fig. 1) (Rasheed et al. 2019a). Life-cycle

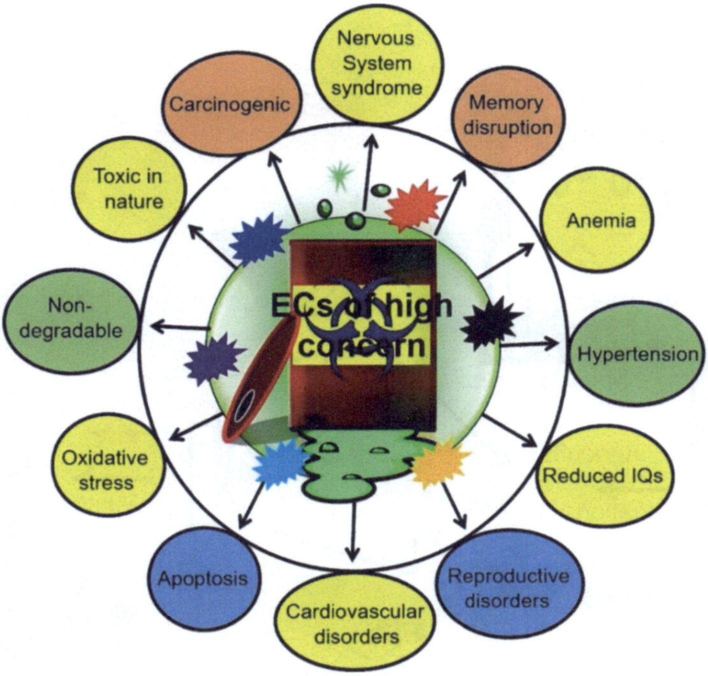

Fig. 1 Major consequences and adverse effects of ECs of high concern on human's health and the environment. Reprinted from Environment International, 122, Rasheed T, Bilal M, Nabeel F, Adeel M & Iqbal HM, Environmentally-related contaminants of high concern: potential sources and analytical modalities for detection, quantification, and treatment, 52–66, copyright © 2018, with permission from Elsevier

distribution of emerging (micro)-pollutants from sources to receptors viewpoint is shown in Fig. 2 (Rasheed et al. 2019a).

In this context, both the European Commission (EC) and the United States Environmental Protection Agency (USEPA) have developed lists and classifications of pollutants of emerging concern, regardless of source and type, as discussed above with suitable examples including EDs (EPA 1979; USEPA). However, there are different classifications of these contaminants based on their usage, origin, and effects, for instance, (1) pesticides, (2) nanomaterials, (3) phthalates, (4) additives to plastics, (5) non-halogenated compounds, (6) personal care products, (7) fluorinated compounds, (8) chlorinated paraffin (flame retardants, sealants, plastic additives), (9) synthetic musks, (10) brominated compounds, (11) phytoestrogens, and (12) pharmaceuticals (Barrios-Estrada et al. 2018b; Bilal and Iqbal 2019a).

Most common methods include spectrophotometric and chromatography techniques to detect and determine these pollutants. However, to identify and quantify with high accuracy over a broad variety of pollutants at the trace level (Marco and Barcelo 1996) remains challenging. Recent researches in monitoring techniques are mainly focused on bioanalytical tools, such as biosensors, which offer advantages

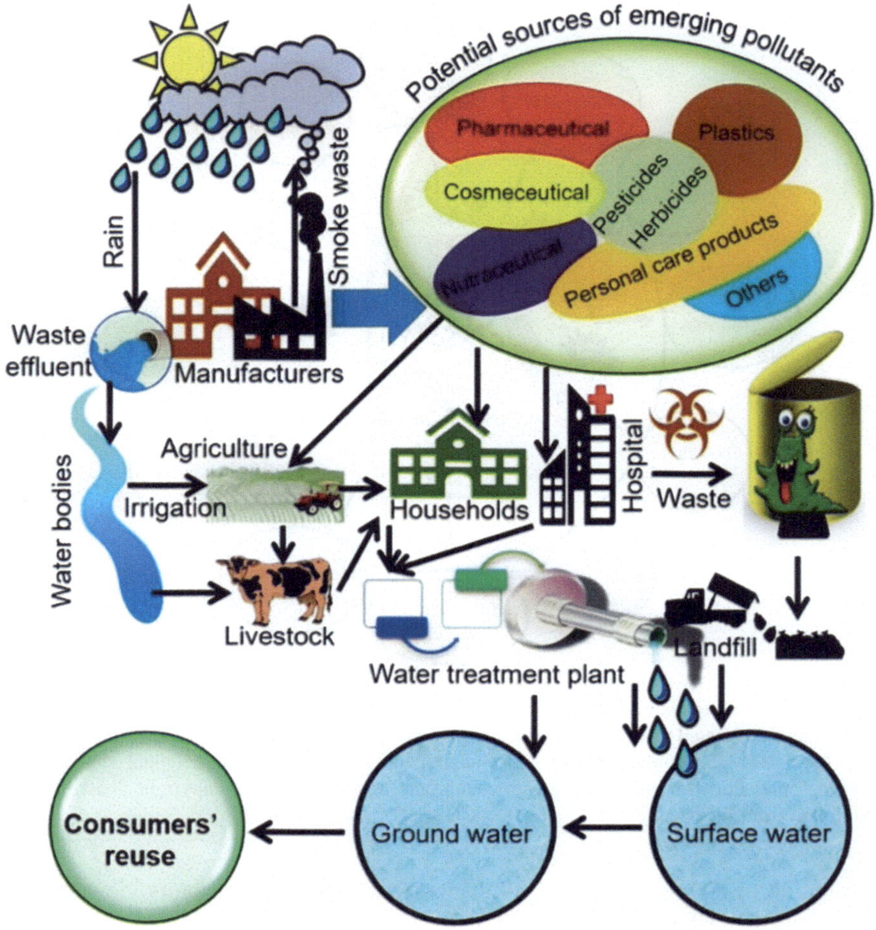

Fig. 2 Life-cycle distribution of emerging (micro)-pollutants from sources to receptors viewpoint. Potential EC sources and pathways of ground and surface water pollution. Reprinted from Environment International, 122, Rasheed T, Bilal M, Nabeel F, Adeel M & Iqbal HM, Environmentally-related contaminants of high concern: potential sources and analytical modalities for detection, quantification, and treatment, 52–66, copyright © 2018, with permission from Elsevier

over classical analytical techniques in terms of selectivity, sensitivity, short assay times, and reduced cost of analysis (Marco and Barcelo 1996). The development and exploitation of enzyme-based biosensing and biocatalytic prototypes has increased in current research activities due to specific and peculiar properties of proteins such as laccase. Enzymes are versatile, efficient, and specific catalysts acting behind all chemical reactions occurring under mild conditions (Cajthaml et al. 2009). Many biosensor research papers have reported previously the detection of phenolic compounds based on several types of enzymes such as tyrosinase (Jaafar et al. 2005; Abdullah et al. 2006a) and horseradish peroxidase (Abdullah et al. 2006b).

Nevertheless, both present some disadvantages, tyrosinase suffer low stability and significant inhibition by reaction products and horseradish peroxidase needs the presence of hydrogen peroxide to complete its catalytic function (Freire et al. 2002). On the other hand, laccase appears as a strong candidate for biosensing and biocatalytic applications providing some specific advantages over other enzymes, like their ability to catalyze electron transfer reactions without additional cofactors and to oxidize phenols and *o*-, *m*-, and *p*-benzenediol compounds in the presence of molecular oxygen with good stability (Munteanu et al. 1998; Baldrian 2006).

Herein, this chapter gives a comprehensive description on ECs and their impact on the environment and human health, especially, those are very soluble in water. In addition, an effort has been made to compile salient information to extend the current knowledge on novel modalities to detect, quantify, and treat ECs efficiently. Risk assessment, detection, prevention, and regulatory perspectives are discussed in the first half. In the second half of the work, useful information is also given on laccase-assisted sensing cues, immobilized laccase-based biocatalytic cues, and their role in the pollution detection and redefining "removal" processes. Toward the end, potential research gaps, future perspectives, and recommendations are given that can help the future studies to present novel potentialities of laccase-assisted biosensing and biocatalytic cues for their proper implementation against a broader spectrum of ECs of high concern.

2 Emerging Contaminants (ECs): A Threatening Concern

ECs are present in the environment, though in pristine and mix forms (which are even more complicated to tackle using a single strategy), and they act synergistically, making it hard to predict their effects on aquatic organisms and human health. Therefore, the relationship between exposure to ECs and health cannot be modeled linearly. EDCs have been associated with disorders such as breast and prostate cancer, malformations in the reproductive tract, cognitive deficiencies, reduced sperm count, obesity, type II diabetes, cardiovascular disease, early puberty, and thyroid disorders (Barrios-Estrada et al. 2018a). Interference of EDCs with the biosynthesis, transportation, metabolism, and exclusion of hormones is thought to be the underlying mechanism of endocrine disruption reducing the level of natural hormones (Mnif et al. 2007; Hashim et al. 2018). Bisphenol A (BPA) has been associated with diabetes, cancer, obesity, and reproductive, renal respiratory, thyroid, developmental immunotoxicity, and autoimmune diseases (Bilal et al. 2018c). BPA binds to estrogen and thyroid receptors, affecting estrogen-sensitive genes and possibly influencing immune cell populations. BPA has shown effects on brain development in rats, and prairie vole Tributyltin compounds promote the development of marine organisms' genitalia (DeWitt and Patisaul 2018). Studies on zebrafish have revealed that the fungicide, imazilil, decreases zebrafish microbiota and affects their hepatic metabolism; zebrafish gather few layers graphene in their gut. Triclosan (TCS) affected the fatty acid synthesis and decreased zebrafish microbiota; and bisphenol A prompted male feminization (Evariste et al. 2019). Benzophenones, the active ingredient in sunscreens, and present in other cosmetics and hygiene products, have shown to alter estrogenic, androgenic, progesterone, and thyroid hormone reception functions in several in vitro and in vivo assays. Glyphosates can harm the liver, kidney, and embryological development. Polyhalogenated aromatic hydrocarbons have immunotoxic effects and lead to increased susceptibility to infections such as otitis media (DeWitt and Patisaul 2018; Pontelli et al. 2019).

Formaldehyde and paraformaldehyde and 1,4-dioxane ($C_4H_8O_2$, dioxane) are commonly found in PCPs, such as shampoo, toothpaste, and mouthwash (Juhász and Marmur 2014). All are potentially known for various types of human health-related vulnerabilities. For instance, the ingestion/consumption of 1,4-dioxane at levels beyond animal's threshold mechanisms can act as a potent carcinogen to cause breast cancer, skin cancer, and liver cancer (Juhász and Marmur 2014). Figure 3 illustrates a schematic representation of potential transmission routes, presence, and human health-related vulnerabilities of 1,4-dioxane (Bilal and Iqbal 2019a). 1,4-Dioxane increases the foaming capabilities of related products; however, it is not a listed component in cosmetic products. 1,4-Dioxane is a by-product contaminant from the manufacturing ethoxylation step, forming other ingredients such as polyethylene, polyethylene glycol, and polyoxyethylene (FDA 2019). Formaldehyde and paraformaldehyde containing/releasing compounds are another important class of broadly employed preservatives in cosmeceuticals and PCPs such as liquid soaps, shampoos, and shower creams/lotions (Lv et al. 2015; Halla et al. 2018).

3 Risk Assessment, Detection, Prevention, and Regulation

Research over localization and measurement of ECs of high concern is increasing in order to assess the risk of such ECs and possibly speed up the policy decision-making regulations (Naidu et al. 2016). So far, an array of methods has been developed and implemented to access the risk assessment of ECs. Among them, a method widely used to assess risks caused by ECs is the calculation of a hazard quotient (HQ). The HQ consists of dividing the estimated concentration of exposure by a toxicity reference value, mainly the threshold at which a chronic risk begins; most times the no-observed-effect concentration (NOEC) or concentration of concern (COC) is used. Potential risks are obtained when the predicted or measured concentration of exposure is higher than the NOEC or COC. If HQ is higher than one, then the level of certainty with which the evaluation was done is further assessed to confirm that there is a possibility of adverse effects. Furthermore, to understand the risk stochastically, probabilistic risk assessments are used where models of the probabilities of reaching an exposure beyond the accepted parameter are used and with enough data can lead to calculations of risk particular to a species (Woodburn et al. 2018).

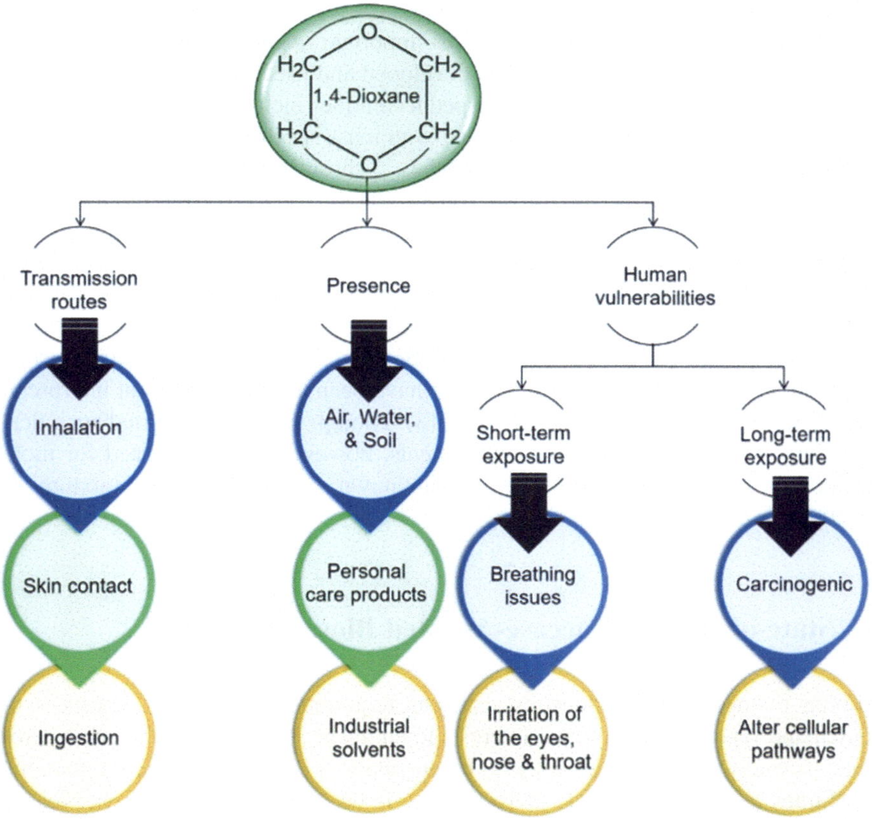

Fig. 3 The chemical structure, potential transmission routes, presence, and human health-related vulnerabilities of 1,4-dioxane. Reprinted from Science of The Total Environment, 670, Bilal M & Iqbal HM, An insight into toxicity and human-health-related adverse consequences of cosmeceuticals—A review, 555–568, copyright © 2019, with permission from Elsevier

Priority ECs are monitored and managed worldwide. The USA has programs funded by USGS, NOAA, USEPA, and the Department of Defense, while the European Union has its corresponding projects (Lapworth et al. 2012). The US Food and Drug Administration demands ecological testing of drug proposals in case it would enter the environment in concentrations higher than 1 ppb. The US Environmental Protection Agency (USEPA) created a Contaminant Candidate List that would be updated every 5 years in order to regulate the compounds' effects on human health and possible endocrine activity in addition to numerous other parameters.

Considering the above-discussed regulatory measures, effective recognition and assessment of ECs have turned into a key scientific task. However, this demands highly sensitive analytical techniques to detect ECs at nanograms per liter (ng/L) scales (Thomaidis et al. 2012). Therefore, the development of highly reliable, efficient, sensitive, fast, and responsive analytical tools for precise monitoring and determination of widespread ECs is utmost important (Agüera et al. 2013). So far, a wide array of analytical tools has been developed and used to detect ECs in various water bodies. Most of the earlier developed techniques include but are not limited to the gas or liquid chromatography coupled with mass spectroscopy. As compared to the simple chromatographic approaches, chromatography coupled mass spectroscopy-based tools have shown more reliability and precision in their detection process. This is mainly because of their high sensitivity, specificity, and selectivity (Agüera et al. 2013), to detect ECs present in complex matrix samples of wastewater or surface water. However, further advanced techniques, e.g., ion trap and triple quadrupole, are even efficient to detect ECs at ng/L scale. In addition, triple quadrupole, linear ion traps quadrupole, quadrupole-time of flight, triple quadrupole, and quadrupole-linear ion trap have been used for structural interpretation of transformed products (Nikolaou 2013). Though the above-reported analytical tools have achieved significant improvements, however, there is a need for more sensitive recognition and detection of unknown compounds or a mixture of several ECs.

4 State-of-the-Art Laccase-Assisted Biosensing Cues

Aiming to meet current industrial catalysts requirements, laccases regardless of source and types are of supreme interests to engineer robust, recyclable, or heterogeneous catalysts. Moreover, enzyme-based catalysis cues are considered potential substitutes to chemical-based catalysis. This is because the use of enzyme-based catalysis offers various catalytically requisite features such as catalytic efficiency, high specificity, selectivity, stability, and mild reaction conditions (Bilal et al. 2017; Zdarta et al. 2018; Bilal and Iqbal 2019b, c). Despite such enormous potentialities, laccases in free forms suffer from some limitations that in turn hinders their industrial exploitability. For instance, free enzyme forms have drawbacks such as low stability under different reaction processes, inhibition, and non-reusability, among others. However, with current biotechnological advancements at large and in protein engineering in particular, catalytic properties that include high catalytic yield, high stability extreme pH and elevated temperatures environment, overall reusability/recyclability, and prolonged shelf-life of enzymes can be improved via enzyme modification and tailoring techniques such as immobilization, genetic engineering approaches, and directed evolution technology (Bilal et al. 2018a; Bilal and Iqbal 2019b; Bilal et al. 2019c). Among various enzyme immobilization approaches, covalent attachment to a support material offers numerous benefits such as enzyme fixation via chemical linkage, minimum leaching, and negligible product contamination issues (Urrutia et al. 2018; Bilal and Iqbal 2019b). Additionally, multipoint covalent immobilization between the support and enzyme also deliver high stabilization properties (Bilal et al. 2019c). Furthermore, the covalent coupling to a solid carrier allows an increase in its resilience against extreme pH or thermal inactivation,

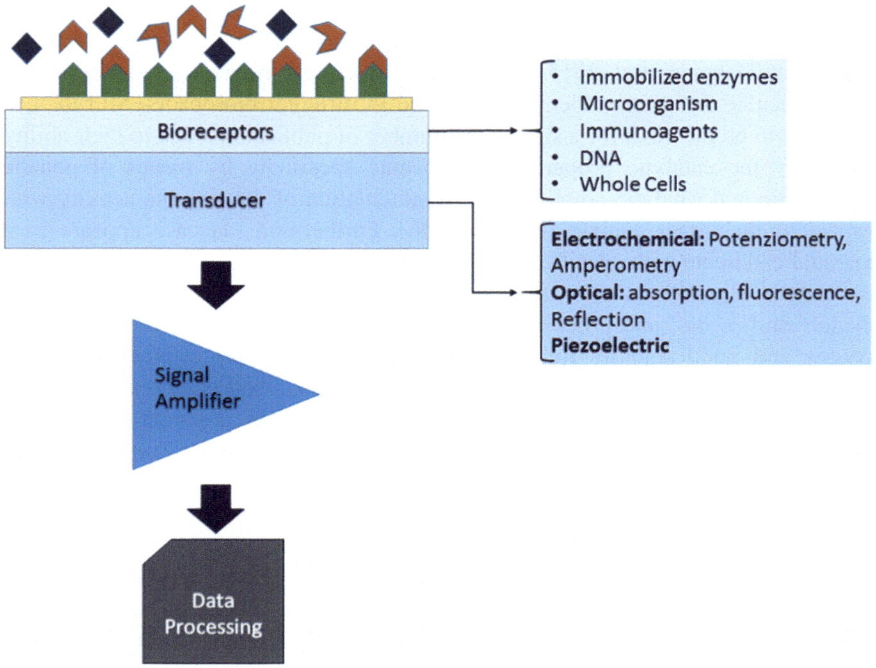

Fig. 4 A simplified schematic representation of various bioreceptors that includes immobilized enzymes and transducer-based biosensing prototype. Modified from Rodríguez-Delgado et al. (2015; https://doi.org/10.1016/j.trac.2015.05.008), an open-access article under the CC-BY license (https://creativecommons.org/licenses/by/4.0/). Copyright © 2015 The Authors. Published by Elsevier

although the immobilization may lead to a significant activity loss of the original enzyme (Durán et al. 2002; Bilal et al. 2017). Enzyme modification through different immobilization strategies is considered as the most effective and straightforward method in green biotechnology allowing for the reuse of laccase, facile recovery, longer half-lives, stabilizing enzyme activity and structure, and diminished proteases activity (Bilal et al. 2018c, b). In addition to catalysis, other useful advantages related to sensing cues can also be obtained from an efficient protocol of immobilization, such as prolonged use of the sensor and anticipated extended storage and working stability (Liu et al. 2006; Rodríguez-Delgado et al. 2015). For such purposes, the immobilization support should be inert, stable, and resistant to mechanical force, also considering shape, distribution, and pore size and expandability. Since stability, selectivity, and activity of an enzyme are obtained by combining immobilization techniques and the proper selection of the support (Fernández-Fernández et al. 2013). For example, the method by which an enzyme is immobilized at an electrode surface is a critical factor to establish an efficient electron transference between the enzyme and the electrode surface (Chen et al. 2015).

Figure 4 illustrates a simplified schematic representation of various bioreceptors that includes immobilized enzymes and transducer-based biosensing prototype (Rodríguez-Delgado et al. 2015). From the historical perspectives, enzymes are the first molecular recognition elements included in different biosensors. Still then, it continues to be the basis for a significant number of publications due to their ability to modify the catalytic properties or substrate specificity by means of genetic engineering and catalytic amplification by modulation of the enzyme activity with respect to the target analyte (Rogers 2006). Furthermore, laccase appears as a potential candidate with specific advantages over other enzymes, such as the ability to catalyze electron transfer reactions without additional cofactors and oxidize phenols and *o*-, *m*-, and *p*-benzenediols compounds in the presence of molecular oxygen and good stability for biosensing applications (Munteanu et al. 1998). Moreover, other characteristics such as good mechanical and electronic properties along with notable ability to transfer electrons make laccase a potent candidate to develop immobilized laccase-based sensing cues. In this context, Fernandes et al. (2008) constructed a biosensor based on laccase immobilized on microspheres of chitosan cross-linked with tripolyphosphate by spray drying method. The developed laccase-based sensor was then used for rutin determination in pharmaceutical formulations, obtaining a bioelectrode with high sensitivity, good reproducibility, low detection, and rapid response (Fernandes et al. 2008). Likewise, Santhiago and Vieira (2007) developed a biosensor based on a carbon paste electrode modified with a fungal laccase from *Aspergillus oryzae* for the determination of L-cysteine in pharmaceutical formulations, performed in the presence of hydroquinone (Santhiago and Vieira 2007). Gupta et al. (2003) used *Coriolus hirsutus* laccase immobilized onto the amine-terminated thiol monolayers on the gold electrode to monitoring catechol (Gupta et al. 2003). Very recently, Kavetskyy et al. (2019) engineered an amperometric laccase biosensor using enzyme-immobilized gold nanoparticles coupling with ureasil polymer as a host matrix. A novel sensitive laccase biosensor using gold nanoparticles and poly-L-arginine to detect catechol in natural water is reported by Maleki et al. (2019). In the same study, the authors have used simple, green, and fast layer-by-layer modification of the glassy carbon electrode which was mainly performed by electrodeposition of gold nanoparticles and then poly-L-arginine, and finally, laccase was covalently bonded to poly-L-arginine using glutaraldehyde. A detailed literature overview of laccase biosensors in terms of analytical characteristics, immobilization type, applications, and other requisite features is provided in Table 1 (Rodríguez-Delgado et al. 2015).

5 Research Gaps and Future Considerations

In spite of current biotechnological advancement in different research sectors including environmental engineering, biocatalysis, biosensors, and others, many unsolved questions are posing a big research gap that must be tackled comprehensively. For instance, the unsatisfactory detection procedures, malpractices,

Table 1 A detailed literature description of laccase biosensors in terms of analytical characteristics, immobilization type, applications, and other requisite features

Laccase sources	Laccase characteristics	Immobilization method	Measurement conditions	Electrode	Analyte	Analytical characteristics	Application	Real Samples	References
Aspergillus oryzae	Commercial laccase; 0.55 units/mg	Encapsulated in microspheres of chitosan cross-linked with tripolyphosphate	Acetate buffer pH 4.0; frequency 30 Hz; pulse amplitude 30 mV; and scan increment 2.0 mV	Printed graphite electrode (PGE), Ag/AgCl reference electrode, and platinum wire as auxiliary electrode	Rutin	Two linear range 0.599 to 3.92 μM 5.82 to 13.1 μM	Pharmaceutical analysis	Pharmaceutical formulations	Fernandes et al. (2008)
Aspergillus oryzae	Commercial laccase genetically modified; Denilite® 800 U/g	Adsorption graphite powder–Nujol–Pt. BMIPF $_6$	Phosphate buffersolution (0.1 M, pH 6.5); frequency 20 Hz; pulse amplitude 80 mV; scan 5.0 mV	Pt-BMI.PF6-laccase, Ag/AgCl reference electrode, and platinum wire as auxiliary electrode	Adrenaline	Linear range 0.999–213 μM	Pharmaceutical analysis	Pharmaceutical formulations	Brondani et al. (2009)
Aspergillus oryzae	Commercial laccase genetically modified; Denilite® 800 U/g	Adsorption graphite powder–Nujol–Pt. BMIPF $_6$	0.1 M acetate buffer solution (pH 5.0). +0.2 V	Pt-BMI.PF6-laccase, Ag/AgCl reference electrode, and platinum wire as auxiliary electrode	Rosmarinic acid	Linear range 0.999–65.4 μM	Food analysis	Plant extract samples	Franzoi et al. (2009b)
Aspergillus oryzae	Commercial laccase 0.29 U/mL	Cross-linked with cyanuric chloride (CC) in chitosan	0.1 M acetate buffersolution (pH 4.0); with frequency 50 Hz; pulse amplitude 100 mV and scan 5.0 mV	Lac nanoparticles-MI.PF_6, and au/Ag electrode	Luteolin	Linear range 0.099–5.825 μM	Food analysis	Chamomile tea samples	Franzoi et al. (2009c)
Aspergillus oryzae	Commercial laccase genetically modified; Denilite® 800 U/g	Adsorption on carbon paste electrode	Performed in the presence of hydroquinone and L-cysteinein 0.1 M phosphate buffer (pH 7.0) at an applied potential of −0.08 V versus Ag/AgCl	Carbon paste electrode, Ag/AgCl reference electrode, and platinum wire as auxiliary electrode	L-cysteine	Linear range 0.197–3.24 mM	Pharmaceutical analysis	Pharmaceutical formulations	Santhiago and Vieira (2007)

(continued)

Table 1 (continued)

Laccase sources	Laccase characteristics	Immobilization method	Measurement conditions	Electrode	Analyte	Analytical characteristics	Application	Real Samples	References
Coriolus hirsutus	Commercial laccase; 40 U/mg	Covalently binding by glutaraldehyde on gold-thiol monolayers	0.1 M acetate buffer at pH 5	Gold electrode, Ag/AgCl reference electrode, and platinum wire as auxiliary electrode	Catechol	Linear range 1 and 400 μM; sensitivity 15 μA/mM	Environmental analysis	Synthetic samples	Gupta et al. (2003)
Trametes versicolor	Commercial laccase	Entrapment using silica spheres asimmobilization matrix	0.10 M PBS solution (pH 5.0)	Lac/Si/MWCNTs/SPE and Ag/AgCl reference electrode	Dopamine	Linear range 1.3 to 85.5 μM; sensitivity 2.787 × 10 3 μA/mM cm	Clinical analysis	Synthetic samples	Li et al. (2012)
Aspergillus oryzae	Commercial laccase genetically modified; Denilite® 800 U/g	Adsorption on cellulose acetate/ $BMI{\cdot}N(Tf)_2$ support.	0.1 M acetatebuffer solution (pH 5.5); frequency 90 Hz, pulse amplitude 100 mV,and scan increment 4.0 mV.	Lac-CA/BMI·N (Tf)2 and Ag/AgCl reference electrode	Methyldopa	Linear range 34.8 to 370.3 μM	Pharmaceutical analysis	Pharmaceutical formulations	Moccelini et al. (2011)
Trametes versicolor	Commercial laccase	Adsorption on multi-walled carbonnanotube-based paste (MWCPE)	47.5 μM of 4-AMP; Britton–Robinson buffer at pH 5; scan rate of 50 mV/s	(MWCPE); Ag/AgClreference electrode and glassy carbon counter electrode	Pirimicarb (4-aminophenol as substrate)	Linear range 0.990–11.5 μM	Food analysis	Vegetables	Oliveira et al. (2013)
Aspergillus oryzae	Commercial laccase; Denilite® II BASE	Adsorption graphite powder–Nujol–ILs 1-butyl-3-methylimidazolium (BMI·Tf 2 N), 1-decyl-3-methylimidazolium (DMI·Tf 2 N) and 1-tetradecyl-3-methylimidazolium (TDMI·Tf 2 N),	0.1 M acetate buffer solution (pH 5.0). +0.2 V	Pt-BMI.PF6-laccase, Ag/AgCl reference electrode, and platinum wire as auxiliary electrode	Rutin detection with BMI·Tf 2 N DMI·Tf 2 N TDMI·Tf 2 N	Linear range 4.77–46.2 μM 5.84–53.6 μM 5.84–53.6 μM	Pharmaceutical analysis	Pharmaceutical formulations	Franzoi et al. (2009a)

Cerrena unicolor	Purified laccase 2.23 mg/ml	Electrolytic deposition under galvanostatic conditions applying current 1 mA	0.1 M phosphate–citrate buffer pH 5.2; scan rate – 100 mV/s	Working platinum electrode; Pt counter electrode and saturated calomel referenceelectrode (SCE)	Hydroquinone	Linear range 2.0–60 μM sensitivity 2.34 ± 0.11 μA/mM	Environmental analysis	Synthetic samples	Jędrychowska et al. (2014)
n.r.	Activity ≥100 units/mg	Entrapment in Nafion matrix	0.1 M acetate buffer solution pH 5.0; scan rate 100 mV/s.	Nafion/laccase/A-CZU-glassy carbon electrode as the working electrode, Pt wirecounter electrode, and Ag/AgCl reference electrode	Catechol	Linear range 0–7 μM	Environmental analysis	Real samples	Chen et al. (2015)
Pleurotus ostreatus	Bi enzyme system lac-peroxidase; crude laccase extracts 15.9 U/mg	Adsorption carbon paste	0.1 M phosphatebuffer solution pH 6.0 to 7.5; scan rate 40 mV/s, potential pulse 50 mV	Nujol/graphite powder laccase/peroxidase as working electrode, Ag/AgCl reference electrode, and platinumauxiliary electrodes	Dopamineadrenaline L-dopa isoprenaline	Linear range 6.6–390 μM 6.1–100 μM 6.7–70 μM 6.2–81 μM	Pharmaceutical analysis	Synthetic samples	Leite et al. (2003)
Cerrena unicolor	Purified laccase; 1.62 mU/10 μl	Adsorption in graphiteelectrode	0.1 M citrate buffer pH 5.5; applied potential: – 50 mV.	Graphite working electrode and Ag/AgCl reference electrode	Caffeic acid Prodelphinidin B3 Epicatechin gallate Catechin Epicatechin	Linear range 1–10 μM 1–10 μM 1–10 μM 4–40 μM 2–60 μM	Food analysis	Synthetic samples	Jarosz-Wilkołazka et al. (2004)

Modified from Rodríguez-Delgado et al., (2015; https://doi.org/10.1016/j.trac.2015.05.008), an open-access article under the CC-BY license (https://creativecommons.org/licenses/by/4.0/). Copyright © 2015 The Authors. Published by Elsevier

$BMIPF_6$, 1-n-butyl-3-methylimidazolium hexafluorophosphate
$BMIBF_4$, 1-n-butyl-3-methylimidazolium tetrafluoroborate
$BMI{\cdot}Tf_2N$, 1-butyl-3-methylimidazolium
$DMI{\cdot}Tf_2N$, 1-decyl-3-methylimidazolium
$TDMI{\cdot}Tf_2N$, 1-tetradecyl-3-methylimidazolium

misconducts, and limitations of in practice technologies greatly affecting the detection and redefining "removal" fate of heavily increasing ECs of high concern. The role of low-risk contaminants in the emergence of new hazardous pollutants should be addressed in future studies. It could be useful to involve the multidisciplinary scientists, technical cooperation, policymakers, and stakeholders to strengthen the EC detection, degradation, and efficient removal at the global level.

From the future development perspective, the use of a synergistic approach could also be a noteworthy solution to define and design impartial tactics to remove and/or degrade environmentally related emerging (micro)-contaminants of high concern. Aiming to get a deeper insight, the influential aspects of numerous sensing devices and removal strategies should be covered in future studies. Another interesting research avenue in the degradation/removal of ECs is the search for novel materials and bio-nano-catalysts following green agenda to establish highly effective and environmentally friendly treatment processes.

6 Concluding Remarks and Outlook

In conclusion, the presence and persistence of environmentally related contaminants in different mediums is a serious ecological concern worldwide. This serious issue is progressively rising with rising industrial development, around the globe. Based on this human-made industrialization and heavy consumption of harsh chemicals during various industrially relevant manufacturing processes, the continuous discharge of toxic compounds with/without partial or insufficient treatments is among a major a cause of environmental contamination. Such careless practices adversely affect the both terrestrial and aquatic systems, alike. To tackle this issue more effectively, many earlier studies and ongoing research efforts have focused on developing new tools aiming to detect, quantify, degrade, and remove a broader range of contaminants from different environments. So far, various strategies have been proposed and developed including laccase-assisted biosensing and bioremediation. A wide array of pollutants has been effectively treated and removed using laccase-based systems. The literature evidenced that EC detection and removal using laccase-based biosensing and biocatalytic cues are promising measures to treat various contaminated matrices. However, owing to any or many reasons, such sensing and catalytic systems are far behind from their real-time and in situ implementation in current industrial sector. Despite several available and in practice solutions to operate bioremediation processes, though using different materials and strategies, numerous challenges are still pending that needs to be addressed. For instance, much critiques are being posed on the (re)-emergence of threatening pollutants, distribution profile of ECs, efficient and "one-pot" detection and degradation fate, and reliable and consistent on-site monitoring of ECs are still outstanding and need to be addressed in the future studies. Future studies must be conducted to optimize the laccase-related parameters for efficient degradation and stringently validated technological solutions. Considering the potent features of laccase, it is equally important to design and engineer laccase-based measuring procedures that can precisely detect and redefine "remove" various contaminants of emerging concern from a broader spectrum. Moreover, a deeper insight and advanced analytical measures are utmost import to understand the degradation pathways involved in the ECs removal following efficient detection by laccase-assisted systems.

Acknowledgments Authors are grateful to their representative institutes and universities for providing the literature services.

Conflict of Interest Authors declare no conflict of interests in any capacity, including competing or financial.

References

Abdullah J, Ahmad M, Heng LY, Karuppiah N, Sidek H (2006b) Stacked films immobilization of MBTH in nafion/sol-gel silicate and horseradish peroxidase in chitosan for the determination of phenolic compounds. Anal Bioanal Chem 386(5):1285–1292

Abdullah J, Ahmad M, Karuppiah N, Heng LY, Sidek H (2006a) Immobilization of tyrosinase in chitosan film for an optical detection of phenol. Sens Actuator B Chem 114(2):604–609

Agüera A, Bueno MJM, Fernández-Alba AR (2013) New trends in the analytical determination of emerging contaminants and their transformation products in environmental waters. Environ Sci Pollut Res 20(6):3496–3515

Baldrian P (2006) Fungal laccases–occurrence and properties. FEMS Microbiol Rev 30(2):215–242

Barrios-Estrada C, de Jesús Rostro-Alanis M, Muñoz-Gutiérrez BD, Iqbal HM, Kannan S, Parra-Saldívar R (2018a) Emergent contaminants: endocrine disruptors and their laccase-assisted degradation–a review. Sci Total Environ 612:1516–1531

Barrios-Estrada C, de Jesús Rostro-Alanis M, Parra AL, Belleville MP, Sanchez-Marcano J, Iqbal HM, Parra-Saldívar R (2018b) Potentialities of active membranes with immobilized laccase for Bisphenol A degradation. Int J Biol Macromol 108:837–844

Bilal M, Adeel M, Rasheed T, Zhao Y, Iqbal HM (2019c) Emerging contaminants of high concern and their enzyme-assisted biodegradation–a review. Environ Int 124:336–353

Bilal M, Asgher M, Cheng H, Yan Y, Iqbal HMN (2019b) Multi-point enzyme immobilization, surface chemistry, and novel platforms: a paradigm shift in biocatalyst design. Crit Rev Biotechnol 39(2):202–219

Bilal M, Asgher M, Parra-Saldivar R, Hu H, Wang W, Zhang X, Iqbal HMN (2017) Immobilized ligninolytic enzymes: an innovative and environmental responsive technology to tackle dye-based industrial pollutants–a review. Sci Total Environ 576:646–659

Bilal M, Iqbal HM (2019a) An insight into toxicity and human-health-related adverse consequences of cosmeceuticals—a review. Sci Total Environ 670:555–568

Bilal M, Iqbal HMN (2019b) Naturally-derived biopolymers: potential platforms for enzyme immobilization. Int J Biol Macromol 130:462–482

Bilal M, Iqbal HMN (2019c) Chemical, physical, and biological coordination: an interplay between materials and enzymes as potential platforms for immobilization. Coord Chem Rev 388:1–23

Bilal M, Iqbal HM, Shuqi G, Hu H, Wang W, Zhang X (2018b) State-of-the-art protein engineering approaches using biological macromolecules: a review from immobilization to implementation viewpoint. Int J Biol Macromol 108:893–901

Bilal M, Rasheed T, Iqbal HM, Yan Y (2018c) Peroxidases-assisted removal of environmentally-related hazardous pollutants with reference to the reaction mechanisms of industrial dyes. Sci Total Environ 644:1–13

Bilal M, Rasheed T, Nabeel F, Iqbal HM, Zhao Y (2019a) Hazardous contaminants in the environment and their laccase-assisted degradation–a review. J Environ Manag 234:253–264
Bilal M, Rasheed T, Zhao Y, Iqbal HM, Cui J (2018a) "Smart" chemistry and its application in peroxidase immobilization using different support materials. Int J Biol Macromol 119:278–290
Brondani D, Scheeren CW, Dupont J, Vieira IC (2009) Biosensor based on platinum nanoparticles dispersed in ionic liquid and laccase for determination of adrenaline. Sens Actuator B Chem 140 (1):252–259
Cajthaml T, Křesinová Z, Svobodová K, Möder M (2009) Biodegradation of endocrine-disrupting compounds and suppression of estrogenic activity by ligninolytic fungi. Chemosphere 75 (6):745–750
Chen X, Li D, Li G, Luo L, Ullah N, Wei Q, Huang F (2015) Facile fabrication of gold nanoparticle on zein ultrafine fibers and their application for catechol biosensor. Appl Surf Sci 328:444–452
DeWitt JC, Patisaul HB (2018) Endocrine disruptors and the developing immune system. Curr Opin Toxicol 10:31–36
Durán N, Rosa MA, D'Annibale A, Gianfreda L (2002) Applications of laccases and tyrosinases (phenoloxidases) immobilized on different supports: a review. Enzym Microb Technol 31 (7):907–931
EC (n.d.) European Commission, "Priority substances under the Water Framework Directive, (2000/60/EC) (WFD)" [Online]. Available from http://ec.europa.eu/environment/water/water-dangersub/pri_substances.htm#list. Last accessed: May 19, 2019
EPA (1979) Water-related environment fate of 129 priority pollutants. United States Environmental Protection Agency II. pp. 412–432
Evariste L, Barret M, Mottier A, Mouchet F, Gauthier L, Pinelli E (2019) Gut microbiota of aquatic organisms: A key endpoint for ecotoxicological studies. Environ Pollut 248:989–999
FDA (2019) 1, 4-Dioxane in cosmetics: A manufacturing byproduct. Available from https://www.fda.gov/cosmetics/productsingredients/potentialcontaminants/ucm101566.htm Last accessed 17 May 2019
Fernandes SC, de Oliveira IRW, Fatibello-Filho O, Spinelli A, Vieira IC (2008) Biosensor based on laccase immobilized on microspheres of chitosan crosslinked with tripolyphosphate. Sens Actuator B Chem 133(1):202–207
Fernández-Fernández M, Sanromán MÁ, Moldes D (2013) Recent developments and applications of immobilized laccase. Biotechnol Adv 31(8):1808–1825
Franzoi AC, Dupont J, Spinelli A, Vieira IC (2009b) Biosensor based on laccase and an ionic liquid for determination of rosmarinic acid in plant extracts. Talanta 77(4):1322–1327
Franzoi AC, Migowski P, Dupont J, Vieira IC (2009a) Development of biosensors containing laccase and imidazolium bis (trifluoromethylsulfonyl) imide ionic liquid for the determination of rutin. Anal Chim Acta 639(1–2):90–95
Franzoi AC, Vieira IC, Dupont J, Scheeren CW, de Oliveira LF (2009c) Biosensor for luteolin based on silver or gold nanoparticles in ionic liquid and laccase immobilized in chitosan modified with cyanuric chloride. Analyst 134(11):2320–2328
Freire RS, Duran N, Kubota LT (2002) Development of a laccase-based flow injection electrochemical biosensor for the determination of phenolic compounds and its application for monitoring remediation of Kraft E1 paper mill effluent. Anal Chim Acta 463(2):229–238
Gavrilescu M (2010) Environmental biotechnology: achievements, opportunities and challenges. Dynam Biochem Proc Biotechnol Molecul Biol 4(1):1–36
Gomes IB, Simões LC, Simões M (2018) The effects of emerging environmental contaminants on *Stenotrophomonas maltophilia* isolated from drinking water in planktonic and sessile states. Sci Total Environ 643:1348–1356
Gupta G, Rajendran V, Atanassov P (2003) Laccase biosensor on monolayer-modified gold electrode. Electroanal Int J Dev Fundamen Pract Asp Electroanal 15(20):1577–1583
Halla N, Fernandes I, Heleno S, Costa P, Boucherit-Otmani Z, Boucherit K, Rodrigues AE, Ferreira ICFR, Barreiro M (2018) Cosmetics preservation: a review on present strategies. Molecules 23 (7):1571

Hashim M, Iqbal R, Afsheen S, Fahad AUM, Asghar U, Daman M, Ghazanfar M (2018) Effect of 1, 4-dioxane on rabbit (*Oryctolagus cuniculus*) reproductive hormonal level and histopathology of testes and ovaries. Pure Appl Biol (PAB) 7(2):466–469
Hernandez-Vargas G, Sosa-Hernández J, Saldarriaga-Hernandez S, Villalba-Rodríguez A, Parra-Saldivar R, Iqbal H (2018) Electrochemical biosensors: a solution to pollution detection with reference to environmental contaminants. Biosensors 8(2):29
Jaafar A, Musa A, Lee YH, Nadarajah K, Hamidah S (2005) The use of immobilized tyrosinase enzyme in sol-gel film for optical phenol detection. Sains Malaysiana 34(2):91–94
Jarosz-Wilkołazka A, Ruzgas T, Gorton L (2004) Use of laccase-modified electrode for amperometric detection of plant flavonoids. Enzym Microb Technol 35(2–3):238–241
Jędrychowska A, Cabaj J, Świst A, Sołoducho J (2014) Electrochemical laccase sensor based on 3-methylthiophene/3-thiopheneacetic acid/bis (3, 4-ethylenedioxythiophene)-N-nonylacridone as a new polymer support. J Electroanal Chem 720:64–70
Juhász MLW, Marmur ES (2014) A review of selected chemical additives in cosmetic products. Dermatol Therap 27(6):317–322
Kavetskyy T, Stasyuk N, Smutok O, Demkiv O, Kukhazh Y, Hoivanovych N, Boev V, Ilcheva V, Petkova T, Gonchar M (2019) Improvement of amperometric laccase biosensor using enzyme-immobilized gold nanoparticles coupling with ureasil polymer as a host matrix. Gold Bull:1–7
Lapworth DJ, Baran N, Stuart ME, Ward RS (2012) Emerging organic contaminants in groundwater: a review of sources, fate and occurrence. Environ Pollut 163:287–303
Leite OD, Lupetti KO, Fatibello-Filho O, Vieira IC, Barbosa Ade M (2003) Synergic effect studies of the bi-enzymatic system laccase–peroxidase in a voltammetric biosensor for catecholamines. Talanta 59(5):889–896
Li Y, Zhang L, Li M, Pan Z, Li D (2012) A disposable biosensor based on immobilization of laccase with silica spheres on the MWCNTs-doped screen-printed electrode. Chem Cent J 6(1):103
Liu Y, Qu X, Guo H, Chen H, Liu B, Dong S (2006) Facile preparation of amperometric laccase biosensor with multifunction based on the matrix of carbon nanotubes–chitosan composite. Biosens Bioelectron 21(12):2195–2201
Lv C, Hou J, Xie W, Cheng H (2015) Investigation on formaldehyde release from preservatives in cosmetics. Int J Cosmet Sci 37(5):474–478
Maleki N, Kashanian S, Nazari M, Shahabadi N (2019) A novel sensitive laccase biosensor using gold nanoparticles and poly L-arginine, to detect catechol in natural water. Biotechnol Appl Biochem 66(4):502–509. https://doi.org/10.1002/bab.1746
Marco MP, Barcelo D (1996) Environmental applications of analytical biosensors. Meas Sci Technol 7(11):1547
Mnif W, Pillon A, Balaguer P, Bartegi A (2007) Endocrine xenoestrogenics disrupters: molecular mechanisms and detection methods. Therapie 62(5):369–386
Moccelini SK, Franzoi AC, Vieira IC, Dupont J, Scheeren CW (2011) A novel support for laccase immobilization: cellulose acetate modified with ionic liquid and application in biosensor for methyldopa detection. Biosens Bioelectron 26(8):3549–3554
Munteanu FD, Lindgren A, Émneus J, Gorton L, Ruzgas T, Csöregi E, Ciucu A, van Huystee RB, Gazaryan IG, Lagrimini LM (1998) Bioelectrochemical monitoring of phenols and aromatic amines in flow injection using novel plant peroxidases. Anal Chem 70(13):2596–2600
Naidu R, Jit J, Kennedy B, Arias V (2016) Emerging contaminant uncertainties and policy: the chicken or the egg conundrum. Chemosphere 154:385–390
Nikolaou A (2013) Pharmaceuticals and related compounds as emerging pollutants in water: analytical aspects. Global NEST J 15(1):1–12
Oliveira TM, Barroso MF, Morais S, de Lima-Neto P, Correia AN, Oliveira MB, Delerue-Matos C (2013) Biosensor based on multi-walled carbon nanotubes paste electrode modified with laccase for pirimicarb pesticide quantification. Talanta 106:137–143
Pontelli RC, Souza MC, Fantucci MZ, de Andrade M, Rocha EM (2019) The role of endocrine disruptors in ocular surface diseases. Med Hypotheses 122:157–164

Rasheed T, Bilal M, Nabeel F, Adeel M, Iqbal HM (2019a) Environmentally-related contaminants of high concern: potential sources and analytical modalities for detection, quantification, and treatment. Environ Int 122:52–66

Rasheed T, Bilal M, Nabeel F, Iqbal HM, Li C, Zhou Y (2018a) Fluorescent sensor based models for the detection of environmentally-related toxic heavy metals. Sci Total Environ 615:476–485

Rasheed T, Li C, Bilal M, Yu C, Iqbal HM (2018b) Potentially toxic elements and environmentally-related pollutants recognition using colorimetric and ratiometric fluorescent probes. Sci Total Environ 640:174–193

Rasheed T, Nabeel F, Adeel M, Bilal M, Iqbal HM (2019b) "Turn-on" fluorescent sensor-based probing of toxic hg (II) and cu (II) with potential intracellular monitoring. Biocatal Agric Biotechnol 17:696–701

Rasheed T, Nabeel F, Bilal M, Zhao Y, Adeel M, Iqbal HM (2019c) Aqueous monitoring of toxic mercury through a rhodamine-based fluorescent sensor. Math Biosci Eng 16(4):1861–1873

Rodríguez-Delgado MM, Alemán-Nava GS, Rodríguez-Delgado JM, Dieck-Assad G, Martínez-Chapa SO, Barceló D, Parra R (2015) Laccase-based biosensors for detection of phenolic compounds. TrAC Trends Anal Chem 74:21–45

Rogers KR (2006) Recent advances in biosensor techniques for environmental monitoring. Anal Chim Acta 568(1–2):222–231

Santhiago M, Vieira IC (2007) L-cysteine determination in pharmaceutical formulations using a biosensor based on laccase from *Aspergillus oryzae*. Sens Actuator B Chem 128(1):279–285

Thomaidis NS, Asimakopoulos AG, Bletsou AA (2012) Emerging contaminants: a tutorial mini-review. Glob NEST J 14(1):72–79

Urrutia P, Bernal C, Wilson L, Illanes A (2018) Use of chitosan heterofunctionality for enzyme immobilization: β-galactosidase immobilization for galacto-oligosaccharide synthesis. Int J Biol Macromol 116:182–193

USEPA (n.d.) United States Environmental Protection Agency. Toxic and Priority Pollutants Under the Clean Water Act. [Online]. Available from https://www.epa.gov/eg/toxic-and-priority-pollutants-under-clean-water-act. Last accessed: May 19, 2019

Woodburn KB, Seston RM, Kim J, Powell DE (2018) Benthic invertebrate exposure and chronic toxicity risk analysis for cyclic volatile methylsiloxanes: comparison of hazard quotient and probabilistic risk assessment approaches. Chemosphere 192:337–347

Zdarta J, Meyer AS, Jesionowski T, Pinelo M (2018) Developments in support materials for immobilization of oxidoreductases: a comprehensive review. Adv Coll Interf Sci 258:1–20

Laccase Engineering by Directed and Computational Evolution

Ivan Mateljak, Bernardo Gomez-Fernandez, and Miguel Alcalde

Contents

Abstract Directed evolution is a powerful strategy to tailor enzymes with improved attributes. The use of laboratory evolution is becoming more refined, whereby computational and experimental approaches are being combined so that more effective libraries can be created, producing enzymes with greater biotechnological potential while notably reducing the demands on screening. This chapter summarizes the most recent findings from our laboratory to tailor fungal high-redox potential laccases by bringing together computational approaches with in vitro and in vivo methods for library creation. We focus on four recent case studies of laccase engineering in which different computational algorithms were applied at both the gene and protein levels:

1. A consensus design to identify consensus ancestral mutations

I. Mateljak · B. Gomez-Fernandez
Department of Biocatalysis, Institute of Catalysis, Madrid, Spain

M. Alcalde (✉)
Department of Biocatalysis, Institute of Catalysis, Madrid, Spain

EvoEnzyme S.L., Madrid, Spain
e-mail: malcalde@icp.csic.es

D. Schlosser (ed.), *Laccases in Bioremediation and Waste Valorisation*, Microbiology Monographs 33, https://doi.org/10.1007/978-3-030-47906-0_8

2. SCHEMA-RASPP computation for homologous in vivo recombination to generate laccase chimeras
3. Computer-guided evolution to enhance the laccase redox potential and the activity toward high-redox potential mediators
4. Ancestral sequence reconstruction for the directed evolution of a resurrected basidiomycete laccase toward initiators

1 Introduction

Laccases (benzenediol: oxygen oxidoreductase, EC 1.10.3.2) are versatile multicopper enzymes that catalyze the oxidation of a wide array of compounds using molecular oxygen as final electron acceptor and releasing water as their only by-product. The catalytic structure of laccases is organized as two centers containing four copper atoms: one Cu at the T1 site, where the substrate is oxidized, and the trinuclear Cu cluster containing one T2 and two T3 coppers, where the reduction of molecular oxygen takes place (Alcalde 2007). Laccases are predominantly found in plants, bacteria, and fungi, and based on equilibrium potentiometric titrations of the T1Cu site, they have been classified as low-, medium-, and high-redox potential, ranging from 370 to 790 mV relative to the NHE—normal hydrogen electrode (Shleev et al. 2005). Due to the high-redox potential at the T1Cu site and their minimal catalytic needs, high-redox potential laccases (HRPLs) from basidiomycete white-rot fungi are particularly relevant in biotechnological contexts, with possible applications in fields as varied as bioremediation, the development of bioelectronic devices (biocathodes and biosensors), organic synthesis, the pulp and paper, textile, and food industries (Mate and Alcalde 2017). Laccases enhance their substrate range by using redox mediators that behave as diffusible electron shuttles between the enzyme and their substrates, enabling their oxidation while overcoming steric hindrance and differences in redox potentials (Morozova et al. 2007). Therefore, the industrial applicability of laccases may be expanded through the use of the laccase-mediator system (LMS).

In recent years, the demonstration of the potential of laccases in a range of biotechnological applications has propelled efforts to engineer laccases, including attempts to adapt them to strict industrial standards. Both fungal and bacterial laccases have been subjected to extensive engineering efforts via rational design and directed evolution (Mate and Alcalde 2015). Considerable research has been carried out to improve or introduce specific enzymatic features, including functional heterologous expression (Bulter et al. 2003; Koschorreck et al. 2009; Mate et al. 2010; Camarero et al. 2011); activity on different substrates (Madzak et al. 2005; Zumarraga et al. 2008; Kataoka et al. 2013; Toscano et al. 2013; Vicente et al. 2016; Pardo et al. 2016; Santiago et al. 2016); resistance to high temperatures (Mollania et al. 2011; Scheiblbrandner et al. 2017) or extreme pH (Torres-Salas et al. 2013); and activity in nonnatural environments like organic solvents (Zumarraga et al. 2007; Rasekh et al. 2014), human blood (Mate et al. 2013), or ionic liquids (Liu et al. 2013). The majority of these adaptations have been achieved through directed evolution, the protein engineering strategy pioneered by Prof. Frances H. Arnold and for which she was awarded the Nobel Prize in Chemistry 2018. Directed evolution is indeed an extremely potent strategy to customize enzymes for different purposes, mimicking the processes of natural evolution in the laboratory. Genetic diversity is created by random mutagenesis and/or the recombination of parental genes. Once the mutant libraries are expressed in a suitable host organism, they can be screened to select for specific biochemical traits, applying a selective pressure that is strictly controlled by the researcher. This procedure can be repeated as many times as necessary until the desired properties are attained.

In recent years, combining directed evolution with computational approaches has accelerated the design of customized molecules. In silico predictions made using computational tools can guide the exploration of the vast protein sequence space, significantly easing the experimental burden while opening new avenues in protein design (Mate and Alcalde 2015; Molina-Espeja et al. 2016). In this chapter, we describe the tricks we have used in our most recent studies where we have combined classic and modern computational strategies with directed evolution in order to engineer HRPLs with novel attributes, from enhancing laccase redox potential (Mateljak et al. 2019a) or recreating the first resurrected basidiomycete laccase (Gomez-Fernandez et al. 2020b) to identifying stabilizing ancestral-consensus mutations via consensus design (Gomez-Fernandez et al. 2020a) or generating a family of laccase chimeras enriched in functional and sequence diversity (Mateljak et al. 2019b).

2 Consensus Design of High-Redox Potential Laccases

Consensus design was first applied by Pantoliano and coworkers (Pantoliano et al. 1989), a theory of the fundamentals of consensus being later drawn up by the group of Steipe (Steipe et al. 1994). This technique is an empirical approach that infers the most conserved amino acid for each position from extant, very similar mesophilic proteins, Fig. 1. To retrieve a consensus sequence from a target protein, the first step is to generate a multiple sequence alignment (MSA) from homologous proteins (Edgar 2004; Larkin et al. 2007), ruling out the sequences that are too short or that lack content relative to the majority of the sequences (Sullivan et al. 2011; Durani and Magliery 2013). The final number of sequences can be as few as 20, although more sequences are required to apply some filters (Lehmann et al. 2002; Jackel et al. 2010; Porebski and Buckle 2016). Once a suitable MSA has been obtained, the most conserved amino acid at each position can be considered the consensus residues, even though not all have the same degree of conservation. There are a number of strategies available to discern the most conserved residues and those with biological significance arising from the bias of databases (Applebaum 1996; Magliery and Regan 2005; Socolich et al. 2005; Polizzi et al. 2006; Sullivan et al. 2012). Once the consensus sequence is defined, it can be synthesized, or the most promising conserved positions can be evaluated by including them as mutations in the target enzyme (Wang et al. 1999; Polizzi et al. 2006; Dai et al. 2007; Khersonsky et al. 2012; Sullivan et al. 2011).

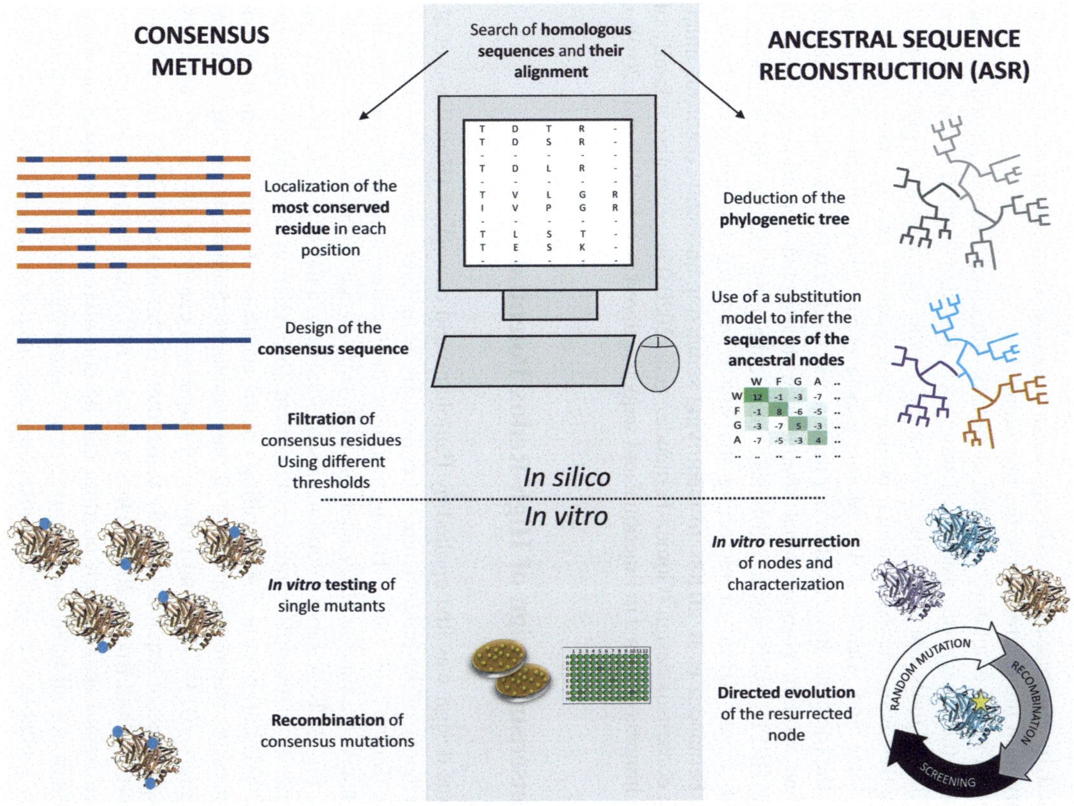

Fig. 1 Comparison between consensus design and ASR. Both methods start from MSAs. **Left:** Consensus sequence is retrieved upon localizing the most conserved residue in each position. Through the application of different metrics, consensus mutations are filtered, in vitro evaluated—individually—and/or recombined in the search of positive epistasis. **Right:** Deduction of the phylogenetic tree by using the MSA and Bayesian/maximum likelihood (ML) method. The combination of the sequences, the phylogenetic tree, and the substitution model (amino acids or nucleotides) give rise to the reconstructed nodes, which are resurrected (i.e., functionally expressed in a modern host), evaluated, and further engineered by directed evolution

Consensus mutations are in many cases stabilizing since they contribute to the progress of molecular evolution by generating a solid backbone that supports the introduction of beneficial yet destabilizing mutations (Lehmann and Wyss 2001; Porebski and Buckle 2016). A good example can be found in the evolution of Kemp eliminase toward the elimination of nonactivated benzisoxazoles, which was initially hampered by the poor stability of the enzyme template (Khersonsky et al. 2012). By inferring some consensus residues and including them in the directed evolution campaign, highly active variants were disclosed. Accordingly, consensus design is an appealing method for protein stabilization. In general, 20–40% of the individual consensus mutations identified have a stabilizing effect (Lehmann et al. 2002; Komor et al. 2012; Magliery 2015). Enzyme engineers obviously wish to improve stability without jeopardizing activity, and the consensus approach has the advantage of overcoming the natural trade-off of activity against stability by inserting stabilizing consensus mutations that do not necessarily dampen activity (Lehmann et al. 2002; Polizzi et al. 2006; Porebski and Buckle 2016). While there are various examples of the use of consensus approach to stabilize proteins (Nikolova et al. 1998; Wirtz and Steipe 1999; Wang et al. 2000; Lehmann et al. 2000; Dai et al. 2007; Zahnd et al. 2007; De Groot and Scott 2007; Bershtein and Tawfik 2008; Paatero et al. 2016), none have focused on laccases.

2.1 OB-1 HRLP Variant: The Starting Point

In the last 3 years, we have used different strategies to explore and improve the stability of a HRPL evolved in the laboratory for expression in yeasts. This variant, termed OB-1, is a highly active, soluble, and stable enzyme, the product of eight rounds of directed evolution (Mate et al. 2010). Indeed, apart from its strong activity and enhanced expression, the OB-1 variant has notable thermostability with a T_{50} value of 73 °C (we defined the T_{50} as the temperature at which the laccase retains 50% of its activity after a 10 min incubation). Attracted by these results, attempts were made to reproduce and rationalize the effect of the secretion mutations on stability and folding of our laccase mutant in silico (Christensen and Kepp 2012). To this end, the FoldX protocol was modified, and as a result, a better predictive method was developed with an error of 1 °C for the T_{50} when compared to the standard FoldX approach that had 3–5 °C accuracy. Encouraged by this data, we wanted to know whether the stability of this already thermostable laccase variant could be pushed even further by different approaches. We first applied a strategy to predict new stabilizing substitutions in OB-1. We scanned flexible loops through a combination of molecular dynamics and an analysis of the *B*-factors, pinpointing ~16% of the protein sequence that could then be explored by random mutagenesis using the MORPHING approach (Gonzalez-Perez et al. 2014). The final double mutant had a 31 min enhancement in the half-life of thermal inactivation at 70 °C ($t_{1/2}$), with an optimum temperature of activity at 75 °C and similar kinetic parameters (Vicente et al. 2019). In view of these promising results, we decided to subject OB-1 to consensus design (Gomez-Fernandez et al. 2020a).

2.2 Consensus Design of the OB-1 Laccase

As there is no perfect method to obtain MSAs, we built three different MSAs, constructing MSA1 and MSA2 with 250 and 1000 sequences from the NCBI database, respectively (Coordinators 2018), and MSA3 with 500 laccase and multicopper oxidase engineering database (LccED) sequences (Sirim et al. 2011). We aligned the three groups of sequences using MUSCLE (available at https://www.ebi.ac.uk/Tools/msa/muscle/), and we then analyzed the resulting MSAs in excel files, following an earlier protocol with minor modifications (Durani and Magliery 2013). In order to obtain a reliable consensus sequence, we calculated the relative entropy (RE) for all possible amino acids at each position (Magliery and Regan 2005; Sullivan et al. 2012). The RE is directly related to the degree of conservation of an amino acid at a position, and it requires a reference distribution. We employed the codon usage of the yeast proteome as the reference distribution (Durani and Magliery 2013), which enhanced the statistical significance. We selected those positions with a RE above the average in the three MSAs, and once these positions had been filtered, we calculated their mutual information (MI). This parameter provides information as to how often two amino acids at different positions are matched in various sequences, revealing a possible interaction between the two (Cronbach 1955; Applebaum 1996). Therefore, a high MI at certain positions means that upon changing a given amino acid, there is a risk of disrupting interactions with other amino acids in the protein. We selected positions with a MI ≤ 0.6 discarding entries with RE > 2.4 (i.e. nearly invariant). By applying these filters, we identified 20 mutations, 18 of which were considered ancestral consensus mutations, as determined by inferring the ancestral node of all basidiomycete laccases (see Sect. 5).

We introduced each of the 20 consensus mutations onto the OB-1 template, subjecting each variant to activity and stability assays. Most of the variants displayed equivalent or higher activity than the parental type, and seven clones had similar or enhanced stability, in good agreement with the success rates of the consensus experiments (Lehmann et al. 2002; Komor et al. 2012; Magliery 2015). The best variant, carrying mutation A240G, increased the $t_{1/2}$ at 70 °C by 1 h relative to OB-1. We then recombined the six most promising mutations in the A240G mutant by site-directed recombination (SDR) in vivo, and when their biochemical properties were analyzed, potential relationships between the consensus mutations were revealed. Ultimately, the best single mutant A240G (named DooKu) was characterized in depth, not only showing notably improved stability but also better catalytic efficiencies for classic laccase substrates and redox mediators, as well as enhanced secretion.

It is worth noting that the main reason why consensus mutations do not affect activity is that the computational search focuses on the surface of the protein and the programs used penalize buried residues (Goldenzweig et al. 2016; Shivange et al. 2016). By using this in-house consensus mutagenesis method that defined the RE and MI thresholds to refine the MSAs, we selected an ancestral consensus mutation (A240G) that notably improved stability, secretion, and activity. This substitution is close to the T2/T3Cu cluster rather than to the surface, which may explain the improvement in activity with all the substrates assayed. Using a method that allows the user to take decisions at each step may enable positions to be identified that may otherwise be overlooked by more straightforward programs. These results demonstrate the potential of a consensus approach to find stabilizing mutations that can be further recombined in other variants. The small libraries and the simple computational requirements make consensus design a reasonable strategy to stabilize laccases without jeopardizing their activity, a key milestone in the engineering of these versatile biocatalysts for industrial applications.

3 SCHEMA-RASPP Computation for Homologous In Vivo Recombination of Stable Laccase Chimeras

Laccase chimeragenesis is an interesting strategy to tailor new laccases with hybrid properties. While directed laccase evolution from single templates has been achieved, it is not common to design chimeric laccases from different orthologues. Indeed, the few reports of laccase chimeras did not yield chimeric proteins with significant sequence and functional diversity (Cusano et al. 2009; Nakagawa et al. 2010; Pardo et al. 2012). This is probably due to the complex structural organization of laccases, with two copper centers linked through a highly conserved HisCysHis tripeptide that allows electrons to transit between the two catalytic sites. Selecting suitable crossover positions in chimeragenesis is paramount to success, and we recently tackled this problem by applying SCHEMA-RASPP structure-guided protein recombination in vivo, a powerful computational method to generate chimeras with high sequence diversity while preserving functionality and stability (Mateljak et al. 2019b). The SCHEMA algorithm identifies protein blocks that can be recombined by selecting suitable crossover locations in contiguous sequences that minimize the SCHEMA energy $< E >$ – the average number of interactions that are broken upon recombination of different parental types (Voigt et al. 2002), Fig. 2. Complementing SCHEMA, the RASPP algorithm helps to set the minimal number of mutations in each protein block and in an entire library, represented as $< m >$ – the average number of mutations in a library relative to the closest parental type (Endelman et al. 2004). By combining SCHEMA and RASPP, the chimeric library is enriched with functional variants with quite diverse sequences. SCHEMA structure-guided recombination has been applied previously to different enzyme families to improve thermostability, substrate specificity, and pH dependence among other traits (e.g., β-lactamases, cellulases, P450 monooxygenases, arginases, and channel rhodopsins), yet it has never been used on laccases (Otey et al. 2004; Meyer et al. 2006;

Otey et al. 2006; Li et al. 2007; Heinzelman et al. 2009, 2010; Romero et al. 2012; Smith et al. 2012; Bedbrook et al. 2017).

Our preferred cloning strategy for directed evolution of eukaryotic genes is based on the high-frequency of homologous DNA recombination of *Saccharomyces cerevisiae*, an approach that we successfully applied to different enzyme systems (Alcalde 2015). We harness *S. cerevisiae* in order to facilitate the cloning and in vivo shuffling of individual computationally predicted SCHEMA blocks, eventually producing functional yet diverse laccase chimeras with inter- and intra-SCHEMA block recombination. In this case study (Mateljak et al. 2019b), OB-1, 3PO from *Pycnoporus cinnabarinus* and Lac3 from basidiomycete *Trametes* sp. strain 30 were used as the points of departure for laccase chimeragenesis. Importantly, all three of these laccases are secreted by *S. cerevisiae* at similar titers (after adjusting the signal peptides), and they shared around 70% sequence identity at the protein level (Mateljak et al. 2017). Considering the need to maintain 40 bp of homology at crossover sites for the in vivo assembly of the SCHEMA blocks within *S. cerevisiae*, we had to write two additional scripts that allowed SCHEMA-RASPP to select crossover positions with larger areas of sequence identity between the three parental laccases. Several computational libraries were created, with four to seven blocks, in an attempt to find suitable crossover spots that preserve protein folding and functionality. We eventually selected a library with four blocks in order to keep a good balance between transformation efficiency and sequence diversity. Indeed, this library was well balanced in terms of homology at the crossover spots, structural conservation at the cupredoxin domains, and good modulation of the Cu-binding sites. Through this strategy we transformed yeast with the computational SCHEMA blocks together with the linearized plasmid. One-pot transformation was sufficient to generate thousands of clones in theory, yielding all the possible combinations of the chimeric laccases. Considering that we have 3 parental types spliced in 4 blocks, we would expect to obtain 81 laccase chimeras, although this number might be expanded by in vivo shuffling within the SCHEMA blocks.

We screened our chimeric library for thermostability and characterized 14 thermostable chimeric laccases. This collection of chimeras had significant sequence diversity, on average differing in 46 amino acids from the closest homologue, Fig. 3. Additional stochastic in vivo shuffling was observed within SCHEMA blocks in four chimeras (i.e., intrablock recombination events), enriching sequence diversity. It is important to highlight that by applying SCHEMA-RASPP recombination in vivo, we generated much more sequence and functional diversity than previous laccase chimeragenesis efforts. Indeed, 12 of the 14 chimeras had a higher $t_{1/2}$ of thermal inactivation at 70 °C than the most stable parent, OB-1 (Fig. 3), of which clone $AAA^{B}C$ particularly stood out, with a fivefold higher $t_{1/2}$ (roughly 120 min) than OB-1. We also investigated the activity of the chimeric family using several laccase substrates and compared them to the closest parent. In order to rationalize the activity of the chimeras, we generated a structural alignment of the T1Cu site catalytic cavity, showing that all four SCHEMA blocks participated in the formation of the catalytic site. Positive, negative, or neutral effects on the oxidation of different substrates were observed depending on how the protein blocks combined with each other. This effect

Fig. 2 SCHEMA approach. Parental proteins are depicted in different colors. After predicting in silico the exact crossover locations (*dashed arrows*), the SCHEMA blocks are randomly recombined to yield a whole set of functional chimeric proteins with combined or improved properties

highlights the potential of SCHEMA-RASPP chimeragenesis, which may alter substrate specificities in promiscuous enzymes by generating functional catalytic sites with different substrate preferences. We also investigated the pH stability of the most active and thermostable chimeras. The thermostable $AAA^{B}C$ clone had higher pH stability in an acidic environment than any of the parental enzymes, retaining 70% of its activity after 72 h at pH 2.0, while its closest OB-1 homologue was fully inactive. A similar effect was observed for CBBC, which displayed eightfold enhancement over its closest parent Lac3 after 72 h at pH 2.0. The enhanced acid stability of these two variants together with their superior thermal stability makes them promising candidates for future industrial applications.

4 Computer-Guided Evolution to Enhance the Laccase Redox Potential and Activity for High-Redox Potential Mediators

The redox potential of the T1Cu site (E_{T1}) crucially influences the overall performance of laccases. It is well-known that electron transfer from the substrate to the T1Cu is the rate-limiting step in laccase catalysis, and the difference between the redox potential of the substrate and the T1Cu is the driving force of reaction, along with the binding event (Jones and Solomon 2015). Despite having very similar T1Cu coordination geometry, redox potential can vary significantly among laccases (Shleev et al. 2004; Hong et al. 2011). Indeed, the complex regulation of laccase's E_{T1} is influenced by many factors, such as hydrophobicity at the T1 site, solvent accessibility, the nature of the axial ligand, the distances between the coordinating histidine and cysteine residues at the T1Cu, the hydrogen bonding to the S(Cys), and stacking and electrostatic interactions in the protein backbone (Palmer et al. 1999; Marshall et al. 2009; Hong et al. 2011; Cambria et al. 2012; Hadt et al. 2012). Furthermore, overall laccase catalysis is not yet fully understood, considering that in addition to substrate oxidation at the T1Cu site and the ensuing intramolecular electron transfer (IET), oxygen is reduced at the trinuclear Cu cluster. Therefore, altering the E_{T1} while preserving the stability and activity of the enzyme, as well as identifying determinants of such modifications, remains a significant challenge to laccase engineers.

The departure point toward higher-redox potential at the T1Cu site was the HRPL OB-1 variant (Mate et al. 2010), employing a strategy in which in vitro directed evolution and computer-aided evolution were run in parallel (Mateljak et al. 2019a). To screen laccase mutant libraries, a reliable screening assay was designed based on two high-redox potential mediators on which laccases hardly act, potassium octacyanomolybdate ($K_4(MoCN)_8$, $E = 780$ mV vs. NHE) and violuric acid ($E = 920$ mV vs. NHE). ABTS was used as a standard non-phenolic mediator substrate that served as a negative control to complement the screening. In the first round of directed evolution, several mutant libraries with different mutational loads were created, and the L3G3 variant was obtained that carried an A162P substitution. The activity of L3G3 improved for both high-redox potential mediators, while its activity on ABTS was dampened slightly. A parallel computational mutagenesis protocol was performed based on ranking sequence positions and sorting them according to the frequency of the beneficial mutations. We screened over 40 positions in the vicinity of T1Cu site by molecular modeling using PELE (protein energy

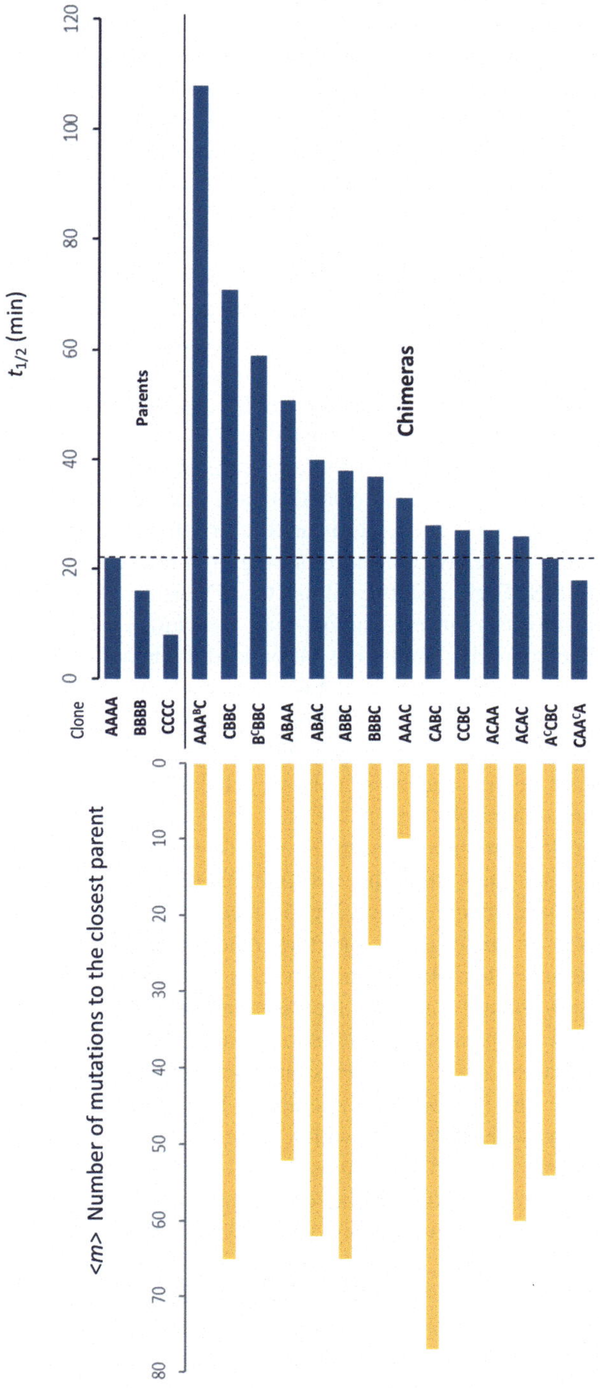

Fig. 3 Family of thermostable chimeric laccases. SCHEMA blocks for OB-1 (AAAA), Lac3 (BBBB), and 3PO (CCCC), with non-computed inserted blocks indicated in *superscript*. The *blue bars* show $t_{1/2}$ of chimeric laccases and parents at 70 °C. *Black vertical dashed line* represents residual activity of most stable parental type. *Yellow bars* are representing amino acid differences between the corresponding chimera and its closest homologue

landscape exploration), a robust algorithm able to score the protein-substrate conformational space (Iglesias et al. 2018). The goal was to reveal positions where mutations enhanced the protein-ligand interaction energy when using an oxidized substrate as a probe, in this case oxidized violuric acid. As a consequence, we found that when the protein-product interaction improves, the overall thermodynamics of the reaction may be enhanced due to a reduction in the electron transfer transition state barrier. After in silico screening, nine positions were identified, of which Ala162 had been uncovered in the previous round of experimental evolution. Of these predicted positions, Ala458 had the most beneficial stabilizing substitutions, and accordingly, Ala162 and Ala458 were subjected to an in vitro round of combinatorial saturation mutagenesis and screening, yielding a double A162V-A458L mutant (named GreeDo). Biochemical and spectro-electrochemical characterization showed this variant to have improved redox potential and enhanced activity for a panel of high-redox potential mediators. Indeed, its E_{T1} was over 50 mV higher than that of the OB-1 laccase (increasing from 740 mV to 790 mV), in the upper limit of redox potential for any laccase reported to date and representing the first time that the E_{T1} of an HRPL had been increased. GreeDo also had improved catalytic efficiency (k_{cat}/K_m) for $K_4Mo(CN)_8$, HBT, and violuric acid (all of them high-redox potential mediators), and its superior behavior in decolorizing high-redox potential dyes was demonstrated using LMS. These results establish GreeDo as a suitable candidate for different LMS applications in bioremediation, pulp bleaching, and other industrial processes. In protein engineering, mutations introduced at the active site often tend to be destabilizing (Bloom et al. 2006), and this is especially true for laccases (Mate and Alcalde 2015). We tested the kinetic thermostability of the GreeDo mutant, and surprisingly its $t_{1/2}$ at 70 °C increased by ~40 min. Moreover, while both OB-1 and GreeDo were stable at neutral and basic pHs, at pH 2.0 the GreeDo variant had 18-fold higher tolerance than OB-1.

The A162V and A458L mutations lie in the vicinity of the T1Cu site, shrinking the catalytic pocket and increasing the hydrophobicity of metal site which is surrounded by bulkier side-chain residues, Fig. 4 Indeed, the hydrophobicity around the metal site seems to be one of major structural determinants of E_{T1}. Embedding the T1Cu of laccase and other blue copper proteins in a more hydrophobic core reduces accessibility to the solvent, with the concomitant stabilization of the less charged Cu (I) ion and the elevation of the redox potential (Battistuzzi et al. 1999; Marshall et al. 2009; Cambria et al. 2012; Hosseinzadeh and Lu 2016; Hosseinzadeh et al. 2016). Moreover, the hydrophobic interaction between Val162 and Leu458 created an energetically beneficial effect, with a more rigid T1Cu environment that positively affects thermal and acid pH stability.

We computationally analyzed the binding of the substrates used, and the A162V and A458L mutations compress the binding cavity around His455 and Asp205, the first electron and proton acceptor, respectively, Fig. 4 Alternatively, the alanine at these two positions in the parental type OB-1 kept the binding pocket open, creating a flat environment around T1Cu. Hence, apart from the enhanced E_{T1}, the formation of more reactive contacts with high-redox potential mediators improves the catalytic constants of this evolved laccase mutant.

In summary, our results stress that the E_{T1} and substrate binding are paramount when assessing the catalytic performance of laccases, both properties that are modulated by a more hydrophobic environment at the T1Cu site.

5 Ancestral Sequence Reconstruction (ASR) for the Directed Evolution of a Resurrected Laccase

Ancestral sequence reconstruction (ASR) involves inferring ancient nodes from the sequences of their modern counterparts. This strategy became feasible at the end of the 1990s when adequate resources became available, which were not accessible when ASR was first hypothesized (Pauling and Zuckerkandl 1963). In addition to the obvious interest in these sequences for the study of natural molecular evolution, the appealing properties that ancient enzymes may display have stimulated the curiosity of protein engineers (Risso et al. 2013; Cox and Gaucher 2014; Whitfield et al. 2015; Babkova et al. 2017; Zakas et al. 2017; Ayuso-Fernández et al. 2017; Gomez-Fernandez et al. 2018). Catalytic and substrate promiscuity, stability in extreme conditions, and improved folding and expression are features often shown by ancestral "resurrected" enzymes. Many of these traits are not inherent to consensus mutations, although they have an evident ancestral origin since they result from residues conserved over time. Although this subject remains open to debate, the ensemble of ancestral-consensus mutations truly defines the unique characteristics of resurrected proteins (Bershtein and Tawfik 2008; Wijma et al. 2013).

To infer the ancestral sequence from a modern enzyme, MSAs must be generated using similar tools as in consensus design, Fig. 1 These MSAs are produced with sequence identity thresholds between 30 and 90% to avoid the inclusion of sequences that are too different or too similar to the query. The outcomes have different magnitudes depending on the number of target protein sequences available, although this does not constitute an impediment to apply this technique (Merkl and Sterner 2016). With a MSA that includes sequences of the same size, and after deleting areas of uncertain alignment, a phylogenetic tree can typically be deduced using either the maximum likelihood (ML) or Bayesian methodology (Yang et al. 1995; Ronquist and Huelsenbeck 2003; Lartillot et al. 2009; Bazinet et al. 2014). Finally, a substitution model based on codons or amino acids can be added to the equation to evaluate the chance of a nucleotide or amino acid being changed to another (Jones et al. 1992; Yang et al. 2000; Whelan and Goldman 2001). Once the sequences of the ancestral nodes are available, they can be resurrected (i.e., functionally expressed in a modern microorganism). Alternatively, the amino acids that differ from an extant version of the protein (ancestral mutations) can be inserted into the modern template or even shuffled to produce ancestral libraries. Indeed, ancestral mutations were recently inserted into a laccase from *Agaricus brasiliensis* that was not expressed by *Pichia pastoris*, and after including 16 ancestral mutations, the

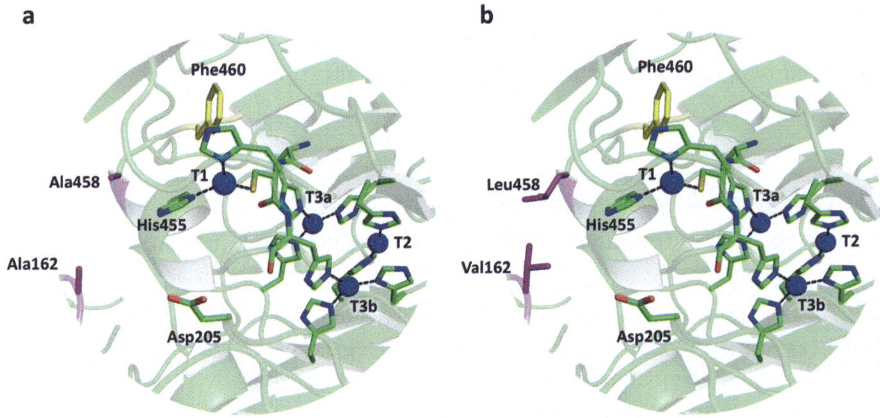

Fig. 4 Location of mutations. Representation of parental OB-1 (**a**) and GreeDo mutant (**b**). A162V and A458L mutations (highlighted as *magenta sticks*) are in the proximity of T1Cu site closing the catalytic pocket. Cooper atoms are shown as *blue spheres* and are coordinated by ten His residues and one Cys depicted as *sticks*. Axial ligand Phe 460 is shown in *yellow sticks*. His 455 and Asp205, the first electron and proton acceptor, respectively, are highlighted as *green sticks*. Model is based on *Trametes trogii* laccase (96% of identity). PDB code 2HRG

laccase generated was successfully produced in this host, and it also showed higher pH and thermal stability (Hamuro et al. 2017).

We consider that the alliance of ASR and directed evolution is very timely, given that the combination of both strategies can lead to the development of a new generation of biocatalysts, harnessing a model to exclude disruptive mutations while enhancing promiscuity and evolvability (Bloom et al. 2006; Cox and Gaucher 2014; Merkl and Sterner 2016; Alcalde 2017; Risso et al. 2018). Indeed, the first proof of concept of the laboratory evolution of a Precambrian enzyme was recently reported. We resurrected several ancestral nodes of the Rubisco from proteobacteria *Rhodospirillum rubrum*, and we subjected both the ancestral and modern Rubiscos to exhaustive directed evolution campaigns aimed at improving stability and evaluating mutational tolerance (Gomez-Fernandez et al. 2018). Encouraged by these results, we recently combined ASR and directed evolution to resurrect and evolve the ancestral node of basidiomycete laccases (Gomez-Fernandez et al. 2020b). We first used the curated MSA1 from the consensus design explained above (see Sect. 2.2) to generate the corresponding phylogenetic tree with MrBayes, while the node sequences were inferred with PAML and an amino acid substitution model. Accordingly, we inferred the sequence of three ancestral nodes dating them back 500–227 million years (at the beginning of the Phanerozoic eon), within the appearance of the common ancestor of some Basidiomycota, in the early Cambrian period (LacAnc95), and the common ancestors of Agaricomycotina and Agaricomycetes, in the Mesozoic era (LacAnc98 and LacAnc100), (Hedges and Kumar 2009). These three nodes were selected for expression in *S. cerevisiae*. Three different signal peptides were employed to enhance expression, all of them evolved versions of the α-factor prepro-leader, and we biochemically characterized the resurrected node 100. This ancestral laccase contained 136 ancestral mutations (73% of identity with OB-1 laccase), dating back ~250–227 million years to the beginning of the Mesozoic era, right after the Permian-Triassic extinction (Nowak et al. 2019). Node 100 had similar kinetic parameters to the modern OB-1 laccase, although it was much more stable at acid pH and its expression was improved ~tenfold.

The warm climate on the earth during the Cambrian period coincided with the most intense evolutionary burst (the Cambrian explosion), with dozens of new species emerging. As such, we were not expecting to find strong differences in terms of thermostability between the resurrected node and the modern laccase, as occurs with many Precambrian proteins (Alcalde 2017 and references herein). Conversely, we hypothesized that improved expression and acid stability are biochemical traits that may confer better adaptation during the biological spark of Basidiomycota expansion during the Cambrian explosion (Gaucher et al. 2008; Hamuro et al. 2017).

We set out to evolve the resurrected laccase to improve its activity against β-diketones, an unusual class of redox mediators that upon oxidation by laccases trigger the polymerization of vinyl monomers. The β-diketone radical itself initiates polymer chain growth by forming part of the polymer, which is why it is referred to as an initiator. That is, after incorporation into the polymer, the β-diketone is no longer available for new cycles of oxidation by the laccase, unlike classical redox mediators whose redox conversion is cyclical (Ikeda et al. 1998; Durand et al. 2000; Hollmann et al. 2008). Initiators are poor laccase substrates so that designing efficient laccases for the oxidation of β-diketones could help to replace the hazardous chemical procedures commonly used in vinyl polymerization reactions (e.g., in the production of polystyrene, polyacrylamide) (Durand et al. 2001; Lizotte and Long 2003; Hollmann et al. 2008). We first developed a high-throughput screening assay for the oxidation of the β-diketone 1,3-cyclopentanedione, and we performed structure-guided evolution. Some hotspot residues were identified and further studied by site-directed recombination in vivo, giving rise to a new P163R-V165R mutant with 160% efficiency in the oxidation of 1,3-cyclopentanedione, a promising point of departure for the enzymatic production of vinyl polymers.

6 Conclusions

The successful case studies discussed in this chapter underline the powerful synergy between computational methodologies and directed evolution in order to tailor new generations of laccases. Computational tools seem to be sufficiently mature as to help protein engineers design smaller yet smarter libraries, significantly reducing the experimental effort required. The incipient incorporation of machine learning to the directed evolution workflow (Saito et al. 2018), along with classical computational methods ranging from the in silico prediction of hotspot residues to de novo enzyme design and/or the generation of ancestral enzymes, will shift the road map of directed evolution in the forthcoming years. In particular, the production of HRPLs tailored *a la carte* at the scale g/Kg seems to be closer than ever, which would allow them to find an outstanding place in the market for many industrial processes (e.g., pulp and paper bleaching, organic synthesis, biosensor design) as well as to aid in environmental concerns (e.g., treatment of industrial wastewaters, removal of PAHs, plastics, and many other xenobiotics).

Acknowledgments This study is based upon work funded by and the Spanish Government projects BIO2013-43407-R-DEWRY and BIO2016-79106-R-Lignolution and by the Bio Based Industries Joint Undertaking under the European Union's Horizon 2020 research and innovation programme under grant agreement n° 886567 (BIZENTE project). B.J.G.F was supported by a FPI national fellowship BES-2014-068887.

References

Alcalde M (2007) Laccases: biological functions, molecular structure and industrial applications. In: Polaina J, MacCaben AP (eds) Industrial enzymes. Springer, Dordrecht, pp 461–476

Alcalde M (2015) Engineering the ligninolytic enzyme consortium. Trends Biotechnol 33 (3):155–162

Alcalde M (2017) When directed evolution met ancestral enzyme resurrection. Microb Biotechnol 10(1):22–24

Applebaum D (1996) Probability and information: an integrated approach. Cambridge University Press, Cambridge

Ayuso-Fernández I, Martínez AT, Ruiz-Dueñas FJ (2017) Experimental recreation of the evolution of lignin-degrading enzymes from the Jurassic to date. Biotechnol Biofuels 10:67–67

Babkova P, Sebestova E, Brezovsky J, Chaloupkova R, Damborsky J (2017) Ancestral Haloalkane Dehalogenases show robustness and unique substrate specificity. Chembiochem 18 (14):1448–1456

Battistuzzi G, Borsari M, Loschi L, Righi F, Sola M (1999) Redox thermodynamics of blue copper proteins. J Am Chem Soc 121(3):501–506

Bazinet AL, Zwickl DJ, Cummings MP (2014) A gateway for phylogenetic analysis powered by grid computing featuring GARLI 2.0. Syst Biol 63(5):812–818

Bedbrook CN, Rice AJ, Yang KK, Ding X, Chen S, LeProust EM, Gradinaru V, Arnold FH (2017) Structure-guided SCHEMA recombination generates diverse chimeric channelrhodopsins. Proc Natl Acad Sci U S A 114(13):E2624

Bershtein S, Tawfik DS (2008) Ohno's model revisited: measuring the frequency of potentially adaptive mutations under various mutational drifts. Mol Biol Evol 25(11):2311–2318

Bloom JD, Labthavikul ST, Otey CR, Arnold FH (2006) Protein stability promotes evolvability. Proc Natl Acad Sci U S A 103(15):5869–5874

Bulter T, Alcalde M, Sieber V, Meinhold P, Schlachtbauer C, Arnold FH (2003) Functional expression of a fungal laccase in *Saccharomyces cerevisiae* by directed evolution. Appl Environ Microbiol 69(2):987–995

Camarero S, Pardo I, Cañas AI, Molina P, Record E, Martínez A, Martínez MJ, Alcalde M (2011) Engineering platforms for the directed evolution of laccase from *Pycnoporus cinnabarinus*. Appl Environ Microbiol 78(5):1370–1384

Cambria MT, Gullotto D, Garavaglia S, Cambria A (2012) In silico study of structural determinants modulating the redox potential of *Rigidoporus lignosus* and other fungal laccases. J Biomol Struct Dyn 30(1):89–101

Christensen NJ, Kepp KP (2012) Accurate stabilities of laccase mutants predicted with a modified FoldX protocol. J Chem Inf Model 52(11):3028–3042
Coordinators NR (2018) Database resources of the National Center for biotechnology information. Nucleic Acids Res 46(D1):D8–D13
Cox VE, Gaucher EA (2014) Engineering proteins by reconstructing evolutionary adaptive paths. In: Gillam EMJ, Copp JN, Ackerley D (eds) Directed evolution library creation: methods and protocols. Springer, New York, pp 353–363
Cronbach LJ (1955) On the non-rational application of information measures in psychology. In: Quastler H (ed) Information theory in psychology problems and methods. Free Press, Glencoe, IL, pp 14–30
Cusano AM, Mekmouche Y, Meglecz E, Tron T (2009) Plasticity of laccase generated by homeologous recombination in yeast. FEBS J 276(19):5471–5480
Dai M, Fisher HE, Temirov J, Kiss C, Phipps ME, Pavlik P, Werner JH, Bradbury AR (2007) The creation of a novel fluorescent protein by guided consensus engineering. Protein Eng Des Sel 20 (2):69–79
De Groot AS, Scott DW (2007) Immunogenicity of protein therapeutics. Trends Immunol 28 (11):482–490
Durand A, Lalot T, Brigodiot M, Maréchal E (2000) Enzyme-mediated initiation of acrylamide polymerization: reaction mechanism. Polymer 41(23):8183–8192
Durand A, Lalot T, Brigodiot M, Maréchal E (2001) Enzyme-mediated radical initiation of acrylamide polymerization: Main characteristics of molecular weight control. Polymer 42 (13):5515–5521
Durani V, Magliery TJ (2013) Protein engineering and stabilization from sequence statistics: variation and covariation analysis. Methods Enzymol 523:237–256
Edgar RC (2004) MUSCLE: multiple sequence alignment with high accuracy and high throughput. Nucleic Acids Res 32(5):1792–1797
Endelman JB, Silberg JJ, Wang Z-G, Arnold FH (2004) Site-directed protein recombination as a shortest-path problem. Protein Eng Des Sel 17(7):589–594
Gaucher EA, Govindarajan S, Ganesh OK (2008) Palaeotemperature trend for Precambrian life inferred from resurrected proteins. Nature 451:704
Goldenzweig A, Goldsmith M, Hill SE, Gertman O, Laurino P, Ashani Y, Dym O, Unger T, Albeck S, Prilusky J, Lieberman RL, Aharoni A, Silman I, Sussman JL, Tawfik DS, Fleishman SJ (2016) Automated structure- and sequence-based Design of Proteins for high bacterial expression and stability. Mol Cell 63(2):337–346
Gomez-Fernandez BJ, Garcia-Ruiz E, Martin-Diaz J, Gomez de Santos P, Santos-Moriano P, Plou FJ, Ballesteros A, Garcia M, Rodriguez M, Risso VA, Sanchez-Ruiz JM, Whitney SM, Alcalde M (2018) Directed -in vitro- evolution of Precambrian and extant Rubiscos. Sci Rep 8(1):5532
Gomez-Fernandez BJ, Risso VA, Sanchez-Ruiz JM, Alcalde M (2020a) Consensus design of an evolved high-redox potential laccase. Front Biotechnol Bioeng 8(354):1–11
Gomez-Fernandez BJ, Risso VA, Rueda A, Sanchez-Ruiz JM, Alcalde M (2020b) Ancestral resurrection and directed evolution of fungal mesozoic laccases. Appl Environ Microbiol 86: e00778-20
Gonzalez-Perez D, Molina-Espeja P, Garcia-Ruiz E, Alcalde M (2014) Mutagenic organized recombination process by homologous in vivo grouping (MORPHING) for directed enzyme evolution. PLoS One 9(3):e90919
Hadt RG, Sun N, Marshall NM, Hodgson KO, Hedman B, Lu Y, Solomon EI (2012) Spectroscopic and DFT studies of second-sphere variants of the type 1 copper site in azurin: covalent and nonlocal electrostatic contributions to reduction potentials. J Am Chem Soc 134 (40):16701–16716
Hamuro Y, Tajima K, Matsumoto-Akanuma A, Sakamoto S, Furukawa R, Yamagishi A, Ohno N, Akanuma S (2017) Characterization of a thermostable mutant of *Agaricus brasiliensis* laccase created by phylogeny-based design. J Biosci Bioeng 124(6):623–629
Hedges SB, Kumar S (2009) The timetree of life. OUP, Oxford

Heinzelman P, Komor R, Kanaan A, Romero P, Yu X, Mohler S, Snow C, Arnold F (2010) Efficient screening of fungal cellobiohydrolase class I enzymes for thermostabilizing sequence blocks by SCHEMA structure-guided recombination. Protein Eng Des Sel 23(11):871–880

Heinzelman P, Snow CD, Wu I, Nguyen C, Villalobos A, Govindarajan S, Minshull J, Arnold FH (2009) A family of thermostable fungal cellulases created by structure-guided recombination. Proc Natl Acad Sci U S A 106(14):5610–5615

Hollmann F, Gumulya Y, Tölle C, Liese A, Thum O (2008) Evaluation of the laccase from *Myceliophthora thermophila* as industrial biocatalyst for polymerization reactions. Macromolecules 41(22):8520–8524

Hong G, Ivnitski DM, Johnson GR, Atanassov P, Pachter R (2011) Design parameters for tuning the type 1 cu multicopper oxidase redox potential: insight from a combination of first principles and empirical molecular dynamics simulations. J Am Chem Soc 133(13):4802–4809

Hosseinzadeh P, Lu Y (2016) Design and fine-tuning redox potentials of metalloproteins involved in electron transfer in bioenergetics. Biochim Biophys Acta Bioenerg 1857(5):557–581

Hosseinzadeh P, Marshall NM, Chacón KN, Yu Y, Nilges MJ, Yee New S, Tashkov SA, Blackburn NJ, Lu Y (2016) Design of a single protein that spans the entire 2-V range of physiological redox potentials. Proc Natl Acad Sci U S A 113(2):262–267

Iglesias J, Saen-oon S, Soliva R, Guallar V (2018) Computational structure-based drug design: predicting target flexibility. Wiley Interdiscip Rev: Comput Mol Sci 8(9):e1367

Ikeda R, Tanaka H, Uyama H, Kobayashi S (1998) Laccase-catalyzed polymerization of acrylamide. Macromol Rapid Commun 19(8):423–425

Jackel C, Bloom JD, Kast P, Arnold FH, Hilvert D (2010) Consensus protein design without phylogenetic bias. J Mol Biol 399(4):541–546

Jones SM, Solomon EI (2015) Electron transfer and reaction mechanism of laccases. Cell Mol Life Sci 72(5):869–883

Jones DT, Taylor WR, Thornton JM (1992) The rapid generation of mutation data matrices from protein sequences. Comput Appl Biosci 8(3):275–282

Kataoka K, Kogi H, Tsujimura S, Sakurai T (2013) Modifications of laccase activities of copper efflux oxidase, CueO by synergistic mutations in the first and second coordination spheres of the type I copper center. Biochem Biophys Res Commun 431(3):393–397

Khersonsky O, Kiss G, Röthlisberger D, Dym O, Albeck S, Houk KN, Baker D, Tawfik DS (2012) Bridging the gaps in design methodologies by evolutionary optimization of the stability and proficiency of designed Kemp eliminase KE59. Proc Natl Acad Sci U S A 109 (26):10358–10363

Komor RS, Romero PA, Xie CB, Arnold FH (2012) Highly thermostable fungal cellobiohydrolase I (Cel7A) engineered using predictive methods. Protein Eng Des Sel 25(12):827–833

Koschorreck K, Schmid RD, Urlacher VB (2009) Improving the functional expression of a *Bacillus licheniformis* laccase by random and site-directed mutagenesis. BMC Biotechnol 9(1):12

Larkin MA, Blackshields G, Brown NP, Chenna R, McGettigan PA, McWilliam H, Valentin F, Wallace IM, Wilm A, Lopez R, Thompson JD, Gibson TJ, Higgins DG (2007) Clustal W and Clustal X version 2.0. Bioinformatics 23(21):2947–2948

Lartillot N, Lepage T, Blanquart S (2009) PhyloBayes 3: a Bayesian software package for phylogenetic reconstruction and molecular dating. Bioinformatics 25(17):2286–2288

Lehmann M, Kostrewa D, Wyss M, Brugger R, D'Arcy A, Pasamontes L, van Loon AP (2000) From DNA sequence to improved functionality: using protein sequence comparisons to rapidly design a thermostable consensus phytase. Protein Eng 13(1):49–57

Lehmann M, Loch C, Middendorf A, Studer D, Lassen SF, Pasamontes L, van Loon AP, Wyss M (2002) The consensus concept for thermostability engineering of proteins: further proof of concept. Protein Eng 15(5):403–411

Lehmann M, Wyss M (2001) Engineering proteins for thermostability: the use of sequence alignments versus rational design and directed evolution. Curr Opin Biotechnol 12(4):371–375

Li Y, Drummond DA, Sawayama AM, Snow CD, Bloom JD, Arnold FH (2007) A diverse family of thermostable cytochrome P450s created by recombination of stabilizing fragments. Nat Biotechnol 25:1051

Liu H, Zhu L, Bocola M, Chen N, Spiess AC, Schwaneberg U (2013) Directed laccase evolution for improved ionic liquid resistance. Green Chem 15(5):1348–1355

Lizotte JR, Long TE (2003) Stable free-radical polymerization of styrene in combination with 2-Vinylnaphthalene initiation. Macromol Chem Phys 204(4):570–576

Madzak C, Mimmi M, Caminade E, Brault A, Baumberger S, Briozzo P, Mougin C, Jolivalt C (2005) Shifting the optimal pH of activity for a laccase from the fungus *Trametes versicolor* by structure-based mutagenesis. Protein Eng Des Sel 19(2):77–84

Magliery TJ (2015) Protein stability: computation, sequence statistics, and new experimental methods. Curr Opin Struct Biol 3:161–168

Magliery TJ, Regan L (2005) Sequence variation in ligand binding sites in proteins. BMC Bioinf 6 (1):240

Marshall NM, Garner DK, Wilson TD, Gao Y-G, Robinson H, Nilges MJ, Lu Y (2009) Rationally tuning the reduction potential of a single cupredoxin beyond the natural range. Nature 462 (7269):113

Mate DM, Alcalde M (2015) Laccase engineering: from rational design to directed evolution. Biotechnol Adv 33(1):25–40

Mate DM, Alcalde M (2017) Laccase: a multi-purpose biocatalyst at the forefront of biotechnology. Microb Biotechnol 10(6):1457–1467

Mate DM, García-Burgos C, García-Ruiz E, Ballesteros AO, Camarero S, Alcalde M (2010) Laboratory evolution of high-redox potential laccases. Chem Biol 17(9):1030–1041

Mate DM, Gonzalez-Perez D, Falk M, Kittl R, Pita M, De Lacey AL, Ludwig R, Shleev S, Alcalde M (2013) Blood tolerant Laccase by directed evolution. Chem Biol 20(2):223–231

Mateljak I, Monza E, Lucas MF, Guallar V, Aleksejeva O, Ludwig R, Leech D, Shleev S, Alcalde M (2019a) Increasing redox potential, redox mediator activity, and stability in a fungal Laccase by computer-guided mutagenesis and directed evolution. ACS Catal 9(5):4561–4572

Mateljak I, Rice A, Yang K, Tron T, Alcalde M (2019b) The generation of Thermostable fungal Laccase chimeras by SCHEMA-RASPP structure-guided recombination in vivo. ACS Synth Biol 8(4):833–843

Mateljak I, Tron T, Alcalde M (2017) Evolved α-factor prepro-leaders for directed laccase evolution in *Saccharomyces cerevisiae*. Microb Biotechnol 10(6):1830–1836

Merkl R, Sterner R (2016) Reconstruction of ancestral enzymes. Perspect Sci 9:17–23

Meyer MM, Hochrein L, Arnold FH (2006) Structure-guided SCHEMA recombination of distantly related β-lactamases. Protein Eng Des Sel 19(12):563–570

Molina-Espeja P, Viña-Gonzalez J, Gomez-Fernandez BJ, Martin-Diaz J, Garcia-Ruiz E, Alcalde M (2016) Beyond the outer limits of nature by directed evolution. Biotechnol Adv 34 (5):754–767

Mollania N, Khajeh K, Ranjbar B, Hosseinkhani S (2011) Enhancement of a bacterial laccase thermostability through directed mutagenesis of a surface loop. Enzym Microb Technol 49 (5):446–452

Morozova OV, Shumakovich GP, Shleev SV, Yaropolov YI (2007) Laccase-mediator systems and their applications: a review. Appl Biochem Microbiol 43(5):523–535

Nakagawa Y, Sakamoto Y, Kikuchi S, Sato T, Yano A (2010) A chimeric laccase with hybrid properties of the parental Lentinula edodes laccases. Microbiol Res 165(5):392–401

Nikolova PV, Henckel J, Lane DP, Fersht AR (1998) Semirational design of active tumor suppressor p53 DNA binding domain with enhanced stability. Proc Natl Acad Sci U S A 95 (25):14675–14680

Nowak H, Schneebeli-Hermann E, Kustatscher E (2019) No mass extinction for land plants at the Permian–Triassic transition. Nat Commun 10(1):384–391

Otey CR, Landwehr M, Endelman JB, Hiraga K, Bloom JD, Arnold FH (2006) Structure-guided recombination creates an artificial family of cytochromes P450. PLoS Biol 4(5):e112

Otey CR, Silberg JJ, Voigt CA, Endelman JB, Bandara G, Arnold FH (2004) Functional evolution and structural conservation in chimeric cytochromes P450. Chem Biol 11(3):309–318

Paatero A, Rosti K, Shkumatov AV, Sele C, Brunello C, Kysenius K, Singha P, Jokinen V, Huttunen H, Kajander T (2016) Crystal structure of an engineered LRRTM2 synaptic adhesion molecule and a model for Neurexin binding. Biochemistry 55(6):914–926

Palmer AE, Randall DW, Xu F, Solomon EI (1999) Spectroscopic studies and electronic structure description of the high potential type 1 copper site in fungal laccase: insight into the effect of the axial ligand. J Am Chem Soc 121(30):7138–7149

Pantoliano MW, Whitlow M, Wood JF, Dodd SW, Hardman KD, Rollence ML, Bryan PN (1989) Large increases in general stability for subtilisin BPN' through incremental changes in the free energy of unfolding. Biochemistry 28(18):7205–7213

Pardo I, Santiago G, Gentili P, Lucas F, Monza E, Medrano FJ, Galli C, Martínez AT, Guallar V, Camarero S (2016) Re-designing the substrate binding pocket of laccase for enhanced oxidation of sinapic acid. Cat Sci Technol 6(11):3900–3910

Pardo I, Vicente AI, Mate DM, Alcalde M, Camarero S (2012) Development of chimeric laccases by directed evolution. Biotechnol Bioeng 109(12):2978–2986

Pauling L, Zuckerkandl E (1963) Molecular "restoration studies" of extinct forms of life. Acta Chem Scand 17:S9–S16

Polizzi KM, Chaparro-Riggers JF, Vazquez-Figueroa E, Bommarius AS (2006) Structure-guided consensus approach to create a more thermostable penicillin G acylase. Biotechnol J 1 (5):531–536

Porebski BT, Buckle AM (2016) Consensus protein design. Protein Eng Des Sel 29(7):245–251

Rasekh B, Khajeh K, Ranjbar B, Mollania N, Almasinia B, Tirandaz H (2014) Protein engineering of laccase to enhance its activity and stability in the presence of organic solvents. Eng Life Sci 14(4):442–448

Risso VA, Gavira JA, Mejia-Carmona DF, Gaucher EA, Sanchez-Ruiz JM (2013) Hyperstability and substrate promiscuity in laboratory resurrections of Precambrian β-lactamases. J Am Chem Soc 135(8):2899–2902

Risso VA, Sanchez-Ruiz JM, Ozkan SB (2018) Biotechnological and protein-engineering implications of ancestral protein resurrection. Curr Opin Struct Biol 51:106–115

Romero PA, Stone E, Lamb C, Chantranupong L, Krause A, Miklos A, Hughes RA, Fechtel B, Ellington AD, Arnold FH, Georgiou G (2012) SCHEMA designed variants of human Arginase I & II reveal sequence elements important to stability and catalysis. ACS Synth Biol 1 (6):221–228

Ronquist F, Huelsenbeck JP (2003) MrBayes 3: bayesian phylogenetic inference under mixed models. Bioinformatics 19(12):1572–1574

Saito Y, Oikawa M, Nakazawa H, Niide T, Kameda T, Tsuda K, Umetsu M (2018) Machine-learning-guided mutagenesis for directed evolution of fluorescent proteins. ACS Synth Biol 7 (9):2014–2022

Santiago G, de Salas F, Lucas MF, Monza E, Acebes S, Martinez ÁT, Camarero S, Guallar V (2016) Computer-aided Laccase engineering: toward biological oxidation of Arylamines. ACS Catal 6(8):5415–5423

Scheiblbrandner S, Breslmayr E, Csarman F, Paukner R, Führer J, Herzog PL, Shleev SV, Osipov EM, Tikhonova TV, Popov VO, Haltrich D, Ludwig R, Kittl R (2017) Evolving stability and pH-dependent activity of the high redox potential *Botrytis aclada* laccase for enzymatic fuel cells. Sci Rep 7(1):13688

Shivange AV, Hoeffken HW, Haefner S, Schwaneberg U (2016) Protein consensus-based surface engineering (ProCoS): a computer-assisted method for directed protein evolution. BioTechniques 61(6):305–314

Shleev SV, Jarosz-Wilkolazka A, Khalunina A, Morozova O, Yaropolov A, Ruzgas T, Gorton L (2005) Direct electron transfer reactions of laccases from different origins on carbon electrodes. Bioelectrochemistry 67(1):115–124

Shleev SV, Morozova OV, Nikitina OV, Gorshina ES, Rusinova TV, Serezhenkov VA, Burbaev DS, Gazaryan IG, Yaropolov AI (2004) Comparison of physico-chemical characteristics of four laccases from different basidiomycetes. Biochimie 86(9):693–703

Sirim D, Wagner F, Wang L, Schmid RD, Pleiss J (2011) The Laccase Engineering Database: a classification and analysis system for laccases and related multicopper oxidases. Database (Oxford) 2011:bar006

Smith MA, Rentmeister A, Snow CD, Wu T, Farrow MF, Mingardon F, Arnold FH (2012) A diverse set of family 48 bacterial glycoside hydrolase cellulases created by structure-guided recombination. FEBS J 279(24):4453–4465

Socolich M, Lockless SW, Russ WP, Lee H, Gardner KH, Ranganathan R (2005) Evolutionary information for specifying a protein fold. Nature 437(7058):512–518

Steipe B, Schiller B, Pluckthun A, Steinbacher S (1994) Sequence statistics reliably predict stabilizing mutations in a protein domain. J Mol Biol 240(3):188–192

Sullivan BJ, Durani V, Magliery TJ (2011) Triosephosphate isomerase by consensus design: dramatic differences in physical properties and activity of related variants. J Mol Biol 413 (1):195–208

Sullivan BJ, Nguyen T, Durani V, Mathur D, Rojas S, Thomas M, Syu T, Magliery TJ (2012) Stabilizing proteins from sequence statistics: the interplay of conservation and correlation in triosephosphate isomerase stability. J Mol Biol 420(4–5):384–399

Torres-Salas P, Mate DM, Ghazi I, Plou FJ, Ballesteros AO, Alcalde M (2013) Widening the pH activity profile of a fungal Laccase by directed evolution. Chembiochem 14(8):934–937

Toscano MD, De Maria L, Lobedanz S, Østergaard LH (2013) Optimization of a small Laccase by active-site redesign. Chembiochem 14(10):1209–1211

Vicente AI, Viña-Gonzalez J, Mateljak I, Monza E, Lucas F, Guallar V, Alcalde M (2019) Enhancing Thermostability by modifying flexible surface loops in an evolved high-redox potential Laccase. AiChE J In press

Vicente AI, Viña-Gonzalez J, Santos-Moriano P, Marquez-Alvarez C, Ballesteros O, Alcalde M (2016) Evolved alkaline fungal laccase secreted by Saccharomyces cerevisiae as useful tool for the synthesis of C-N heteropolymeric dye. J Mol Catal B Enzym 134:323–330

Voigt CA, Martinez C, Wang Z-G, Mayo SL, Arnold FH (2002) Protein building blocks preserved by recombination. Nat Struct Biol 9:553

Wang Q, Buckle AM, Fersht AR (2000) Stabilization of GroEL minichaperones by core and surface mutations. J Mol Biol 298(5):917–926

Wang Q, Buckle AM, Foster NW, Johnson CM, Fersht AR (1999) Design of highly stable functional GroEL minichaperones. Protein Sci 8(10):2186–2193

Whelan S, Goldman N (2001) A general empirical model of protein evolution derived from multiple protein families using a maximum-likelihood approach. Mol Biol Evol 18(5):691–699

Whitfield JH, Zhang WH, Herde MK, Clifton BE, Radziejewski J, Janovjak H, Henneberger C, Jackson CJ (2015) Construction of a robust and sensitive arginine biosensor through ancestral protein reconstruction. Protein Sci 24(9):1412–1422

Wijma HJ, Floor RJ, Janssen DB (2013) Structure-and sequence-analysis inspired engineering of proteins for enhanced thermostability. Curr Opin Struct Biol 23(4):588–594

Wirtz P, Steipe B (1999) Intrabody construction and expression III: engineering hyperstable V (H) domains. Protein Sci 8(11):2245–2250

Yang Z, Kumar S, Nei M (1995) A new method of inference of ancestral nucleotide and amino acid sequences. Genetics 141(4):1641–1650

Yang Z, Nielsen R, Goldman N, Pedersen AM (2000) Codon-substitution models for heterogeneous selection pressure at amino acid sites. Genetics 155(1):431–449

Zahnd C, Wyler E, Schwenk JM, Steiner D, Lawrence MC, McKern NM, Pecorari F, Ward CW, Joos TO, Plückthun A (2007) A designed Ankyrin repeat protein evolved to Picomolar affinity to Her2. J Mol Biol 369(4):1015–1028

Zakas PM, Brown HC, Knight K, Meeks SL, Spencer HT, Gaucher EA, Doering CB (2017) Enhancing the pharmaceutical properties of protein drugs by ancestral sequence reconstruction. Nat Biotechnol 35(1):35–37

Zumarraga M, Bulter T, Shleev S, Polaina J, Martínez-Arias A, Plou FJ, Ballesteros A, Alcalde M (2007) In vitro evolution of a fungal laccase in high concentrations of organic cosolvents. Chem Biol 14(9):1052–1064

Zumarraga M, Camarero S, Shleev S, Martínez-Arias A, Ballesteros A, Plou FJ, Alcalde M (2008) Altering the laccase functionality by in vivo assembly of mutant libraries with different mutational spectra. Proteins: Struct Funct Bioinf 71(1):250–260

Laccases from Extremophiles

Yordanis Pérez-Llano, Heilyn Pérez Soler, Axel Ramírez Olivano, Jorge Luis Folch-Mallol, Hubert Cabana, and Ramón Alberto Batista-García

Contents

Abstract Extremophilic microorganisms usually inhabit extreme ecosystems and show optimal growth under extreme conditions. Extreme conditions are defined as

Y. Pérez-Llano · R. A. Batista-García (✉)
Centro de Investigación en Dinámica Celular. Instituto de Investigaciones en Ciencias Básicas y Aplicadas. Universidad Autónoma del Estado de Morelos, Morelos, Cuernavaca, Mexico
e-mail: rabg@uaem.mx

H. P. Soler · A. R. Olivano
Centro de Investigación en Dinámica Celular. Instituto de Investigaciones en Ciencias Básicas y Aplicadas. Universidad Autónoma del Estado de Morelos, Morelos, Cuernavaca, Mexico

Facultad de Ciencias Químicas e Ingeniería. Universidad Autónoma del Estado de Morelos, Morelos, Cuernavaca, Mexico

J. L. Folch-Mallol
Centro de Investigación en Biotecnología. Universidad Autónoma del Estado de Morelos, Morelos, Cuernavaca, Mexico

H. Cabana
University of Sherbrooke Water Research Group (USWRG), Department of Civil and Building Engineering, University of Sherbrooke, Sherbrooke, QC, Canada

D. Schlosser (ed.), *Laccases in Bioremediation and Waste Valorisation*,
Microbiology Monographs 33, https://doi.org/10.1007/978-3-030-47906-0_9

those far from the optimal conditions for human life. Thus, ecosystems that present very low or high temperature, pH values, or radiation index, habitats containing a NaCl concentration higher than 1.0 M and deep sea, among others, are considered extreme ecosystems. Derived from these physicochemical conditions, extreme ecosystems have an interesting biodiversity that could represent an attractive source of novel extremozymes. The use of many methods to ensure the bioprospection is needed to find robust enzymes with industrial applications. In this sense, dependent and independent culture approaches are successfully used to identify and characterize novel biocatalysts as laccases. Laccases are copper-containing oxides enzymes found in many organisms (plants, bacteria, fungi, insects, oysters), and they have a key role in the lignin deconstruction. These enzymes catalyze the monoelectronic oxidation of a huge range of substrates. Especially, microbial laccases have been extensively studied since structural and functional views. These oxidoreductases are frequently extracellular proteins, and the interest in their applications has significantly grown in recent years. Currently, laccase-based industrial processes as green-clean bioprocesses are very attractive for the biotechnology. Thus, the search for new genes encoding laccases is an important objective to supply the increasing demand of the bioprocess-based industry. In this chapter, a discussion of the structure and application of laccases will be presented, with an emphasis in extremophilic laccases. Also, a revision related to biotechnological applications using extremophilic laccases will be addressed.

1 Introduction

One of the main concerns in current biotechnology industries is to increase efficiency and competitiveness in the face of challenges imposed by the well-established chemistry industry. In some industrial fields, shifting the framework toward biotechnological processes is not yet affordable (Chen and Jiang 2018). Even though biotechnological solutions promise to be a "green" alternative, the truth is that current methods have still many shortcomings that should be addressed to effectively introduce biological treatments in the market. Among others, biotechnological processes:

- Rely on high-energy consumption (for sterilization, optimal temperature stabilization, aeration, agitation).
- Have a considerable production of wastewater.
- Are prone to microbial contamination.
- Are more time-consuming due to constraints imposed by microorganism growth rates.
- Are not easily adapted to continuous flow as many fermentations are best held on batch.
- Are not easily automated as many variables require expert assessment.
- Broth media is difficult to recycle.

- Product separation and purification are lengthy and expensive.
- Require high infrastructure investment.
- Require highly trained personnel.

In recent years, extremophiles have arisen as a promising tool to overcome many of these difficulties. Imposing extreme conditions such as high salinity, low or high pH, low temperatures, or high solute concentration can significantly reduce the possibilities of microbial contamination in the processes. Using extreme conditions would eliminate the requirement of sterility and the associated cost (Chen and Jiang 2018). Also, since there is a low risk of contamination, batch fermentations are no longer a necessity, and continuous flow processes could be designed. Altogether, these changes could considerably reduce the cost of infrastructure. Then, the main challenge becomes screening for extremophile microorganisms and their enzymes and understanding their mechanisms to thrive in extreme environments.

Extreme ecosystems (i.e., salterns, stromatolites, acidic hot springs, hydrothermal vents, alkaliphilic lakes, and acid water of mines) could harbor an enormous diversity of unknown laccase-like enzymes. Because our knowledge about laccase genes in extreme ecosystems is still very limited, we even cannot visualize the great biotechnological potential that extremophile laccase-like enzymes could represent. Considering that industrial applications demand large-scale enzyme productions, laccases isolated from extremophiles need to be heterologously expressed in different hosts. Currently, the heterologous overexpression is considered the main bottleneck for the laccase industry. There are serious restrictions with the expression levels of many extremophile laccases in classical industrial hosts such as *Escherichia coli, Pichia pastoris, Saccharomyces cerevisiae, and Aspergillus niger.* Particularly "rare laccase genes" isolated from extremophile microorganisms impose serious limitations for the protein heterologous expression, because the codon usage and the promoter regulation can negatively impacting on recombinant laccase production in different hosts. In addition, in many cases posttranslational modifications are also needed to obtain functional extremophile laccases. Thus, new industrial heterologous systems with a particular focus on extremophile enzymes (i.e., laccases) are needed to increase the chances of producing sufficient levels of extremozymes. As we later discuss, after recombinant laccases are successfully produced, they are useful for a huge range of biotechnological applications. The biochemical properties of the extremophile laccases allow their application under critical conditions such as high or low temperature, huge range of pH, among others.

2 Extremophile Microorganisms

Extremophiles are a heterogeneous group of microorganisms growing under extreme conditions. Extreme environments are characterized by very particular physicochemical conditions; for example, very high or low temperature, over 1 atm hydrostatic pressure, extremely acid or alkaline pH values, high radiation index, severe nutrient limitations, hypersalinity and low water activity, among others.

According to the environmental requirements for these microorganisms to grow optimally, they can be classified as thermophiles or hyperthermophiles (when >40 °C is required), psychrophile (when <15 °C is required), acidophiles or alkaliphiles (when acid or alkaline pH values are required), halophiles (when >0.5 M NaCl is required), and barophiles (when >1 atm is required). Other classifications include the term xerophiles for microorganisms requiring low water activity (<0.8) at optimum growth. Organisms that tolerate these adverse conditions but have their optimal growth rate under standard conditions are defined as extremotolerant.

Interestingly, many environments show a combination of these conditions. For example, deep oceans and deep-sea sediments combine high pressure, high salinity, and low temperature; hydrothermal vents have high temperature and high pressure with a low content of organic matter; acid mine drainages are characterized by low pH values (<3), high heavy metal concentrations, and low nutrient availability. All these habitats are classified as poly-extreme, and their endogenous microbiota is considered poly-extremophilic.

Extreme environments are widely distributed around the world. In addition to those previously mentioned, these include volcanoes, arid and hyperarid zones, hypersaline soils, salterns, deserts, alkaline or acid lakes, and frozen habitats, among others. Another type of extreme environments has attracted our attention in recent years. Certain habitats are classified as extreme due to high levels of environmental pollution, which imposes several restrictions for life. These environments are derived from anthropogenic activities such as industrialization and are also colonized by a huge microbial biodiversity that can withstand high levels of toxic compounds.

Extremophiles include microorganisms belonging to the three domains of life, Archaea, Bacteria, and Eukarya. Figure 1 displays a phylogeny representing extremophilic microorganisms of different genera and their extremophilic characteristics. Historically, Archaea and Bacteria have been largely studied as extremophile microorganisms. However, the study of extremophilic fungi has been recently addressed (since the last 20 years) (Gunde-Cimerman et al. 2003). Extremophiles have emerged as a unique opportunity to analyze the biology in the limits of life in a planetary context (Merino et al. 2019). Frequently, the study of extremophilic microorganisms has been related to the origin of life, as well as with the potential search of life in alternative biospaces. Over the past several decades, the search for life in planets or celestial bodies different than Earth has been a human obsession. Extremophiles are biological systems that offer a great opportunity to analyze biological behaviors in conditions that could simulate other planetary bodies. In this context, the Atacama Desert has been the classical environment since its soil composition mimics the Mars' surface. This is one of the oldest, arid, and most hostile deserts on Earth with <1 mm of rain per year. For many years it was considered as uninhabited territory due to its poly-extreme features. However, the study of its autochthone microbiota that includes archaeal, bacteria, fungi, protozoa, algae, and viruses has allowed establishing "the basis for a new perspective for life at this poly-extreme territory" (cited from Gómez-Silva 2018).

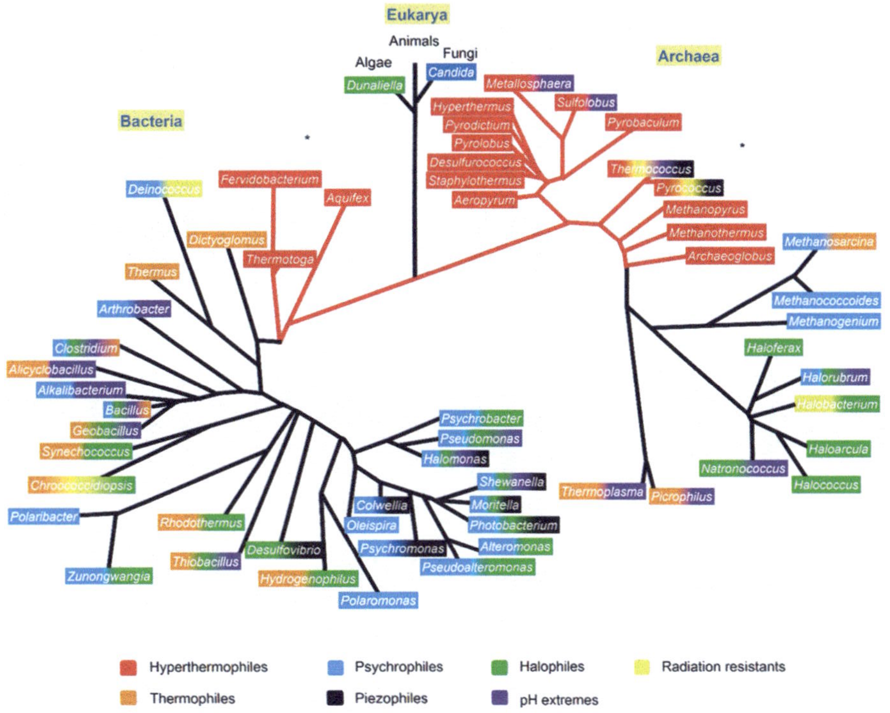

Fig. 1 Phylogenetic tree showing the extremophiles and the resistant characteristics that appear in at least one species of each genera, identified with the color code. Reprinted from Leal G, Ferreira D, and Vermelho A, Marine extremophiles. A source of hydrolases for biotechnological applications. Mar Drugs 2015, 13 (4), 1925–1965. Copyright © 2015 the authors; licensee MDPI, Basel, Switzerland. Open-access article distributed under the terms and conditions of the Creative Commons Attribution license (http://creativecommons.org/licenses/by/4.0/)

For modern science, extremophiles are fascinating microbes because they can successfully live under conditions in which human life would be remotely imaginable. In contrast with our anthropocentric point of view, the Earth harbors many niches that represent no conventional habitable ecosystems. However, all of these are absolutely colonize by microorganisms with a huge diversity of molecular and cellular adaptations to survive under extreme conditions (Singh et al. 2019a, b). These microbes usually exhibit strong metabolic plasticity at the genomic level that reflects their potentialities to implement biochemical strategies to maintain the metabolic homeostasis under extreme conditions.

3 Structural Basis for Extremotolerance

An important aspect of using extremophilic strains for biotechnological processes is not only profiting from their optimal growth rate at extreme conditions but also ensuring the maximal activity of the desired enzymatic treatment. A strain which grows in acidic pH is of no use if the target enzyme required for the process is not produced or has suboptimal kinetic parameters under these conditions. Therefore, the characterization of enzyme activity at extreme conditions is also critical.

In general, halophilic enzymes that are stable at high salt concentrations can be used after or concurrently with chemical processes that use salts, acids, or bases. Psychrophilic enzymes can be applied in situations where the components of a reaction are heat sensitive. Thermophilic enzymes not only are stable under high temperatures but in many cases are also highly resistant to proteases, chaotropic agents, detergents, and organic solvents. All these adaptations of extremozymes could give a more versatile toolbox to couple biotechnological processes with the already efficient and established physicochemical processes in the industry.

The factors that influence the stability of extremozymes include changes in the distribution of hydrophobic patches and charges in the solvent-exposed surface of the protein, reduction in the size of loops and number of cavities, increase in the extent of secondary structure, and more extensive networks of ionic interactions stabilizing the tertiary structure as well as multimeric protein conformation.

For example, comparison of crystallographic structure between thermophilic and mesophilic proteins has revealed an increase of networks of ionic interactions in the former, which can have a stabilizing effect over a much longer range than hydrophobic interactions. Also, site-directed mutagenesis studies have shown that introducing mutations to achieve a more rigid core protein can also increase the stability at high temperatures. Finally, thermophilic enzymes have a lower occurrence of labile residues such as cysteine, asparagine, and aspartate.

Among laccases, several characteristics are proposed to contribute to thermal stability. One of the first identified characteristics is the high glycosylation content in some thermostable laccases, e.g., from *Botrytis cinerea* (49% glycosylation) (Slomczynski et al. 1995), *Ceriporiopsis subvermispora* (Fukushima and Kirk 1995), *Postia placenta* (40% glycosylation) (An et al. 2015), and *Pycnoporus sanguineus* (Dantán-González et al. 2008). For example, six and four N-glycosylation positions were predicted in the two laccases (PPLCC1 and PPLCC2, respectively) of *Postia placenta* (An et al. 2015). While PPLCC1 showed a high thermostability, PPLCC2 was drastically sensible at the temperature. Thus, the glycosylation extent of PPLCC1 (40%) vs 4.7% presents in PPLCC2 might partially explain the higher thermostability of PPLCC1 than PPLCC2 (An et al. 2015). Also, directed evolution of laccases from *P. cinnabarinus* (PcL) (Camarero et al. 2012) and *M. thermophila* (MtL) (Bulter et al. 2003; Zumárraga et al. 2007) showed that mutants with increased stability were hyperglycosylated, with sugar moieties contributing around 50% of their molecular weight. However, other studies suggest that the thermal stability of laccases is not explained exclusively by glycosylation (Koschorreck et al. 2008; Marques De Souza and Peralta 2003). In fact, as with other thermophilic proteins, the crystal structure of the *B. subtilis* CotA shows higher hydrophobic interaction between domains when compared with other monomeric LMCO (laccase-like multicopper oxidase) and laccases and reduced flexibility due to the increased content of proline residues (Enguita et al. 2003).

Contrary to proteins from thermophiles, psychrophilic enzyme adaptations must ensure the required flexibility for the catalytic mechanisms to occur. Therefore, the activity of cold-active enzymes often relies on decreasing the hydrophobicity in the core of the protein and decreasing the number of hydrogen bonds, the ion interaction networks or salt bridges, and the overall charge of the protein. Also, these enzymes must remain active at low water activity, i.e., when the water molecules are less available to interact with proteins as they are forming structured ice lattices. In these conditions, the enzymes must bind more tightly the water to ensure that there are multiple hydration layers supporting enzyme function.

At high salinity, water is sequestered in hydrated ionic structures, similarly as happens in ice structures, limiting the availability of free water molecules for protein hydration. Halophilic enzymes generally have numerous acidic residues located on the surface which increases the negative surface charge and therefore the hydration of the protein. The acidic residues on the surface of halophilic enzymes also reduce the hydrophobicity and promote electrostatic repulsion of the proteins. Additionally, genome-wide analysis of proteins from halophiles has shown a reduction in the number of lysine or larger hydrophobic residues (phenylalanine, isoleucine, and leucine) at the protein surface, consistent with a reduction in surface hydrophobicity and increased flexibility. These properties prevent the "salting-out" effect, i.e., the aggregation of proteins at high salt concentrations.

4 Extremophiles as a Source of Laccases: Some Examples

Extremophile microbes are an attractive source of enzymes for metabolizing recalcitrant compounds such as polycyclic aromatic hydrocarbons, pesticides, pharmaceutical compounds, polysaccharides, lignocellulose, and resins, among others. Enzymes produced by extremophiles are usually extremozymes as they display robust biochemical properties and are functionally active in extreme conditions (Singh et al. 2019a, b). From a biotechnological point of view, these enzymes are very attractive for industries due to their downstream applications as biocatalysts to produce fine chemicals such as biopolymers, biomaterials, and biofuels (Raddadi et al. 2015).

Extreme environments have been routinely explored to discover extremozymes. One example is the bioprospection of thermophiles and hyperthermophiles as a novel source of polysaccharide-degrading enzymes such as amylases, pullulanases, cellulases, hemicellulases, esterases, and pectinases. Some of these lignocellulolytic proteins displayed stability at temperatures higher than 70 °C (Elleuche et al. 2015). High-temperature biotechnological processes demand thermozymes, which should be functional at high temperatures (>60 °C). Functional-based and sequencing-based metagenomics have been successfully driven to describe new thermostable enzymes. Figure 2 shows a schematic representation to identify genes encoding thermozymes.

Microorganisms inhabiting marine ecosystems, especially those found in deep sea, also constitute an interesting source of enzymes. Marine extremophiles include psychrophiles, halophiles, and piezophiles, and their enzymes are functionally active under harsh physical-chemical conditions (Leal et al. 2015). According to this, hydrolases from prokaryotes isolated from these environments have been largely investigated resulting in the characterization of several glycoside hydrolases useful for industrial applications.

The biobased industry is considered one of the best solutions to face global challenges: climate change, agriculture production, generation of new antibiotics and biomaterials, etc. The progress of this industry will positively impact on bioprocesses and will allow developing biological processes to convert biomass to high-value products (Krüger et al. 2018). Biobased economy, as also known, requires sustainable and high efficient eco-friendly technologies based on enzymes. Extreme habitats represent an excellent niche for this industry, since various sectors could be benefited with biocatalysts derived from ecosystems such as hot spring, salt lakes, polar sea ice, etc. Figure 3 shows the contribution of extremophilic microorganisms and their enzymes to biobased industry.

In the search for laccases from extremophiles, many efforts have been made. Miyazaki (2005) purified and characterized a hyperthermophilic laccase obtained from the thermophilic bacterium *Thermus thermophilus* HB27. This enzyme displayed an optimum temperature of 92 °C and a high thermostability (>14 h at 80 °C). The proteins showed activity on the canonical laccase substrates: guaiacol and 2,2′-azino-bis(3-ethylbenzothiazoline-6-sulfonate). This laccase was ranked in 2005 as the most thermophilic laccase reported.

Recently, a new thermophile laccase was reported from *Thermus* sp., and its capacity to delignify lignocellulosic biomass was investigated. This enzyme shows activity on different laccase substrates such as 2,2′-azino-di-(3-ethylbenzothiazoline sulfonate) (ABTS) and 2,6-dimethoxyphenol (DMP). Also, the protein exhibited a high thermostability because retained 80% of its activity was exposed 16 h at 70 °C and revealed a copper-dependence activity (Navas et al. 2019). Moreover, this

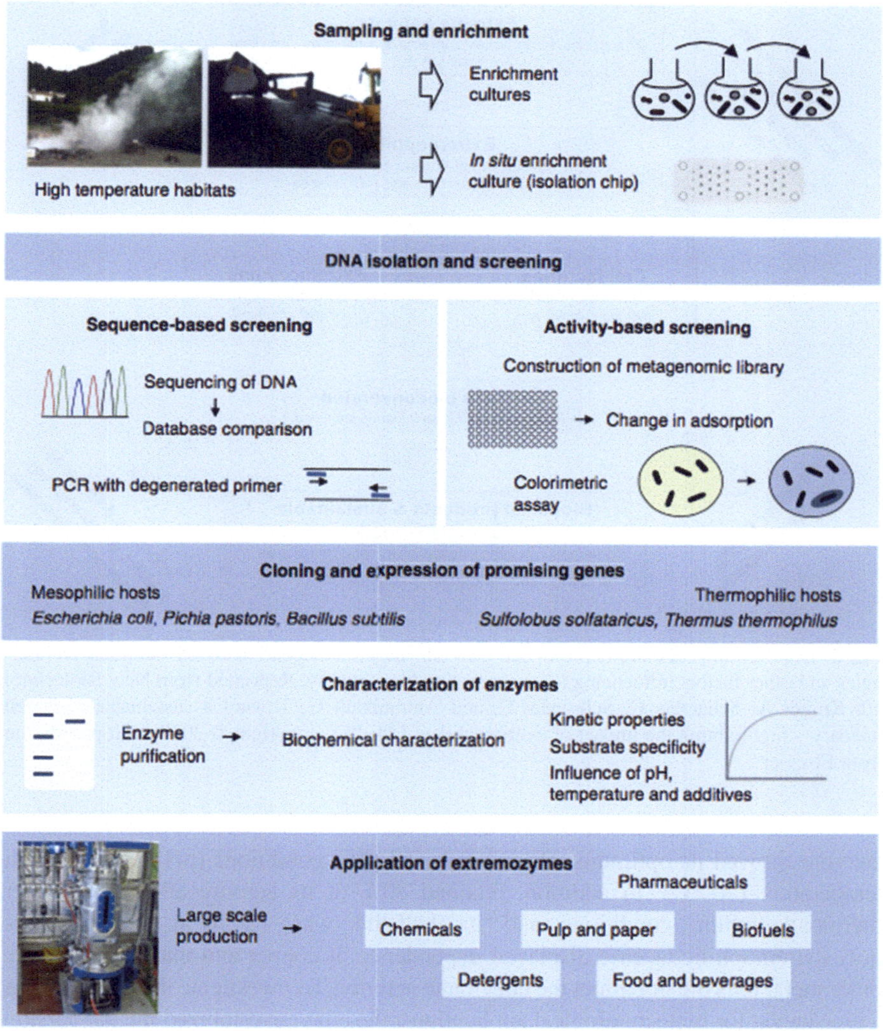

Fig. 2 Schematic representation of high-throughput screenings for genes encoding thermozymes, cloning, and expression followed by biochemical characterization of biocatalysts and subsequent large-scale production for suitable industrial applications. Reprinted from Curr Opin Microbiol 25, Elleuche S, Schäfers C, Blank S, Schröder C, and Antranikian G, Exploration of extremophiles for high-temperature biotechnological processes, 113–119, copyright © 2015, with permission from Elsevier

laccase evidenced biotechnological potentialities for delignification of *Eucalyptus* biomass.

A novel laccase from thermoalkaliphilic bacterium *Caldalkalibacillus thermarum* able to degrade a lignin model compound was described (Ghatge et al. 2018). This

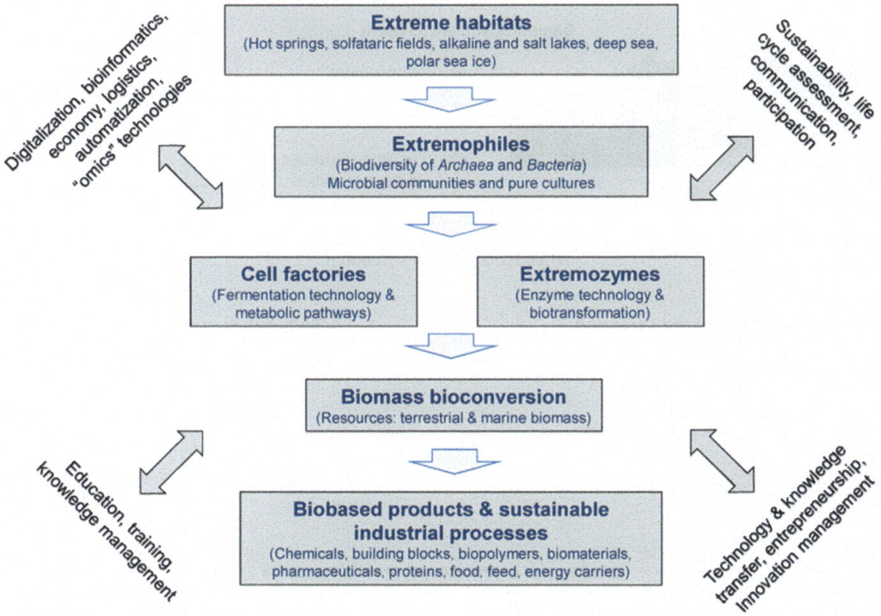

Fig. 3 Contributions of extremophiles and their enzymes to bioeconomy. Cross-sectional technologies and other factors influencing a new emerging bioeconomy. Reprinted from New Biotechnol 40, Krüger A, Schaefers C, Schroeder C, and Antranikian G, Toward a sustainable biobased industry – highlighting the impact of extremophiles, 144–153, copyright © 2018, with permission from Elsevier

enzyme showed its optimum activity under alkaline conditions (pH = 8) and high temperature (70 °C). In addition, retained 80% of its activity after 24 h under thermoalkaliphilic conditions and its half-life was 12 h at 90 °C. This poly-extremophilic laccase displayed dependence of copper and manganese, while other metal ions did not affect the enzymatic activity. To investigate the properties as a biocatalyst for biotechnological applications, the enzyme catalyzed the conversion of guaiacylglycerol-B-guaiacyl ether (GGGE) to C5-C5 biphenyl tetramer, and the mechanisms for dimerization of a dimeric lignin model compound GGGE were proposed as Fig. 4 represents. The laccase of *C. thermarum* oxidized a wide range of substrates (2,6-DMP, ABTS, catechol, caffeic acid, ferulic acid, and phloroglucinol), which indicates that this laccase is a versatile enzyme.

Other thermotolerant bacterial laccases have been also characterized. For example, McoA was described as a thermostable laccase from the bacterium *Aquifex aeolicus* because it showed thermostability at 90 °C for 5 h (Fernándes et al. 2007). McoA was annotated as a multicopper oxidase and presents higher specificity for cuprous and ferrous ions than for aromatic compounds. According to this, McoA was classified as a metallo-oxidase. *Bacillus* has been also described as a bacteria genus with high potential to produce thermotolerant enzymes. *Bacillus* sp. PC-3 was

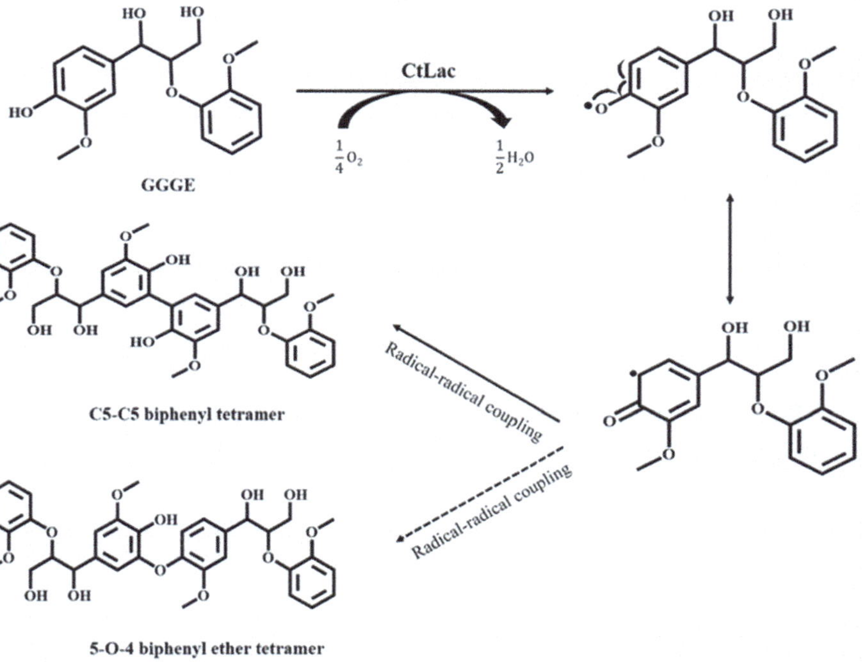

Fig. 4 Proposed mechanism of laccase for dimerization of a dimeric lignin model compound GGGE. Reprinted by permission from Springer Nature: Springer, Appl Microbiol Biotechnol 102 (9), 4075–4086, A novel laccase from thermoalkaliphilic bacterium *Caldalkalibacillus thermarum* strain TA2.A1 able to catalyze dimerization of a lignin model compound, Ghatge S, Yang Y, Song WY, Kim TY, and Hur HG, copyright © Springer-Verlag GmbH Germany, part of Springer Nature 2018

reported as a novel thermophilic bacterium that excretes a thermostable laccase with a significant activity at a wider pH range (Sharma et al. 2018). In addition, *Anoxybacillus* sp. was also reported as a new thermophilic bacterium with hyperthermostable alkaline laccase activity (Al-balawi et al. 2017).

But not only thermophilic laccases have been studied. Halophilic bacterial laccases are also very attractive for biotechnological applications. With this in mind, a laccase produced by *Aquisalibacillus elongatus* was investigated resulting that the purified enzymes showed high stability against organic solvents, salts, metals, inhibitors, and surfactants. The enzyme is active in a huge range of pH (from 5 to 10) and revealed to be useful for industrial applications related to lignin deconstruction (Rezaei et al. 2017).

From Arctic marine environments, 13 laccase-positive *Psychrobacter* species were identified (Moghadam et al. 2016). It was demonstrated that some laccases are located on plasmids while others in the chromosomes of these psychrophilic bacteria. The laccase multicopper oxide exhibited activity on ABTS and guaiacol, and the enzymes also operated by Mg^{2+} and Ca^{2+} addiction. In addition, *Cladosporium tenuissimum* was classified as psychrotolerant bacterium that produces extracellular laccase activity at low temperature (Dhakar and Pandey 2016).

Also, laccases with high pH tolerance range have been also studied in the extremophile bacteria *Thioalkalivibrio* sp. (Ausec et al. 2015). This genus seems to be a potential source of robust bacterial laccases with a high tolerance to temperature and pH (Ausec et al. 2015). On other hand, *Geobacillus thermopakistaniensis* isolated from hot spring in Pakistan produces extracellular laccase isoforms highly stable to several halides (NaF, NaCl, NaBr) and organic solvents (methanol, ethanol, propanol, isopropanol) (Basheer et al. 2017).

5 Metagenomics as a Powerful Method for Laccase Bioprospection

Over the last two decades, omic technologies have changed the paradigm of enzyme bioprospection from extreme environments. Metagenomic-based approaches have allowed to explore a wide variety of different ecosystems and exploit their microbial biodiversity. Laccase bioprospection from extreme environments has gained attention since novel oxidoreductases could be discovered using both dependent and independent culture approaches. Figure 5 displays a general approach that could be useful for laccase bioprospection. Transcriptomic and proteomic can be considered as powerful strategies to discover laccases from extremophiles. On the other hand, functional metagenomics allows exploring genomes to identify genes encoding laccases. Independent culture methods could result in the discovery of new laccases with robust properties over a wider range of environmental conditions. However, from our point of view, one of the limitations for functional-based metagenomic screening is the activity-based method. Typically, these are based on the degradation of a specific substrate, and generally, a color change around the metagenomic clone is used to identify the positive clone screened. For laccases, substrates should be carefully selected to increase the success rate. PCR-based strategies also are an alternative to amplify laccase genes from a metagenome isolated from extreme environments. To do so, three different strategies could be used as Fig. 6 displays. PCR-based methods have been poorly employed to detect laccases from extreme ecosystems.

A novel polyphenol oxidase actives on a huge range of substrates (syringaldazine, 2,6 dimethoxyphenol, veratryl alcohol, guaiacol, tetramethylbenzidine, 4-methoxybenzyl alcohol, 2,2-azino-bis(3-ethylbenzothiazoline-6-sulfonic acid, phenol red) were characterized from metagenomic libraries obtained from bovine rumen microbiota (Beloqui et al. 2006). This enzyme showed activity in pH from 3.5 to 9.0 and was characterized as a new family of polyphenol oxidases. Other alkaline laccase was identified from a metagenomic library obtained from mangrove soil (Ye et al. 2010). However, laccase detection from functional-based screening is very difficult. In this sense, during the functional analysis of 52,000 metagenomic clones

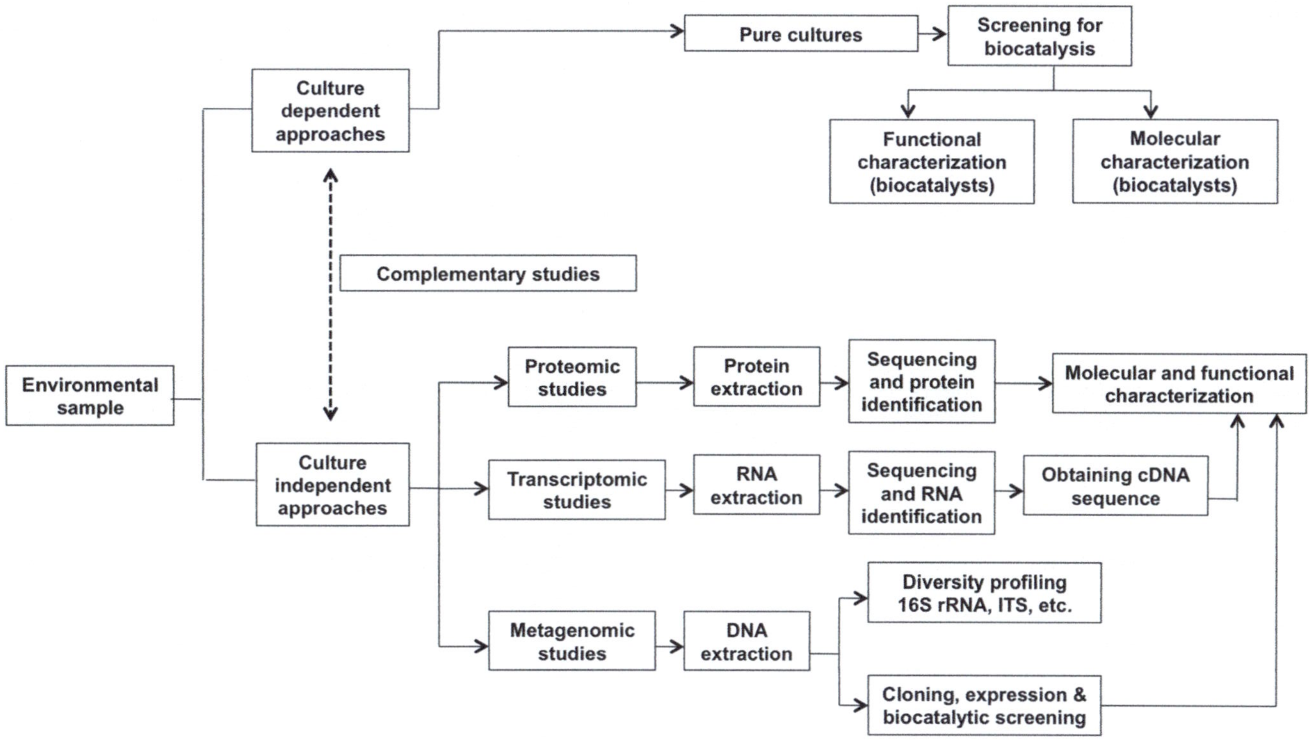

Fig. 5 Culture-dependent and culture-independent approaches for biocatalytic screenings. Reprinted from Batista RA, del Rayo M, Talia P, Jackson SA, O'Leary ND, Dobson AD, and Folch-Mallol JL, From lignocellulosic metagenomes to lignocellulolytic genes: trends, challenges, and future prospects, Biofuel Bioprod Bior 10(6), 864–882, with permission from John Wiley & Sons (copyright © 2016 Society of Chemical Industry and John Wiley & Sons, Ltd)

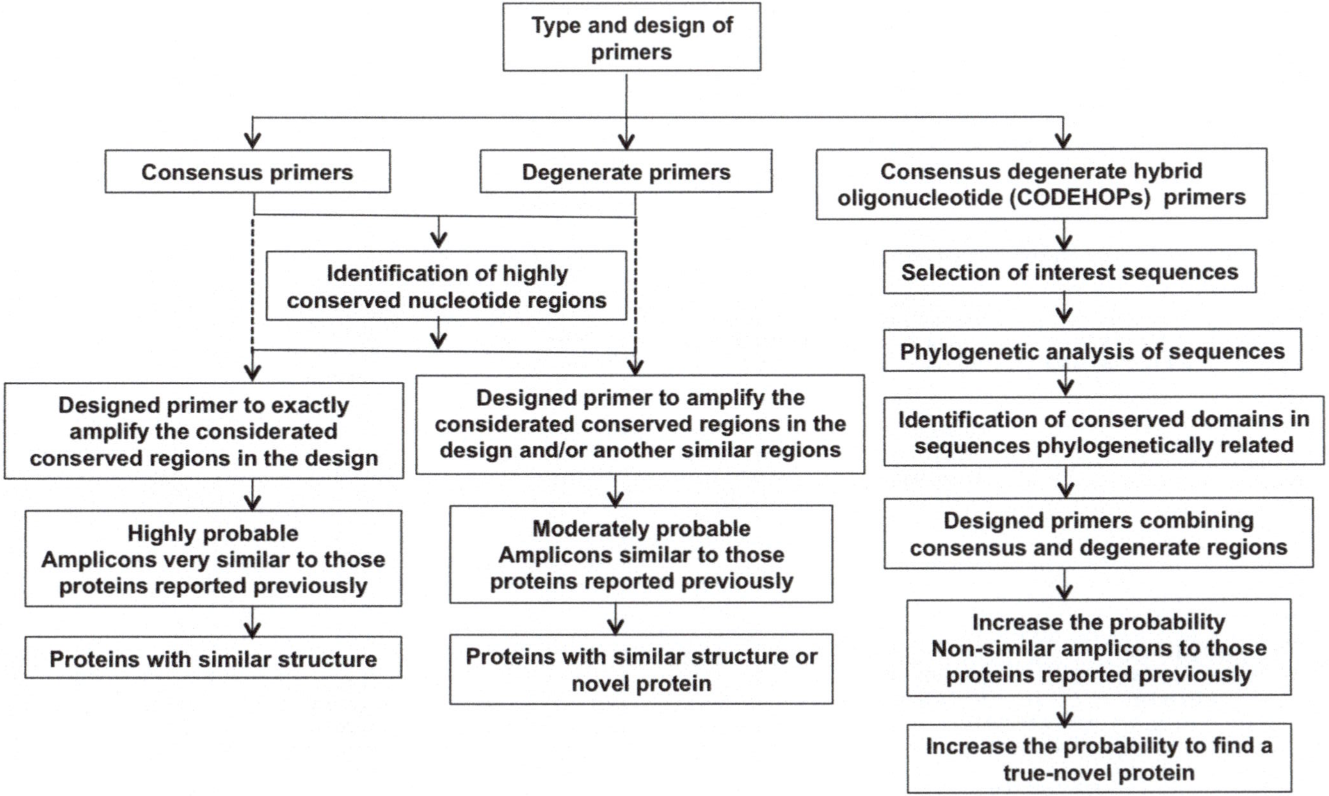

Fig. 6 Consensus, degenerate, and CODEHOPs primers for functional screenings. Reprinted from Batista RA, del Rayo M, Talia P, Jackson SA, O'Leary ND, Dobson AD, and Folch-Mallol JL, From lignocellulosic metagenomes to lignocellulolytic genes: trends, challenges, and future prospects, Biofuel Bioprod Bior 10(6), 864–882, with permission from John Wiley & Sons (copyright © 2016 Society of Chemical Industry and John Wiley & Sons, Ltd)

obtained from Antarctic DNA, any laccase activity was detected, while lipases, esterases, and cellulases were detected (Ferrés et al. 2015).

In addition to activity-based screening, sequence-screening strategies have been used for laccase detection. Using this methodology a new bacterial laccase was characterized from marine environments in South China Sea. This enzyme was overexpressed in *E. coli* and is active on syringaldazine and in high temperature. Moreover, this marine-derived laccase showed an alkaline activity, which is enhanced by chloride addiction (Fang et al. 2012). Genome mining is also useful for laccase bioprospection. It has demonstrated when an alkaline and thermostable laccase Ssl1 from *Streptomyces sviceus* was characterized (Gunne and Urlacher 2012). This phenol oxidase is a small laccase, and its activity is not affected by addiction of detergents and organic solvents. Metagenomic approaches have also derived in the crystallization of laccase. For example, a putative two-domain-type laccase detected from a metagenomic screening was characterized and structurally studied by X-ray analysis (Komori et al. 2009).

6 Pushing the Boundaries of Natural Laccases

As previously seen, there is great potential in screening microorganisms from diverse environments to obtain novel catalysts for industrial applications. Nevertheless, natural enzymes are not specifically suited to perform in settings with all the requirements of industrial processes, as the evolutionary pressures that have driven their changes were oriented to ensure fitness in a biological environment. As our understanding of the molecular dynamics and behavior of protein increases, new and innovative approaches to improve enzyme function and properties are gaining relevance. The engineering of protein structure and function can increase and expand enzyme applications in the industry. Obtaining engineered enzymes is generally achieved by two different strategies: directed evolution or rational design. Other strategies are the result of the combination of these methods.

Directed evolution is one of the most powerful approaches for improvement of enzyme function. The rationale of this method is to mimic the effect of evolution by applying a selective pressure through several generations of a protein population. In each generation, mutations are (randomly or selectively) introduced in the gene and evaluated to assess if the resulting protein can pass to the next generation. The major advantage of this method is that prior knowledge of the protein structure or molecular interactions is not required. The limitations though are the technical availability of a high-throughput screening method that can help select the proteins with the desired enhanced property and the high combinatorial number of mutations that must be screened if the sequence space is not constrained.

The method of directed evolution has been implemented both in vitro and in silico with very good results. In vitro strategies generally rely on the construction of gene libraries that are expressed in an appropriate host. Mutations can be introduced randomly by using physical or chemical mutagenic treatments or specifically by using several molecular biology techniques. To simulate the process in silico, an initial protein structure must be obtained or modeled, and the screening methods rely on evaluating protein structural properties that are previously known to correlate with protein function.

Many of the factors that determine enzyme stability and extremophile performance have been studied and are partially understood. Therefore, several attempts have been made to apply that knowledge for obtaining engineered proteins that display desirable characteristics. In the case of laccases, many efforts have been oriented to increase the redox potential of the enzymes in order to improve their catalytic properties, but some studies have been also pursued an increase in optimal temperature and stability. For example, Mateljak et al. (2019) screened a library of laccase variants derived by multiple strategies of recombination and mutation of an original gene from a basidiomycete strain (PM1). The natural laccase produced by this strain had an optimal enzymatic activity at 80 °C (Coll et al. 1993a, b), ensuring a good starting point for improvement. The library was subjected to sequential rounds of computationally guided mutagenesis and directed evolution, and the resulting laccase product not only showed increased redox potential but also had increased stability to thermal denaturation (near 3-fold increase in the halftime of the enzyme activity at 70 °C) and pH (around 18-fold increase in stability at pH 2 after 96 h) when compared to a previously obtained mutant. One interesting finding in this study was that the introduction of hydrophobic residues that "sheltered" the copper catalytic site from solvent exposure increased the stability of the enzymatic activity. This phenomenon was also observed in some natural laccase with high thermal stability, suggesting this could be a structural feature of thermophilic/thermotolerant laccases (Mateljak et al. 2019). In another example, Sheng et al. (2017) used the methodology of *Bacillus subtilis* spore display to screen laccase mutants with increased pH stability, reaching a 62-fold increase in half lifetime at pH 4 when compared with the wild-type laccase.

In contrast to directed evolution, the rational design method is based on the understanding of the enzyme structure-function relationship. In this case, targeted mutations are oriented to change a structural property of the protein that is already known to be involved in its function. In the case of laccases, an example of site-directed mutagenesis showed that changes in a loop of a native laccase from *Bacillus* HR03 increased thermal stability and indirectly improved kinetic parameters, apparently by the stabilizing effect of the mutations (Mollania et al. 2011).

Figure 7 illustrates all the functional properties of laccases that have been engineered with particular industrial purposes in mind. All these examples show that improving the function and stability of laccases is a promising field of research that will enable obtaining more efficient biocatalysts with improved stability to be used in industrial processes.

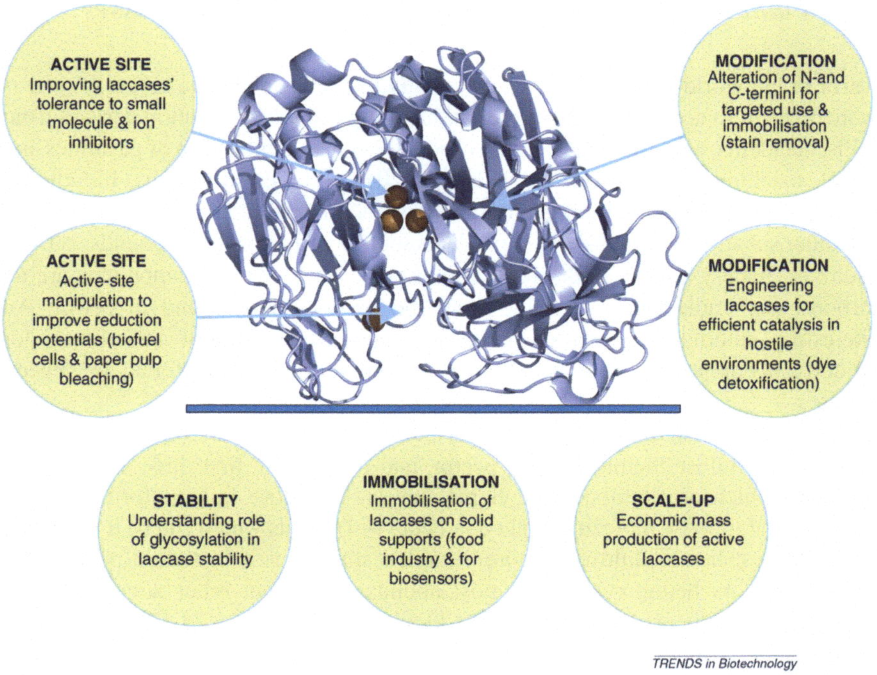

Fig. 7 Designer laccases. Modified, mass-produced laccases with improved stability, activity, and specificity will emerge tailor-made for disparate industrial purpose. Reprinted from Trends Biotechnol 28(2), Rodgers CJ, Blanford CF, Giddens SR, Skamnioti P, Armstrong FA, and Gurr SJ, Designer laccases: a vogue for high-potential fungal enzymes? 63–72, copyright © 2010, with permission from Elsevier

7 Biotechnological Applications of Extremophilic Laccases

Extremophilic enzymes usually withstand harsh conditions (pH, salinity, high or low temperatures, etc.), so many have been proven to show advantages for biotechnological processes, which mesophilic enzymes cannot withstand (Chandra et al. 2017). These enzymes could perform better in several applications in environmental, industrial, medical, chemical, pharmaceutical, textile, agriculture, food or animal feed, and biofuels industries (Ibrahim and Ma 2017). Besides, an important increase in the global market demand of enzymes has been registered, and it was estimated to be USD 4.61 billion in 2016 and is projected to reach USD 6.30 billion by 2022 (Industrial Enzymes Market 2019).

Here we will review some aspects of the biotechnological uses of extremophilic laccases.

7.1 Plant Biomass Decomposition for Biorefinery Processes

Petrol in the world is becoming scarce or found at deep-sea wells, so every time is more difficult to acquire. Furthermore, oil handling provokes pollution derived from spills, accidents, and use of oil-derived products such as combustion of gasolines and diesel, machinery wasted oil, etc. (Thomas 2017). For these reasons, biorefinery technology has been developing to obtain biofuels (mostly bioethanol and biodiesel) and other substitutes for oil-derived products such as waterproof compounds, cement additives, battery components, etc. (Parajuli et al. 2015). These compounds are all carbon-based molecules that come from the hydrocarbons found in petrol. An interesting alternative carbon source to manufacture current substitutes for oil-derived compounds is lignocellulose, a complex composite found in vegetal biomass.

The cell wall in plants and algae is composed mainly of four carbon-based polymers: cellulose, hemicellulose, pectin, and lignin. The first three are polysaccharides, which are imbibed in a lignin matrix. Cellulose is a uniform polymer composed of glucose molecules linked in beta-(1–4) bonds which can be hydrolyzed by cellulases; hemicellulose is more complex since it is a branched polymer of different sugars being xylose the most abundant one, but other sugars such as arabinose, and glucose, or organic acids like acetic or ferulic acid can be linked to the structure. Different enzymes are then needed to obtain the monomeric sugars or acids from hemicellulose. Pectin is mostly composed of linear chains of alpha-(1–4) linked D-galacturonic acid (Toushik et al. 2017). As can be known, lignocellulose is a great feedstock from which many different carbohydrates can be obtained (Toushik et al. 2017; Guo et al. 2018; Lange 2018).

However, lignin is a polyphenolic polymer that hampers the access of enzymes to the polysaccharides found in plant biomass. This polymer contains different types of bonding among its aromatic rings, mainly C–O–C bonds (β-O-4/4′, α-O-4/4′, α/γ-O-γ, 4-O-5/5′, etc.) and C–C bonds (5–5/5′, β-β, β-1, β-5, etc.), which constitute complex linkages difficult to degrade by enzymes (Zakzeski et al. 2010; Zhou et al. 2011).

Fortunately, a certain type of microorganisms (bacteria and fungi) is able to oxidize and thus depolymerize lignin to gain access to the polysaccharide matrix. This process is achieved by the secretion of oxidative enzymes such as peroxidases, monooxygenases, and most importantly laccases (Roth and Spiess 2015).

In biorefinery applications are necessary to pre-treat the biomass to separate the components. Currently, pre-treatments for biomass are carried out using physical (steam explosion, high temperatures) or chemical (acid or alkaline treatments, ozonolysis, etc.) procedures (Barrera et al. 2016). An alternative is to use enzymes, and laccase has been used to depolymerize lignin in order to get access to the polysaccharides, which are then saccharified, by cellulases and hemicellulases to obtain sugar-rich syrups that can be fermented by yeasts or bacteria to obtain bioethanol (Molina et al. 2014). Most laccases are optimally active at low pHs and high temperatures, but this provokes that activity is lost rapidly due to enzyme inactivation (see Table 1 in Baldrian 2006). It is then important that these enzymes can withstand extreme conditions because then pre-treatments using laccase in combination with chemical or physical pre-treatments could result in a beneficial synergistic mode of obtaining biomass components. Several reports have shown promising results on the use of extremophilic laccases to enhance carbohydrate recovery from biomass (Brander et al. 2014; Chen et al. 2012; Rezaie et al. 2017; Guerriero et al. 2015).

7.2 *Pulp and Paper Industry*

Paper is produced from wood and consists mainly of cellulose, so it is necessary to remove the lignin from the wood pulp. Currently, oxygen chlorine-based delignification, oxygen delignification, or bleaching procedures are used to treat the wood pulp in order to manufacture paper (Carter et al. 1997; Ramesh et al. 2017). However, these methods cause environmental concerns because of the chemical wastes that are generated (e.g., hydrogen peroxide or chlorine derivatives). A more environmentally friendly alternative is the use of enzymes, which besides maintain better integrity of cellulose due to its specificity (Yang et al. 2018a, b). However, there are still challenges regarding costs of production (Ramesh et al. 2017). So a cost-effective, environmentally friendly and specific process could be achieved using extremophilic enzymes, especially laccase, which specifically removes lignin, preserving the carbohydrate structure of wood pulp (Rezaie et al. 2017; Navas et al. 2019).

7.3 *Textile Industry*

The dyes used in the textile industry are synthetic and thus many times toxic and recalcitrant due to its chemical structure, which renders them resistant to decolorization by chemicals such as detergents, high temperatures, and reactive oxidative species such as H_2O_2 or light (Chandra et al. 2017). Besides, the dying process itself is carried out in conditions in which mesophilic enzymes would denature (high temperatures, acidic pH, high NaCl concentrations, organic products, etc.). So, wastewaters from the textile industry are a problem because they can have catastrophic effects in rivers, lakes, and other water bodies because they reduce light income to the system and reduce oxygen solubility, and many of them are toxic due to heavy metals in their structure, or they can generate amines which can cause mutations and are carcinogenic.

For this reason, wastewater from textile industries must be treated before discharging, although currently most of it goes untreated to the ecosystems. Traditionally, chemical treatments have been used for dye degradation such as ozonification, photocatalytic degradation, oxidation by Fenton's reagent, etc. Also, physical methods such as adsorption or filtration are used. However, a large hope is placed in biodegradation of industrial dyes by extremophilic microorganisms, especially those that can produce laccases (Amoozegar et al. 2015). Since extremophilic laccases can withstand harsh conditions, they can be combined with other methods that require low pH values and high temperatures or metal ions in coexistence of inorganic solvents, for example (Hayat et al. 2015; Zhuo et al. 2015; Singh et al. 2019a, b). Being unspecific oxidases, extremophilic laccases are promising enzymes for synthetic dye decolorization (Rodríguez-Couto 2019). For example, a thermostable laccase from the basidiomycete fungus *Pycnoporus sanguineus* was heterologously expressed in *Trichoderma atroviride* achieving a substantial amount of azo dyes decolorization (Balcázar et al. 2016).

7.4 Bioremediation

Currently, a plethora of xenobiotic compounds is released into the biosphere both from natural and artificial sources. Natural xenobiotic compounds are mainly heavy metals (Ortiz et al. 2018) and polycyclic aromatic hydrocarbons (PAHs) from incomplete wood combustion or volcanic emanations (Hussein and Mansour 2016). Anthropogenic sources of xenobiotics also include PHAs from the oil and gas industry, together with pesticides, industrial dyes, explosives, endocrine disruptors, and pharmaceutical products. These compounds are found in air, water, and soils, so their impact in living forms is disastrous (Giuliani et al. 2015; Hussein and Mansour 2016). Laccase promiscuity toward its substrates has been widely used to treat xenobiotic pollution, especially in soil and water, but also in the air (Sharma et al. 2018; Vergara et al. 2018). An advantage of laccases is that they can use low-molecular-weight mediators to oxidize compounds with nonoptimal redox potential for the enzyme, so they have proved to degrade a huge variety of xenobiotic compounds (Sharma et al. 2018).

For the treatment of xenobiotic compounds, enzyme immobilization is a method that has resulted in a more efficient process to cope with enzyme stability and re-usage. Several substrates have been used to immobilize enzymes that can now be used in columns or aqueous beds to treat polluted waters (Zhong et al. 2017; Zheng et al. 2016). An advantage of this technique is that natural substrates as porous rocks or natural fibers can be used to immobilize the laccases, and then recycling results in low-cost strategies (Abdel et al. 2013).

A problem to consider in which extremophile laccases could help is that many of the aforementioned xenobiotic compounds have very limited solubility in water. This fact can reduce the concentration at which many PAHs, pesticides endocrine disruptors, etc. can be oxidized by these kinds of enzymes. To overcome this difficulty, organic solvents could be used to prepare solutions in which the xenobiotic compound could be more concentrated, but to achieve this, the laccase must be robust (Fig. 8). Extremophilic laccases fulfill this criterion (Lončar et al. 2016; Yang et al. 2018a, b).

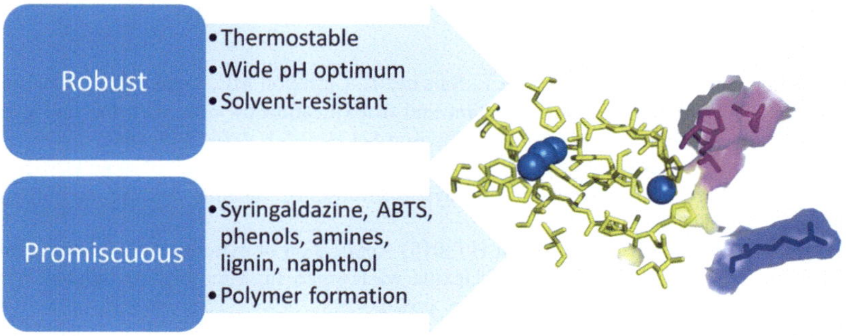

Fig. 8 Biochemical properties and substrates exhibited for several laccases. Reprinted from J Mol Catal B Enzym 134, Lončar N, Božić N, and Vujčić Z, Expression and characterization of a thermostable organic solvent-tolerant laccase from *Bacillus licheniformis* ATCC 9945a, 390–395, copyright © 2016, with permission from Elsevier

8 Application: Perspectives

Stability, half-time life, amount of activity, and cost are factors that still hamper the use of enzymes for biotechnological purposes. To look for commercial applications which are cost rendering, the temperature-dependent catalytic properties of laccases are most important. Extremophilic organisms have been selected to produce enzymes with robust characteristics regarding thermal stability, organic solvent resistance, performance in the presence of metals or high salt conditions, etc. Thus, these organisms represent an invaluable source of enzymes which, if properly isolated and characterized, could bridge the cost-limiting effect of denaturation of mesophilic enzymes. Metagenomic prospection is also a most promising tool for the isolation of extremophilic laccases without the need of culturing extremophile organisms, which precisely by their nature could be an impairment (Batista et al. 2016). An important bottleneck for obtaining industrially cost-effective enzymes is laccase expression. Usually, yeasts have been a choice for laccase expression since they are single-celled organisms that proliferate in simple media and have a rapid doubling time. However, although some promising reports are arising using *Pichia pastoris* (Zerva et al. 2019), best results are obtained with filamentous fungi as host for the heterologous laccase genes (Wikee et al. 2019; Wang et al. 2017). Although laccases were originally discovered in plants, expression in these organisms has proved to be difficult (De-Jesús et al. 2019). Despite some limitations, extremophilic laccases still represent the best option to use for biotechnological applications.

References

Abdel S, Abdel M, Folch JL, Perezgasga L, Sánchez E, Castrejón ML, Ortiz ML (2013) Optimization of methyl parathion biodegradation and detoxification by cells with the opd gene in suspension or immobilized on tezontle. J Environ Sci Health B 48(6):449–461

Al-balawi T, Wood A, Solis A, Cooper T, Barabote R (2017) *Anoxybacillus* sp. strain UARK-01, a new thermophilic soil bacterium with hyperthermostable alkaline laccase activity. Curr Microbiol 74:762–771

Amoozegar MA, Mehrshad M, Akhoondi H (2015) Application of extremophilic microorganisms in decolorization and biodegradation of textile wastewater. In: Microbial degradation of synthetic dyes in wastewaters. Springer, Cham, pp 267–295

An H, Xiao T, Fan H, Wei D (2015) Molecular characterization of a novel thermostable laccase PPLCC2 from the brown rot fungus *Postia placenta* MAD-698-R. Electron J Biotechnol 18:451–458. https://doi.org/10.1016/j.ejbt.2015.09.004

Ausec L, Crnigoj M, Snajder M, Poklar N, Mandic I (2015) Characterization of a novel high-pH-tolerant laccase-like multicopper oxidase and its sequence diversity in *Thioalkalivibrio* sp. Appl Microbiol Biotechnol 99:9987–9999

Balcázar E, Méndez LH, Batista RA, Esquivel U, Ayala M, Kumar VV, Savary O, Cabana H, Herrera A, Folch JL (2016) Xenobiotic compounds degradation by heterologous expression of a *Trametes sanguineus* laccase in *Trichoderma atroviride*. PLoS One 11(2):e0147997. https://doi.org/10.1371/journal.pone.0147997

Baldrian P (2006) Fungal laccases–occurrence and properties. FEMS Microbiol Rev 30 (2):215–242. https://doi.org/10.1111/j.1574-4976.2005.00010.x

Barrera I, Guzmán N, Peña E, Vázquez T, Cerón R, Folch J, Salazar J, Aburto J (2016) Ozonolysis of alkaline lignin and sugarcane bagasse: structural changes and their effect on saccharification. Biomass Bioenergy 94:67–172

Basheer S, Rashdi N, Ashraf R, Sohail M, Siddiqui M, Imanaka T, Aktar R (2017) Identification of a novel copper-activated and halide tolerant laccase in *Geobacillus thermopakistaniensis*. Extremophiles 21:563–571

Batista RA, del Rayo M, Talia P, Jackson SA, O'Leary ND, Dobson AD, Folch-Mallol JL (2016) From lignocellulosic metagenomes to lignocellulolytic genes: trends, challenges and future prospects. Biofuels Bioprod Biorefin 10(6):864–882

Beloqui A, Pita M, Polaina J, Martínez-Arias A, Golyshina O, Zumárraga M, Yakimov M, García-Arellano H, Alcalde M, Fernández V, Elborough K, Andreu J, Ballesteros A, Plou F, Timmis K, Ferrer M, Golyshin P (2006) Novel polyphenol oxidase mined from a metagenome expression library of bovine rumen. Biochemical properties, structural analysis, and phylogenetic relationships. J Biol Chem 281:22933–22942

Brander S, Mikkelsen JD, Kepp KP (2014) Characterization of an alkali- and halide-resistant laccase expressed in *E. coli*: cot a from *Bacillus clausii*. PLoS One 9(6):e99402. https://doi.org/10.1371/journal.pone.0099402

Bulter T, Alcalde M, Sieber V, Meinhold P, Schlachtbauer C, Arnold FH (2003) Functional expression of a fungal laccase in *Saccharomyces cerevisiae* by directed evolution. Appl Environ Microbiol 69:987–995. https://doi.org/10.1128/AEM.69.2.987

Camarero S, Pardo I, Cañas AI, Molina P, Record E, Martínez AT, Martínez MJ, Alcalde M (2012) Engineering platforms for directed evolution of laccase from *Pycnoporus cinnabarinus*. Appl Environ Microbiol 78:1370–1384. https://doi.org/10.1128/aem.07530-11

Carter DN, McKenzie DG, Johnson AP, Idner K (1997) Performance parameters of oxygen delignification. Tappi J (USA), Erratum 81(3):12

Chandra R, Kumar V, Yadav S (2017) Extremophilic ligninolytic enzymes. In: Extremophilic enzymatic processing of lignocellulosic feedstocks to bioenergy. Springer, Cham, pp 115–154

Chen GQ, Jiang XR (2018) Next generation industrial biotechnology based on extremophilic bacteria. Curr Opin Biotechnol 50:94–100. https://doi.org/10.1016/j.copbio.2017.11.016

Chen SC, Wu PH, Su YC, Wen TN, Wei YS, Wang NC, Hsu CA, Wang HJA, Shyur LF (2012) Biochemical characterization of a novel laccase from the basidiomycete fungus *Cerrena* sp. WR1. Protein Eng Des Sel 25(11):761–769. https://doi.org/10.1093/protein/gzs082

Coll PM, Fernandez-Abalos JM, Villanueva JR, Santamaria R, Perez P (1993a) Purification and characterization of a phenoloxidase (laccase) from the lignin-degrading basidiomycete PM1 (CECT 2971). Appl Environ Microbiol 59(8):2607–2613

Coll PM, Tabernero CARLOS, Santamaría RAMON, Pérez PILAR (1993b) Characterization and structural analysis of the laccase I gene from the newly isolated ligninolytic basidiomycete PM1 (CECT 2971). Appl Environ Microbiol 59(12):4129–4135

Dantán-González E, Vite-Vallejo O, Martínez-Anaya C, Méndez-Sánchez M, González MC, Palomares LA, Folch-Mallol J (2008) Production of two novel laccase isoforms by a thermotolerant strain of *Pycnoporus sanguineus* isolated from an oil-polluted tropical habitat. Int Microbiol 11:163–169. https://doi.org/10.2436/20.1501.01.56

De-Jesús R, Folch JL, Dubrovsky JG (2019) Transgenic callus of *Nicotiana glauca* stably expressing a fungal laccase maintains its growth in presence of organic contaminants. Plant Cell Tiss Org 138:311–324. https://doi.org/10.1007/s11240-019-01626-2

Dhakar K, Pandey A (2016) Extracellular laccase from a newly isolated psychrotolerant strain of *Cladosporium tenuissimum* (NFCCI 2608). Proc Natl Acad Sci India 83:685–690

Elleuche S, Schäfers C, Blank S, Schröder C, Antranikian G (2015) Exploration of extremophiles for high temperature biotechnological processes. Curr Opin Microb 25:113–119

Enguita FJ, Martins LO, Henriques AO, Carrondo MA (2003) Crystal structure of a bacterial endospore coat component. J Biol Chem 278:19416–19425. https://doi.org/10.1074/jbc.m301251200

Fang Z, Li T, Wang Q, Zhang X, Peng H, Fang W, Hong Y, Ge H, Xiao Y (2012) A bacterial laccase from marine microbial metagenome exhibiting chloride tolerance and dye decolorization ability. Appl Microbiol Biotechnol 89:1103–1110

Fernándes AT, Soares CM, Pereira MM, Huber R, Grass G, Martins LO (2007) A robust metallo-oxidase from the hyperthermophilic bacterium *Aquifex aeolicus*. FEBS J 274(11):2683–2694

Ferrés I, Amarelle V, Noya F, Fabiano E (2015) Construction and screening of a functional metagenomic library to identify novel enzymes produced by Antarctic bacteria. Adv Pol Sci 26:96–101

Fukushima Y, Kirk TK (1995) Laccase component of the *Ceriporiopsis subvermispora* lignin-degrading system. Appl Environ Microbiol 61:872–876

Ghatge S, Yang Y, Song WY, Kim TY, Hur HG (2018) A novel laccase from thermoalkaliphilic bacterium *Caldalkalibacillus thermarum* strain TA2. A1 able to catalyze dimerization of a lignin model compound. Appl Microb Biotech 102(9):4075–4086

Giuliani S, Piazza R, El Moumni B, Polo FP, Vecchiato M, Romano S, Zambon S, Frignani M, Bellucci LG (2015) Recognizing different impacts of human and natural sources on the spatial distribution and temporal trends of PAHs and PCBs (including PCB-11) in sediments of the Nador lagoon (Morocco). Sci Tot Environ 526:346–357

Gómez-Silva B (2018) Lithobiontic life: "atacama rocks are well and alive". Antonie Van Leeuwenhoek 111(8):1333–1343

Guerriero G, Hausman JF, Strauss J, Ertan H, Siddiqui KS (2015) Destructuring plant biomass: focus on fungal and extremophilic cell wall hydrolases. Plant Sci 234:180–193

Gunde-Cimerman N, Zalar P, Petrovic U, Plemenitas A (2003) Fungi in salterns. In: Ventosa A (ed) Halophilic microorganisms. Springer, Berlin, Heidelberg, pp 103–113. https://doi.org/10.1007/978-3-662-07656-9_7

Gunne M, Urlacher V (2012) Characterization of the alkaline laccase Ssl1 from *Streptomyces sviceus* with unusual properties discovered by genome mining. PLoS One 7(12):e52360

Guo H, Wang XD, Lee DJ (2018) Proteomic researches for lignocellulose-degrading enzymes: a mini-review. Bioresour Technol 265:532–541

Hayat H, Mahmood Q, Pervez A, Bhatti ZA, Baig SA (2015) Comparative decolorization of dyes in textile wastewater using biological and chemical treatment. Sep Purif Technol 154:149–153

Hussein AS, Mansour MS (2016) A review on polycyclic aromatic hydrocarbons: source, environmental impact, effect on human health and remediation. Egyp J Petrol 25(1):107–123

Ibrahim N, Ma K (2017) Industrial applications of thermostable enzymes from extremophilic microorganisms. Curr Biochem Engin 4(2):75–98

Industrial Enzymes Market (2019). Available from: www.marketsandmarkets.com/MarketReports/IndustrilEnzyme. Accessed:29 May 2019

Komori H, Miyazaki K, Higuchi Y (2009) Crystallization and preliminary X-ray diffraction analysis of a putative two-domain-type laccase from a metagenome. Cryst Com 65:264–266

Koschorreck K, Richter SM, Ene AB, Roduner E, Schmid RD, Urlacher VB (2008) Cloning and characterization of a new laccase from *Bacillus licheniformis* catalyzing dimerization of phenolic acids. Appl Microbiol Biotechnol 79:217–224. https://doi.org/10.1007/s00253-008-1417-2

Krüger A, Schaefers C, Schroeder C, Antranikian G (2018) Towards a sustainable biobased industry–highlighting the impact of extremophiles. New Biotechnol 40:144–153

Lange JP (2018) Lignocellulose liquefaction to biocrude: a tutorial review. ChemSusChem 11 (6):997–1014

Leal G, Ferreira D, Vermelho A (2015) Marine extremophiles. A source of hydrolases for biotechnological applications. Mar Drugs 13:1925–1965. https://doi.org/10.3390/md13041925

Lončar N, Božić N, Vujčić Z (2016) Expression and characterization of a thermostable organic solvent-tolerant laccase from *Bacillus licheniformis* ATCC 9945a. J Mol Catal B: Enzymatic 134:390–395

Marques De Souza CG, Peralta RM (2003) Purification and characterization of the main laccase produced by the white-rot fungus *Pleurotus pulmonarius* on wheat bran solid state medium. J Basic Microbiol 43:278–286. https://doi.org/10.1002/jobm.200390031

Mateljak I, Monza E, Lucas MF, Guallar V, Aleksejeva O, Ludwig R, Leech D, Shleev S, Alcalde M (2019) Increasing redox potential, redox mediator activity, and stability in a fungal laccase by computer-guided mutagenesis and directed evolution. ACS Catal 9:4561–4572. https://doi.org/10.1021/acscatal.9b00531

Merino N, Aronson HS, Bojanova DP, Feyhl-Buska J, Wong ML, Zhang S, Giovannelli D (2019) Living at the extremes: extremophiles and the limits of life in a planetary context. Front Microbiol 10:780

Miyazaki K (2005) A hyperthermophilic laccase from *Thermus thermophilus* HB27. Extremophiles 9(6):415–425

Moghadam M, Albersmeier A, Winkler A, Cimmino L, Rise K, Frank M, Kalinowski J, Ruckert C, Wentxel A, Lale R (2016) Isolation and genome sequencing of four artic marine *Psychrobacter* strains exhibiting multicopper oxidase activity. BMC Gen 17:117–129

Molina C, Sánchez A, Serafín-Muñoz A, Folch-Mallol J (2014) Optimization of enzymatic saccharification of wheat straw in a micro-scale system by response surface methodology. Rev Mex de Ingen Quím 13(3):765–768

Mollania N, Khajeh K, Ranjbar B, Hosseinkhani S (2011) Enhancement of a bacterial laccase thermostability through directed mutagenesis of a surface loop. Enzym Microb Technol 49:446–452. https://doi.org/10.1016/j.enzmictec.2011.08.001

Navas LE, Martínez FD, Taverna ME, Fetherolf MM, Eltis LD, Nicolau V, Estenoz D, Campos E, Benintende GB, Berretta FB (2019) A thermostable laccase from *Thermus* sp. 2.9 and its potential for delignification of eucalyptus biomass. AMB Express 9(1):24

Ortiz ML, Mussali P, Sánchez-Salinas E, Tovar-Sánchez E (2018) Mining and mine tailings: characterization, impacts, ecology and bioremediation strategies. In: Fuentes MS, Colin VL, Saez JM (eds) Strategies for bioremediation of organic and inorganic pollutants. CRC Press, Boca Raton, pp 190–214

Parajuli R, Dalgaard T, Jørgensen U, Adamsen APS, Knudsen MT, Birkved M, Gylling M, Schjørring JK (2015) Biorefining in the prevailing energy and materials crisis: a review of sustainable pathways for biorefinery value chains and sustainability assessment methodologies. Renew Sustain Energy Rev 43:244–263

Raddadi N, Cherif A, Daffonchio D, Neifar M, Fava F (2015) Biotechnological applications of extremophiles, extremozymes and extremolytes. Appl Microbiol Biotechnol 99(19):7907–7913

Ramesh P, Krishna Prasad R, Baskar R (2017) Enzymatic and chemical delignification of Kraft wood pulp: optimization & sequential studies. Environ Eng Manag J 6(11):2497–2504

Rezaei S, Shahverdi AR, Faramarzi MA (2017) Isolation, one-step affinity purification, and characterization of a polyextremotolerant laccase from the halophilic bacterium *Aquisalibacillus elongatus* and its application in the delignification of sugar beet pulp. Bioresour Technol 230:67–75

Rezaie R, Rezaei S, Jafari N, Forootanfar H, Khoshayand MR, Faramarzi MA (2017) Delignification and detoxification of peanut shell bio-waste using an extremely halophilic laccase from an *Aquisalibacillus elongatus* isolate. Extremophiles 21(6):993–1004. https://doi.org/10.1007/s00792-017-0958-7

Rodríguez-Couto S (2019) Fungal Laccase: a versatile enzyme for biotechnological applications. In: Recent advancement in white biotechnology through fungi. Springer, Cham, pp 429–457

Roth S, Spiess AC (2015) Laccases for biorefinery applications: a critical review on challenges and perspectives. Bioproc Biosyst Engin 38(12):2285–2313. https://doi.org/10.1007/s00449-015-1475-7

Sharma V, Ayothiraman S, Dhakshinamoorthy V (2018) Production of highly thermo-tolerant laccase from novel thermophilic bacterium Bacillus sp. PC-3 and its application in functionalization of chitosan film. J Biosci Bioeng 127(6):672–678

Sheng S, Jia H, Topiol S, Farinas ET (2017) Engineering cot a laccase for acidic pH stability using *Bacillus subtilis* spore display. J Microbiol Biotechnol 27:507–513. https://doi.org/10.4014/jmb.1608.08026

Singh P, Jain K, Desai C, Tiwari O, Madamwar D (2019b) Microbial community dynamics of extremophiles/extreme environment. In: Microbial diversity in the genomic era. Academic Press, pp 323–332

Singh G, Singh S, Kaur K, Kumar S, Sharma A, Sharma P (2019a) Thermo and halo tolerant laccase from *Bacillus* sp. SS4: evaluation for its industrial usefulness. J Gen Appl Microbiol 65 (1):26–33. https://doi.org/10.2323/jgam.2018.04.002

Slomczynski D, Nakas JP, Tanenbaum SW (1995) Production and characterization of laccase from *Botrytis cinerea* 61-34. Appl Environ Microbiol 61:907–912

Thomas CS (2017) Petroleum and coal proven reserves: the case for coal and the demise of OPEC. In: Stopping climate change: the case for hydrogen and coal. Springer, Cham, pp 35–40

Toushik SH, Lee KT, Lee JS, Kim KS (2017) Functional applications of lignocellulolytic enzymes in the fruit and vegetable processing industries. J Food Sci 82(3):585–593

Vergara A, Revah S, Moreno P, Scott F (2018) Biofiltration of volatile organic compounds using fungi and its conceptual and mathematical modeling. Biotechnol Adv 36(4):1079–1093

Wang R, Yang J, Zhang G, Chao Y, Li Z, Ye Q, Qian S (2017) Co-expression of beta-glucosidase and laccase in *Trichoderma reesei* by random insertion with enhanced filter paper activity. Mol Biotechnol 59(8):353–364. https://doi.org/10.1007/s12033-017-0018-7

Wikee S, Hatton J, Turbé D, Yann A, Yann M, Daou M, Lomascolo A, Kumar A, Lumyong S, Sciara G, Faulds CB, Record E (2019) Characterization and dye decolorization potential of two laccases from the marine-derived fungus *Pestalotiopsis* sp. Intern J Mol Sci 20(8):1864

Yang X, Berthold F, Berglund LA (2018a) Preserving cellulose structure: delignified wood fibers for paper structures of high strength and transparency. Biomacromolecules 19(7):3020–3029. https://doi.org/10.1021/acs.biomac.8b00585

Yang Q, Zhang M, Zhang M, Wang C, Liu Y, Fan X, Li H (2018b) Characterization of a novel, cold-adapted, and thermostable laccase-like enzyme with high tolerance for organic solvents and salt and potent dye decolorization ability, derived from a marine metagenomic library. Front Microbiol 9:2998. https://doi.org/10.3389/fmicb.2018.02998

Ye M, Li G, Liang W, Liu Y (2010) Molecular cloning and characterization of a novel metagenome-derived multicopper oxidase with alkaline laccase activity and high soluble expression. Appl Microbiol Biotechnol 87:1023–1031

Zakzeski J, Bruijnincx PC, Jongerius AL, Weckhuysen BM (2010) The catalytic valorization of lignin for the production of renewable chemicals. Chem Rev 110(6):3552–3599

Zerva A, Koutroufini E, Kostopoulou I, Detsi A, Topakas E (2019) A novel thermophilic laccase-like multicopper oxidase from *Thermothelomyces thermophila* and its application in the oxidative cyclization of 2′, 3, 4-trihydroxychalcone. New Biotechnol 49:10–18

Zheng F, Cui BK, Wu XJ, Meng G, Liu HX, Si J (2016) Immobilization of laccase onto chitosan beads to enhance its capability to degrade synthetic dyes. Int Biodeterior Biodegradation 110:69–78

Zhong Z, Pang S, Wu Y, Jiang S, Ouyang J (2017) Synthesis and characterization of mesoporous cu–MOF for laccase immobilization. J Chem Technol Biotechnol 92(7):1841–1847. https://doi.org/10.1002/jctb.5189

Zhou CH, Xia X, Lin CX, Tong DS, Beltramini J (2011) Catalytic conversion of lignocellulosic biomass to fine chemicals and fuels. Chem Soc Rev 40(11):5588–5617

Zhuo R, He F, Zhang X, Yang Y (2015) Characterization of a yeast recombinant laccase rLAC-EN3-1 and its application in decolorizing synthetic dye with the coexistence of metal ions and organic solvents. Biochem Eng J 93:63–72

Zumárraga M, Bulter T, Shleev S, Polaina J, Martínez-Arias A, Plou FJ, Ballesteros A, Alcalde M (2007) In vitro evolution of a fungal laccase in high concentrations of organic cosolvents. Chem Biol 14:1052–1064. https://doi.org/10.1016/j.chembiol.2007.08.010

Correction to: Laccases in the Context of Potentially Cooperating Enzymes

Kamahldin Haghbeen and Dietmar Schlosser

Correction to: Chapter 4 in: D. Schlosser (ed.), *Laccases in Bioremediation and Waste Valorisation*, Microbiology Monographs 33, https://doi.org/10.1007/978-3-030-47906-0_4

The original version of Chapter 4 was inadvertently published with incorrect last name of the author. The name "Dietmar Schlossser" has now been corrected to "Dietmar Schlosser".

The updated online version of this chapter can be found at
https://doi.org/10.1007/978-3-030-47906-0_4

D. Schlosser (ed.), *Laccases in Bioremediation and Waste Valorisation*,
Microbiology Monographs 33, https://doi.org/10.1007/978-3-030-47906-0_10

The manufacturer's authorised representative in the EU is Springer Nature Customer Service Centre GmbH, Europaplatz 3, 69115 Heidelberg, Germany. If you have any concerns regarding our products, please contact ProductSafety@springernature.com

Printed and bound by CPI Group (UK) Ltd, Croydon, CR0 4YY
30/09/2026
02225492-0015